Communications in Computer and Information Science   **2308**

Series Editors

Gang Li, *School of Information Technology, Deakin University, Burwood, VIC, Australia*

Joaquim Filipe, *Polytechnic Institute of Setúbal, Setúbal, Portugal*

Zhiwei Xu, *Chinese Academy of Sciences, Beijing, China*

Rationale

The CCIS series is devoted to the publication of proceedings of computer science conferences. Its aim is to efficiently disseminate original research results in informatics in printed and electronic form. While the focus is on publication of peer-reviewed full papers presenting mature work, inclusion of reviewed short papers reporting on work in progress is welcome, too. Besides globally relevant meetings with internationally representative program committees guaranteeing a strict peer-reviewing and paper selection process, conferences run by societies or of high regional or national relevance are also considered for publication.

Topics

The topical scope of CCIS spans the entire spectrum of informatics ranging from foundational topics in the theory of computing to information and communications science and technology and a broad variety of interdisciplinary application fields.

Information for Volume Editors and Authors

Publication in CCIS is free of charge. No royalties are paid, however, we offer registered conference participants temporary free access to the online version of the conference proceedings on SpringerLink (http://link.springer.com) by means of an http referrer from the conference website and/or a number of complimentary printed copies, as specified in the official acceptance email of the event.

CCIS proceedings can be published in time for distribution at conferences or as post-proceedings, and delivered in the form of printed books and/or electronically as USBs and/or e-content licenses for accessing proceedings at SpringerLink. Furthermore, CCIS proceedings are included in the CCIS electronic book series hosted in the SpringerLink digital library at http://link.springer.com/bookseries/7899. Conferences publishing in CCIS are allowed to use Online Conference Service (OCS) for managing the whole proceedings lifecycle (from submission and reviewing to preparing for publication) free of charge.

Publication process

The language of publication is exclusively English. Authors publishing in CCIS have to sign the Springer CCIS copyright transfer form, however, they are free to use their material published in CCIS for substantially changed, more elaborate subsequent publications elsewhere. For the preparation of the camera-ready papers/files, authors have to strictly adhere to the Springer CCIS Authors' Instructions and are strongly encouraged to use the CCIS LaTeX style files or templates.

Abstracting/Indexing

CCIS is abstracted/indexed in DBLP, Google Scholar, EI-Compendex, Mathematical Reviews, SCImago, Scopus. CCIS volumes are also submitted for the inclusion in ISI Proceedings.

How to start

To start the evaluation of your proposal for inclusion in the CCIS series, please send an e-mail to ccis@springer.com.

Anish Gupta · Michael Hinchey · Zeev Zalevsky
Editors

Artificial Intelligence and Its Applications

First International Conference, ICAIA 2023
Pune, India, December 18–19, 2023
Proceedings, Part II

 Springer

Editors
Anish Gupta
iNurture Education Solutions Pvt. Ltd.
Bengaluru, Karnataka, India

Michael Hinchey
University of Limerick
Limerick, Ireland

Zeev Zalevsky
Bar-llan University
Ramat Gan, Israel

ISSN 1865-0929   ISSN 1865-0937 (electronic)
Communications in Computer and Information Science
ISBN 978-3-031-84393-8   ISBN 978-3-031-84394-5 (eBook)
https://doi.org/10.1007/978-3-031-84394-5

This Springer imprint is published by the registered company Springer Nature Switzerland AG
The registered company address is: Gewerbestrasse 11, 6330 Cham, Switzerland

If disposing of this product, please recycle the paper.

Preface

Welcome to the second volume of proceedings of the inaugural International Conference on Artificial Intelligence and its Applications (ICAIA 2023), hosted by Ajeenkya DY Patil University in Pune in collaboration with iNurture Education Solutions. This volume presents a comprehensive collection of research and innovation showcased by scholars and practitioners from around the world. Featuring 20 insightful papers across 300 pages, this compilation captures the breadth and depth of the discussions that characterized the conference.

Held from December 18th to 19th, 2023, ICAIA 2023 aimed to be a premier forum for exchanging knowledge and insights on the applications of artificial intelligence. The conference focused on three primary domains: [1] Use of AI in Health Care, [2] Machine Vision and Image Processing, and [3] Automated and Digital Manufacturing Systems. This event convened academics, industry experts, and enthusiasts to explore pioneering research, technological advancements, and practical applications in these critical areas. It not only stimulated intellectual dialogue but also provided valuable networking opportunities, enabling participants to forge new collaborations and explore innovative research avenues.

At ICAIA 2023, artificial intelligence took center stage, reflecting its transformative potential across the featured domains. The papers presented in this volume underscore the conference's commitment to advancing knowledge and harnessing AI to address contemporary challenges and unlock new opportunities.

All submissions to ICAIA 2023 underwent a rigorous double-blind peer review process, with each paper evaluated by a minimum of three reviewers from the conference's international Program Committee. This committee, composed of experts aligned with the conference's themes, played a critical role in ensuring the quality and relevance of the contributions. Their reviews were essential not only for selecting papers for inclusion but also for refining the work of authors and guiding the Program Chairs in highlighting exceptional contributions.

Ajeenkya DY Patil University, our esteemed host, represents a tradition of excellence in education and innovation. Established under the Maharashtra Government Act of 2015, the university is a leader in fostering progressive learning and scholarly exploration, with a strong commitment to supporting creativity and academic advancement through its diverse programs and cutting-edge facilities.

As you explore the contents of this volume, we invite you to engage with the diverse ideas and insights presented by our contributors. Whether you are an academician, researcher, industry professional, or AI enthusiast, we hope these papers will offer inspiration and open new avenues for exploration.

We extend our deepest gratitude to all the authors, reviewers, organizers, and participants whose contributions made ICAIA 2023 a success. Your dedication and scholarly efforts are a testament to the vibrancy and dynamism of the AI research community.

We look forward to your engagement with Volume 2 of the conference proceedings and anticipate the continued growth and impact of AI research in the future.

**Editorial Team

International Conference on Artificial Intelligence and its Applications (ICAIA 2023)**

December 2023

Anish Gupta
Michael Hinchey
Zeev Zalevsky

Organization

General Chairs

Vinay Goyal	iNurture Education
Manas Paul	iNurture Education
Gokuldev	iNurture Education
Parameshachari B. D.	NITTE Meenakshi Institute of Technology, India

Program Committee Chairs

Anish Gupta	iNurture Education Solutions, India
Michael Gerard Hinchey	University of Limerick, Ireland
Zeev Zalevsky	Bar-llan University, Israel

Steering Committee

Michael Gerard Hinchey	University of Limerick, Ireland
Zeev Zalevsky	Bar-llan University, Israel
Anish Gupta	iNurture Education Solutions, India

Program Committee

Digital Chairs

Supriya Bhosle	Ajeenkya DY Patil University, India
Lovenish Sharma	Ajeenkya DY Patil University, India
Swati More	Ajeenkya DY Patil University, India
Shiv Preet	iNurture Education Solutions, India
Faisal Firdous	iNurture Education Solutions, India
Nitin Thapliyal	iNurture Education Solutions, India

Publicity Chairs

Sukhesh Kohtari	Ajeenkya DY Patil University, India
Isha Sood	Ajeenkya DY Patil University, India
Rahul Borate	Ajeenkya DY Patil University, India
Sahil Kansal	iNurture Education Solutions, India
Ravi Kumar Sharma	iNurture Education Solutions, India
Rashi Chaudhary	iNurture Education Solutions, India
Rani	iNurture Education Solutions, India
Yogesh Kumar	iNurture Education Solutions, India
Kalidas S.	iNurture Education Solutions, India
Swati Rehal	iNurture Education Solutions, India
P. Sivaprakash	iNurture Education Solutions, India
Mohammed Mallick	iNurture Education Solutions, India
V. Kanimozhi	iNurture Education Solutions, India

Session Chairs

Saurabh	JSPM University, India
Ankita Aggarwal	Ajeenkya DY Patil University, India
Himanshu Amritlal Patel	Ajeenkya DY Patil University, India
Parmeshwari Anand	Ajeenkya DY Patil University, India
Asif Ali	iNurture Education Solutions, India

Review Chairs

Aishwaraya	Ajeenkya DY Patil University, India
Peeyush Pareek	Ajeenkya DY Patil University, India
Prini Rastogi	Ajeenkya DY Patil University, India
Rajdavinder Singh Boparai	iNurture Education Solutions, India
Suman Saurabh Sarkar	iNurture Education Solutions, India
Shikha Sharma Sarkar	iNurture Education Solutions, India
Hemanth Sharma	iNurture Education Solutions, India
Neeraj Sharma	iNurture Education Solutions, India

Hospitality Chairs

Chandan Parsad	Ajeenkya DY Patil University, India
Ranjana Singh	Ajeenkya DY Patil University, India
Vishal Badgujar	Ajeenkya DY Patil University, India
Bhavesh Shah	iNurture Education Solutions, India
Lokesh Sahu	iNurture Education Solutions, India

Smriti Jain	iNurture Education Solutions, India
Lakshita Mandpe	iNurture Education Solutions, India
Nitish Patil	iNurture Education Solutions, India

Finance Chairs

Rohit Sharma	Ajeenkya DY Patil University, India
Vishal Sain	iNurture Education Solutions, India
Mahendra Kumar Meena	iNurture Education Solutions, India

Registration Chairs

Ankush Patil	Ajeenkya DY Patil University, India
Minhaj Khan	Ajeenkya DY Patil University, India
Himanshu Sharma	iNurture Education Solutions, India
Ghufran khan	iNurture Education Solutions, India
Aditya Tripathi	iNurture Education Solutions, India
Arshad Ali	iNurture Education Solutions, India
Praful Saxena	iNurture Education Solutions, India
Mohd Salman	iNurture Education Solutions, India

Additional Reviewers

Shreyanth S.	Birla Institute of Technology and Science, India
Niveditha S.	Rajalakshmi Engineering College, India
Parveen Singla	Chandigarh Engineering College, India
Rupali Santosh Kalekar	Tilak Maharashtra Vidyapeeth University, India
Rashmi Tiwari	Nagindas Khandwala College, India
Shubham Mahajan	Ajeenkya DY Patil University, India
Khusbhu Khandait	Ajeenkya DY Patil University, India
Pragati Hiwarkar	Tilak Maharashtra University, India
Ankita Aggarwal	Ajeenkya DY Patil University, India
Mohamed Abouhawwash	Michigan State University, USA
Jyotir Moy Chatterjee	Graphic Era University, India
Shakir Khan	Imam University, Saudi Arabia
Gourav Kumar	Central University of Jammu, India
Arjun Puri	Chitkara University, India
Davinder Paul Singh	Pandit Deendayal Energy University, India
Sachin Kumar Gupta	Central University of Jammu, India
Pushpalata Shivaram Patil	MIT ADT University, India

Chandan Prasad	JSPM's Rajarshi Shahu College of Engineering, India
Amit Kumar	Chitkara University, India
Nitin Mittal	Shri Vishwakarma Skill University, India
Kiran Jot Singh	Chandigarh University, India
Arun Kumar	Galgotias University, India
Mayur Mali	Ajeenkya DY Patil University, India
Surbhi Kaul	Ajeenkya DY Patil University, India
Lalit Kumar Tyagi	GL Bajaj, India
Anand Srivastava	TCS, Noida, India
Dhiraj Gupta	GNIOT, India
Saurabh Mittal	JSPM University, India
Krishna Gupta	APEX University, India
Shikha Sharma	APEX University, India
Prerana Vyas	APEX University, India
Vaibhav Jain	SAGE University Bhopal, India
Smriti Jain	SAGE University Bhopal, India
Bhavesh Shah	SAGE University Bhopal, India
Bhoopendra Dwivedy	Galgotias University, India
Vibhash Singh Sisodia	Shri Bhawani Niketan Institute of Technology and Management, India
Sachin Yadav	Shree Dhanvantary College of Engineering and Technology, India
Sukhdeep Kaur	Chandigarh Engineering College, India
Pooja Sahni	Chandigarh Engineering College, India
Mohit Srivastava	Chandigarh Engineering College, India
Rinkesh Mittal	Chandigarh Engineering College, India
Sumit Singh	Galgotia University, India
Rohit Anand	G.B. Pant DSEU Okhla-1 Campus, India
Nitin Saluja	Chitkara University, India
Anupma Gupta	Chitkara University, India
Harvinder Kang	Panjab University, India
Hardeep Singh	SVIET, India

Contents – Part II

Contents – Part I

Machine Learning Algorithms for Identifying Spam Emails

Ajmeera Kiran[1], Mudassir Khan[2(✉)] [ID], J. Chinna Babu[3], and B. P. Santosh Kumar[4]

[1] Department of Computer Science and Engineering, MLR Institute of Technology, Hyderabad, Telangana 500043, India
[2] Department of Computer Science, College of Science and Arts Tanumah, King Khalid University, Abha, Saudi Arabia
mudassirkhan12@gmail.com
[3] Department of ECE, Annamacharya Institute of Technology and Sciences, Rajampet, AP, India
[4] Department of ECE, YSR Engineering College, YV University, Proddatur, AP, India

Abstract. Emails are utilised in practically all spheres of today's society, from the professional world to the academic sphere. Ham and spam are the two subcategories that may be found within emails. Email spam, also known as junk email or unwelcome email, is a sort of email that may be used to cause harm to any user by wasting his or her time, using an excessive amount of computing resources, and stealing important information. The proportion of unsolicited emails is rising at an alarming rate day by day. Predicting the value of a company's stock is difficult for academics, investors, and analysts. The majority of people are interested in learning about stock prices in order to enhance their own finances. Long-Short-Term Memory (LSTM) is the abbreviation for the time series notation. In today's market, a stock trading system needs to adhere to this paradigm and combine KNN and LSTM in order to achieve higher levels of accuracy in its models. The majority of people in today's world improve their financial situations by trading on the stock market. When this doesn't work, individuals' resort to criminal behaviour. The two equities are compared using this procedure. In order to solve the problems that the pure KNN method was having with distance metrics, the suggested model uses an optimised version of the KNN technique. The fact that the majority of test data in the present KNN is focused on focal points has no impact on the stock prediction because all of the qualities are connected. The Programme places a greater emphasis on data points that are favorable rather than ones that are centered. The KNN distance probability is determined by an optimization procedure. The present iteration of the KNN model demands a significant amount of memory in addition to other resources so that it can compute the distance between each data point and the test point. By removing duplicates, the procedure eliminates the need to double-check the records. The procedure is sped up by reducing the number of iterations. The effectiveness of the model is demonstrated by comparisons with standard classifiers.

Keywords: Random Forest · Machine Learning · Naïve Bayes · Support Vector Machine

A. Gupta et al. (Eds.): ICAIA 2023, CCIS 2308, pp. 1–14, 2025.
https://doi.org/10.1007/978-3-031-84394-5_1

1 Introduction

"Email spam" is the practice of sending unsolicited commercial or promotional emails to a group of subscribers via electronic mail. This practice is also known as "spam." It is a sign that a person has not granted their consent to receive emails in a certain category when that person receives emails in that category that they have not requested. There has been a general increase tendency in the use of spam emails since the beginning of the previous decade. This trend started in the previous decade. On the internet, spam has become a big problem that needs to be addressed. It is a waste of both the space in the storage system and the time as well as the speed at which messages can be transmitted. It's possible that using an automated email filter is the most effective way to detect spam, but in today's world, spammers can easily circumvent all of these spam filtering applications by using relatively simple methods. Therefore, using an automated email filter might not be the most effective way to detect spam.

The vast majority of spam that emanated from specific email addresses could be manually blocked prior to these latter years. The process of identifying spam will make use of machine learning as its primary methodology. Text analysis, white and blacklists of domain names, and community-primarily based techniques are some of the primary approaches that have been taken in recent years toward the direction of junk mail filtering. In recent years, there has been a significant increase in the amount of junk mail that has been received. One strategy that is extensively utilised in the fight against spam is the analysis of the text included within emails. [Further citation is required] A great number of solutions can either be installed on the server or bought by consumers, and both options are available.

When an attacker sends a message or email that is irrelevant and undesirable to a large number of recipients, this is an example of spam [3]. Spam may be distributed by email or any other channel that allows for the sharing of information. As a result, a very high level of concern for the safety of the email system is required. Emails marked as spam could include malware such as viruses, worms, or Trojan horses. The majority of the time, attackers will employ this strategy to lure people towards online services. (They may send spam emails that contain attachments with the multiple-file extension, packed URLs that drive the user to harmful and spamming websites, and ultimately result in some form of data or financial fraud as well as identity theft [4, 5]. A great number of email service providers give their users the ability to create keyword-based filters that automatically filter incoming messages. However, this strategy is not very useful due to the fact that it is difficult and users do not want to customise their emails, which leads to spammers attacking their email accounts.

An example of email spam is the sending of unsolicited emails or emails that contain material that is intended for commercial purposes to a list of subscribers. This type of spam is also widely referred to as electronic mail spam. When an individual receives emails that they have not requested, it indicates that the individual has not provided their authorization to receive those emails. Since the beginning of the previous decade, an increasing number of people have been sending unsolicited commercial emails known as spam. On the internet, the issue of spam has grown to become one of great importance. The sending and receiving of spam constitutes a waste of both space and time. Even while automatic email filtering is most likely the most efficient strategy for combating spam, it is likely that spammers of today will quickly develop ways to get through all of these tools. A few years ago, it was possible to manually block the vast majority of spam that emanated from specific email addresses. This was accomplished by using a combination of filters and blacklists. The detection of spam will be accomplished by the application of a technique known as machine learning. Three of the most important tactics that have lately been applied closer to junk mail filtering include text analysis, whitelists and blacklists of domain names, and community-based procedures. Text analysis is a type of anti-spam method that makes considerable use of the content of emails as one of its primary weapons against spam. It is possible to implement any one of a number of different solutions, based on the specifications that have been laid forth by either the server or the buyer. One of the algorithms that is used in these procedures that is particularly well-known is called the naive Bayes algorithm. If false positives arise, it may be difficult to reject sends that are mostly based on content evaluation because of the possibility of a false positive. Customers and businesses alike normally do not want any crucial communications to be lost in any circumstance. It is highly possible that the boycott method was the very first one attempted for classifying spam into its various subtypes.

According to estimates made by specialists in social networking, around forty percent of accounts on social networking sites are used to send spam [8]. (e spammers utilize popular social networking technologies to target certain segments, review pages, or fan pages in order to send hidden links in the text to pornographic or other product sites aimed to sell something from fake accounts. These sites may contain explicit content or other content designed to sell anything. (e spam emails that are sent to the same type of persons or organizations typically contain the same types of recurring highlights. Through further investigation of these highlights, it is possible to improve the detection of emails of this kind. We are able to sort emails into those that are spam and those that are not spam by utilizing artificial intelligence (AI) [9]. (This problem may be solved by employing feature extraction to obtain information from the headers, subjects, and bodies of the messages. Following the extraction of these data based on their characteristics, we are able to classify them as either spam or ham. Learning-based classifiers [10] are now widely employed for spam identification in today's online environments. In learning-based categorization, the detection method begins with the presumption that spam emails possess a unique collection of characteristics that set them apart from legitimate emails [11]. The task of identifying spam in models that are based on learning is made more difficult by the presence of many different components. (These reasons include the subjectivity of spam, concept drift, challenges with language, burdensome processing, and text delay.

The intention is to reply to each and every transmission, with the exception of messages sent from local or electronic mail ids [14–16]. issued a demand for the establishment of a boycott. Because more recent geographic areas have joined the category of spamming domain names, the efficacy of this strategy is not as high as it once was when it was originally deployed. This is because more recent geographic areas have joined the category of spamming domain names. The "white list technique" is a system that accepts e-mails only from domain names and addresses that have been publicly whitelisted, while placing all other e-mails in a queue with a significantly lower priority than the ones accepted from the "whitelisted" domain names and addresses. In some circles, this strategy is also referred to as the "white list method." This strategy is particularly successful in circumstances in which the sender answers to a request from a "junk mail filtering system" by confirming their identification. In other words, the sender has been asked to verify their identity by the system. The terms "ham" and "spam" are defined by Wikipedia as "the use of electronic messaging networks to distribute undesired bulk communications, including mass advertisement, dangerous links, and other such things." Both "the use of electronic messaging networks to transmit undesired bulk communications" (often known as "spam") and "ham" are instances of this practise. Unsolicited communications are ones that you receive from various sources despite not having made a request to do so in order to do so. As a result, you ought to presume that the email is spam if the sender of the message is someone you are not familiar with. When consumers are in the process of downloading any free services or software, or when they are upgrading the programme, they are typically ignorant that they have just signed up for such mailers. This might happen when people are upgrading the programme. Approximately about the year 2001, Spam Bayes came up with the phrase "Ham," which may be defined as "Emails that are not generally solicited and are not tagged spam." It is possible to filter out spam using domain names that have been checked out and are recognised as having a high level of credibility. When it comes to differentiating spam emails from non-spam emails and when it comes to classifying emails, "the category of spam emails is very crucial." "The huge body may utilize this method to differentiate between valid emails, which would comprise exclusively of emails they intended to receive,"

The proposed method analyzed the performance with KNN, Random Forest, Decision Tree and SVM algorithms. The proposed approach achieves an accuracy of 82%, recall of 62%, and precision of 71% correspondingly using KNN, the random forest algorithm method acquires accuracy of 96%, recall of 98%, and precision of 92% and decision algorithm achieves an accuracy of 91%, a recall of 78%, and a precision of 84% respectively.

The remainder of this paper is organised as follows: Sect. 2 reviews the related work on available anti-spam filters, focusing on learning-based and novel alternative approaches. Section 3 presents the implementation and results. Section 4 concludes the paper and suggests directions of further research.

2   Related Works

Suryawanshi, Shubhang, Goswami, Anurag & Patil, Pramod. (2019). et al.: It is recommended that when conversing by email, you make use of the most recent developments that have been made in both the hardware and the software. On the other hand, sending emails that are not wanted has a negative impact on the communication [1]. At this time, there is a pressing requirement for the identification and categorization of spam in email. Models are being built as part of the continuing study to assist in the identification and classification of spam that is sent over email. We have made use of a wide variety of machine learning classifiers, including Naive Bayes (NB),

SVM (Support Vector Machines), KNN (K Nearest Neighbour), bagging and boosting (Adaboost), and ensemble classifiers with a voting mechanism. These classifiers have helped us make accurate predictions. The email spam dataset is used to evaluate and test classifiers. This dataset can be accessed in the UCI Machine Learning repository as well as on the Kaggle website. Accuracy can be evaluated using a variety of metrics, some of which include Accuracy Score, F measure, Recall, Precision, Support, and ROC [2]. [Support] is another metric that can be used to evaluate accuracy. An ensemble classifier that makes use of a voting mechanism has been determined to be the most effective classification approach based on the preliminary findings. It has a high overall degree of accuracy and produces relatively few false positives compared to other methods as per the discussed by authors in [10].

Karim, A., Azam, S., Shanmugam, B., Krishnan, K., &Alazab, M. (2019). et. al: The constantly expanding problem of phishing emails, which are also known as spam, spear phishing, or malware that is spread through spam, has led in a demand for dependable and intelligent anti-spam e-mail filters [3]. This article is a summary of a more in-depth literature study that focuses on AI and ML algorithms for intelligent spam email identification. This survey article [4] provides an overview of the methodologies that, in our opinion, have the potential to be helpful in the process of developing effective countermeasures. For the purpose of our investigation, we zeroed in on the following four structural aspects of email that lend themselves well to in-depth examination: The headers (A) include information about the email's initial origin, intermediate destinations, and ultimate destination. This information includes the IP addresses of the Mail Transfer Agents (MTA) for each sender and receiver. Headers also provide information about the email's routing. Footers (B) Include information about the sender of the email, as well as its initial destination, intermediate destinations, and final destination. Footers (C) are used to provide information about the sender and recipient of the email. The SMTP (Simple Mail Transfer Protocol) Envelope, which includes the originating source and destination domains of the users as well as the names of the persons who are exchanging mail. The initial component of the SMTP data, which is displayed in the vast majority of email clients and includes information such as the sender, recipient, date, and subject of the message. (D) Both the attachment and the email content are included in the SMTP data for the second segment. Articles relevant to each new intelligent technique were tracked down, analysed, and condensed into a summary based on the overall number of papers and their level of application [6]. [Citation needed] This body of work brings to light a variety of intriguing findings, difficulties, and research concerns. This comprehensive analysis opens the path for future research efforts that will address theoretical and empirical elements of the detection of intelligent spam email.

Saleh et al. [17] give a survey on intelligent spam email detection. (They examine several security threats associated with emails, particularly spam emails, the scope of spam analysis, and various ways for detecting and filtering spam, including both machine learning and non-machine learning approaches. (They get to the conclusion that there is widespread application of supervised learning [18] algorithms for the identification of email spam. (They claim that the accuracy and consistency of supervised methods are the reasons for their widespread use, which accounts for the large prevalence of supervised learning. (They also spoke about multialgorithm frameworks and came to the conclusion that multialgorithm frameworks are more effective than a single algorithm. (They discovered that practically all of the research work that leverages the content of emails for the identification of spam, most specifically phishing emails, relies on word-based categorization or clustering techniques.

The methods of learning-based email spam filtering are enumerated by Blanzieri and Bryl [2, 19]. They tackled the spam problem and offered a review of several learning-based spam filtering methods in this work. (They discuss numerous characteristics of unsolicited emails known as spam. This research looked at the consequences that spam emails have had on a variety of distinct domains. This study also discusses a variety of ethical and economic problems that are associated with spam. (e antispam technique that is widely used, as well as learning-based filtering, has reached a mature stage. (e frequently used filters are based on several categorization strategies that are applied to various components of email messages in order to filter out unwanted content. (According to the findings of this study, the Naive Bayes classifier has a unique place among the many different types of learning algorithms that are utilised for spam filtering. It works at a wonderful rate and is really straightforward, and it produces very precise results.

Bhuiyan et al. [20] provide an overview of various ways that are currently used to filter out email spam. (They review a number of different spam filtering techniques and compile information on the correctness of various parameters of various proposed systems by analysing a large number of processes. (They discuss how all of the existing approaches for screening spam emails are effective and efficient. Some people have achieved their goals, while others are looking into new strategies to improve their level of precision in their performance. Even if they are all effective, there are still certain problems with the methods of spam filtering, which is the major worry of researchers. (They are attempting to develop a spam filtering mechanism for the next generation that is capable of understanding enormous quantities of multimedia data and screening spam emails. (They came to the conclusion that the majority of email spam filtering was carried out using the Naive Bayes and SVM algorithms. In order to validate the efficacy of the spam filtration models, it is possible to train these models using a variety of datasets, such as "ECML" and the UCI dataset [21].

A survey of deep learning techniques used in intrusion detection systems and spam detection datasets was published by Ferrag et al. [20, 22–24]. (They examined many different detection techniques that were based on deep learning models and analysed how successful those models were. (They conducted an analysis of 35 well-known cyber datasets and separated them into seven groups before doing so. (these categories include those that are based on Internet traffic, network traffic, Interanet traffic, electrical network traffic, virtual private network traffic, andriod app traffic, and Internet of Things traffic. A list of acronyms that have been used in this article, together with their meanings, may be found in Table 1. Definition of the Acronym KNN: K-closest neighbours NN: nearest neighbours Neural networks SVM Machine à vecteurs de soutien MLP Neural network with many layers of multilayer perceptrons ECML (European Conference on Artificial Intelligence and Machine Learning). Density-based spatial clustering of applications with noise abbreviated as DBSCAN ELM Machines capable of extreme learning tree of AD Various iterations of the decision tree Protection of Information and Communication Networks three datasets based on traffic, as well as devices connected to the Internet [25–28]. (They came to the conclusion that deep learning models have the potential to outperform classic machine learning and lexical models when it comes to the identification of intrusions and spam.

A survey of supervised machine learning algorithms for screening spam emails is presented by Vyas et al. [20]. (They came to the conclusion that the Naive Bayes approach gives faster results and reasonable precision compared to every other method out of all the strategies that were studied. SVM and ID3 give more precision than Naive Bayes, however it takes significantly more effort to build a system using these two methods. There is a compromise to be made between exactness and timeliness [30, 31]. (They came to the conclusion that choosing a learning algorithm is strongly dependent on the circumstances, as well as the needed level of accuracy and amount of time. (They suggest that in the future, all aspects of the email should be taken into consideration in order to develop a more comprehensive framework for screening spam.

Email has emerged as one of the most practical and cost-effective channels of communication for business and governmental users in the current world, according to research conducted by K. Agarwal and T. Kumar et al. [5]. This was discovered as a result of the broad availability of internet connections. Both K. Agarwal and T. Kumar made contributions to the research that was carried out for the purpose of this study. When it comes to managing official documents and communicating vital pieces of information with one another, the technique of choice for the great majority of individuals living in today's society is the use of email. On the other hand, much like how a coin has two sides, many individuals take advantage of how simple it is to use this kind of communication by sending unwanted and unneeded bulk emails to other individuals. This is comparable to how a coin has two sides. This is comparable to the way that there is a similarity between the two sides of a coin. These unwanted emails are known as spam emails, and they are the cause of problems that the average user is obliged to deal with, such as excessive mailbox memory utilisation and the difficulty in discriminating between important emails and undesired, useless emails. Spam emails are sent by unsolicited third parties. These problems can be traced back to unwanted emails known as spam. As a result of this, there is a requirement for a method that is not dependent on any one else and is able to filter out spam emails that include meaningless data. The removal of the stuff that is not required can accomplish this goal. In order to spot spam email, this research uses a method called naive bayes, which is based on the principles of machine learning, and an algorithm called PSO, which is based on the ideas of computational intelligence. Together, these two methods are known as the PSO approach. The methodologies of machine learning and computational intelligence, respectively, provide as the conceptual underpinnings for each of these approaches. During this stage, the text of the email is analysed, and then, using the Naive Bayes algorithm, it is determined whether or not the email contains spam. PSO is taken into account for the overall optimisation of the parameters that are used in the NB approach [7] because to the stochastic distribution and swarm behaviour properties that it contains. This is the case since PSO exhibits these traits. This is due to the fact that PSO possesses both of these characteristics. For the purposes of the experiment, the Ling spam dataset is taken into consideration, and its overall performance in terms of precision, recall, f-measure, and accuracy is assessed and graded. According to the results of the evaluation, the performance of the PSO method is significantly better than that of the individual NB technique [8, 9].

This section gives an overview of three primary approaches to machine learning that are currently in use for spam filtering. We take a look at a variety of articles, analyses the strategies that have been suggested, and evaluate the issues that spam detection and filtration systems face. (This article also focuses on the benefits and drawbacks of the proposed methods for spam identification and filtration, which have never been addressed in the past.

2.1 Module Description

Load Dataset or collecting data: Our model's accuracy is directly proportional to the quantity as well as the quality of the data that you submit to it. The data set should be loaded by utilising the pandas read csv() method. The result of this stage is often a representation of the data that will be used for training in the following step(s).

Split Data Set: Gather information and get it organised so it can be used in the training. Eliminate everything that could possibly need it (remove duplicates, correct errors, deal with missing values, normalization, data type conversions, etc.) Since the data have been randomised, the impacts of the particular sequence in which we collected and/or otherwise prepared our data have been removed from consideration. Distribution at random of the data Make use of data visualisation to assist in the identification of pertinent correlations between variables or class imbalances (bias warning!). Alternatively, carry out alternative forms of exploratory analysis. There was a separation made between the training sets and the evaluation ones. The set of data was partitioned into two different categories. The first one is a train data test, while the second one is a test data set to be used during training.

Train Data Set: The majority of the work that goes into machine learning is focused on the process of training the model. During this period, the majority of the "learning" that takes place takes place. The fit technique will be used to train our data set, and train data set will be used to train the fit method.

Data set for testing The data set that has been provided will be utilised as the test subject for the algorithm. Predict data set: A precise prediction of the outcomes can be obtained by the use of the Predict() method.

3 Implementation

3.1 System Design

Our model was trained using a huge number of different classifiers, and each of those classifiers was subsequently examined and compared to one another to ensure the best level of accuracy. Following the completion of the evaluation, the findings of each classifier will be made available to the user. When all classifiers have finished their work and reported their findings to the user, the user can then compare the result with other results to determine whether or not the data in question is "ham" or "spam." When exhibiting the results of each classification, graphs and tables will be utilised so that the information may be comprehended in a more straightforward manner. The dataset is utilised

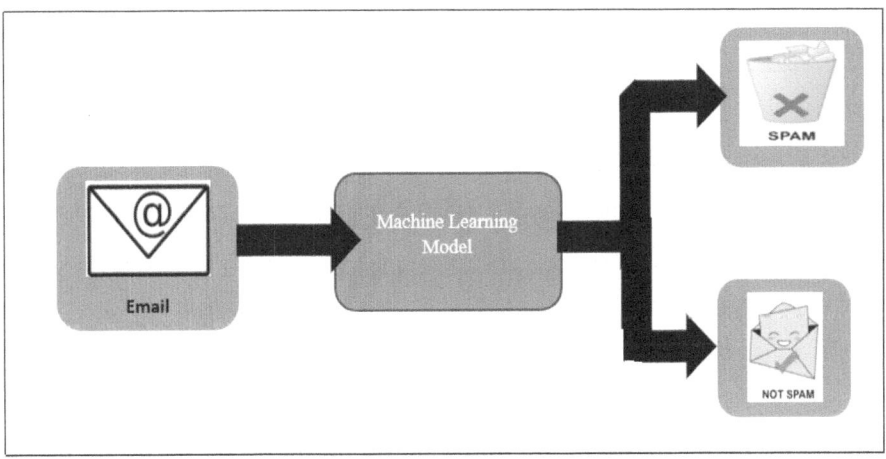

Fig. 1. The Architecture of the Proposed Method

for training after being obtained from the "Kaggle" website and being downloaded. The categorization system for passive aggressive behaviour is a significant improvement over its predecessors in terms of its level of accuracy (Fig. 1).

3.2 Results and Analysis

When a new message is delivered to the inbox, an export of that message will be done to the dataset using the format that is presented further down in this paragraph. A decision will be made regarding whether or not this message is regarded to be spam [25]. When a new message is delivered to the inbox, an export of that message will be done to the dataset using the format that is presented further down in this paragraph. A decision will be made regarding whether or not this message is regarded to be spam. There was a noticeable level of disparity in the outcomes produced by each of the four distinct algorithms that we investigated [26]. The databases both contain information on words that are classified as either "spam" or "ham," according to their respective names. All of the true positives, true negatives, false positives, and false negatives are factored into the calculation of the precision, recall, and accuracy graphs with their associated values. This ensures that the results are as accurate as possible by the researches [27, 28].

Table 1 Comparison between KNN, Random Forest, Decision Tree and SVM

ML Algorithms	Accuracy	Recall	Precision
KNN	82	62	71
Rondom Forest	96	98	92
Decision Tree	91	78	84

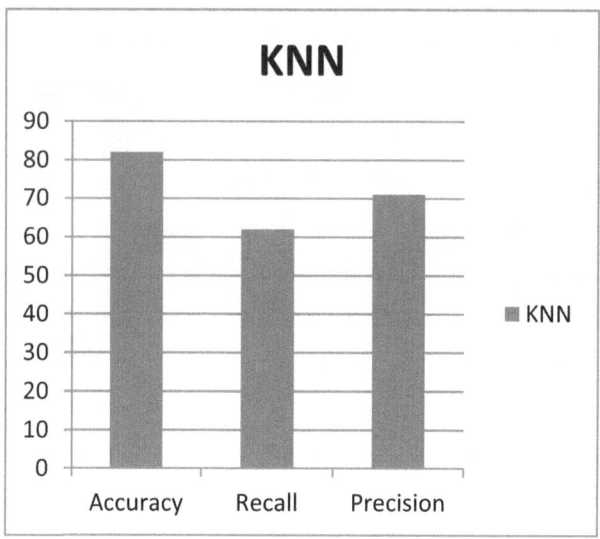

Fig. 2. Comparison Spam mail Using KNN Algorithm

Figure 2 shows the graph that was produced for the comparative examination of accuracy, recall, and precision using KNN algorithms. On the other hand, it has been established that the KNN approach achieves an accuracy of 82%, recall of 62%, and precision of 71% correspondingly.

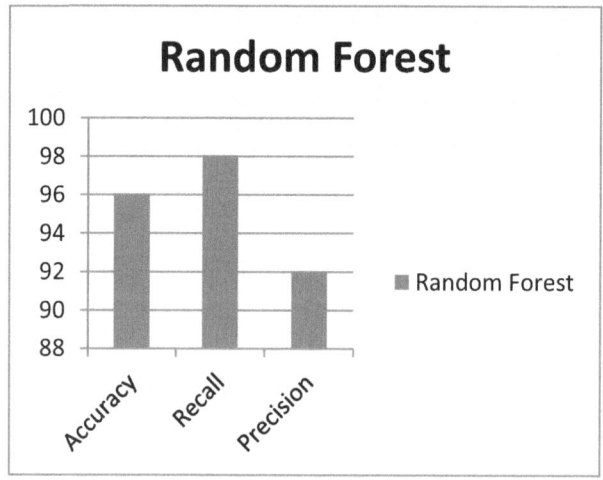

Fig. 3. Random Forest

The Graph is drawn for comparative analysis of accuracy, recall and precision using random forest algorithms are plotted in Fig. 3. However, it is observed that the random forest algorithm method acquires accuracy of 96%, recall of 98%, and precision of 92% respectively.

Fig. 4. Decision Tree

Figure 4 shows the graph that was generated for the comparative examination of accuracy, recall, and precision utilizing decision algorithm. In spite of this, it has been determined that the decision algorithm achieves an accuracy of 91%, a recall of 78%, and a precision of 84% correspondingly.

4 Conclusion

In the past twenty years, the identification and filtering of spam have attracted the interest of a substantial portion of the scientific community. (e reason why there is a lot of study being done in this field is because of its costly and significant influence in many different scenarios, such as bogus reviews and customer behavior. (e survey includes a variety of machine learning strategies and models that have been suggested by various researchers to detect and filter spam in email systems and IoT platforms. (e research put them into categories such as supervised learning, unsupervised learning, reinforcement learning, and so on. (e research makes a comparison of these methodologies and presents a synopsis of the key takeaways from each of the groups. People who aren't aware of these cons are the ones that spammers go for, and the way they do this is by swiftly generating bogus profiles and email accounts for those folks. Those individuals are then targeted by the spammers. They send spam emails that make it appear as though they came from a real individual. As a consequence of this, having the ability to recognize spam emails that have fraudulent content is absolutely necessary. This objective will be achieved by the project through the application of a variety of machine learning methodologies. In this article, we will explore various machine learning techniques, and then we will demonstrate how to apply these algorithms to our data sets. After that, the algorithm that demonstrates the highest level of both precision and accuracy is chosen as the one that is the most effective at identifying spam in email. The findings of this research endeavor offer a huge prospect for growth in a number of different areas. The following is a list of additional improvements that could be made: "It is possible to eliminate spam by using domain names that have been checked and validated as the foundation for doing so." "The spam email classification is extremely important in determining which emails are spam and which emails are not spam," "This method can be used by the vast body in order to differentiate decent emails, which are simply the emails they wish to receive," it is written.

References

1. Suryawanshi, S., Goswami, A., Patil, P.: Email spam detection: an empirical comparative study of different ML and ensemble classifiers 69–74 (2019). https://doi.org/10.1109/IAC C48062.2019.8971582.
2. Kiran, A., Vasumathi, D.: A comprehensive survey on privacy preservation algorithms in data mining. In: Proceedings of the 2017 IEEE International Conference on Computational Intelligence and Computing Research (ICCIC-2017), 978-1-5090-6621-6/17/$31.00 ©2017 IEEE (2017).
3. Karim, A., Azam, S., Shanmugam, B., Kannoorpatti, K., Alazab, M.: A comprehensive survey for intelligent spam email detection. IEEE Access 7, 168261–168295 (2019). https://doi.org/ 10.1109/ACCESS.2019.2954791
4. Kiran, A., Shirisha, N.: K-anonymization approach for privacy preservation using data perturbation techniques in data mining. Mater. Todays (2022). https://doi.org/10.1016/J.Matpr. 2022.05.117
5. Agarwal, K., Kumar, T.: Email spam detection using integrated approach of naïve bayes and particle swarm optimization. In: 2018 Second International Conference on Intelligent Computing and Control Systems (ICICCS) , pp. 685–690. Madurai, India (2018)
6. Harisinghaney, A., Dixit, A., Gupta, S., Arora, A.: Text and image-based spam email classification using KNN, Naïve Bayes and reverse DBSCAN algorithm. In: 2014 International Conference on Optimization, Reliabilty, and Information Technology (ICROIT), pp.153–155. IEEE (2014)
7. Mohamad, M., Selamat. A.: "An evaluation on t he efficiency of hybrid feature selection in spam email classification. In: Computer, Communications, and Control Technology (I4CT), 2015 International Conference on, pp. 227–231. IEEE (2015)
8. Kiran, A., Devara, V.: Optimal privacy preserving technique over big data analytics using oppositional fruit fly algorithm. Recent Adv. Comput. Sci. Commun. 13(2), 283–295 (2020). https://doi.org/10.2174/2213275911666181119113913
9. Shradhanjali, T.V.: E-mail spam detection and classification using SVM and feature extraction. Int. J. Adv. Reas. Ideas Innov. Technol. (2017). Khan, M., Ansari, M.D.: Security and privacy issue of big data over the cloud computing: a comprehensive analysis. Int. J. Recent Technol. Eng. 413–417 (2019).
10. Faris, H., Al-Zoubi, A.M., Heidari, A.A., et al.: An intelligent system for spam detection and identification of the most relevant features based on evolutionary random weight networks. Inform. Fusion 48, 67–83 (2019)

11. Ferrag, M.A., Maglaras, L., Moschoyiannis, S., Janicke, H.: Deep learning for cyber security intrusion detection: approaches, datasets, and comparative study. J. Inform. Secur. Appl. **50**, 102419 (2020)
12. Alghoul, A., Al Ajrami, S., Al Jarousha, G., Harb, G., Abu-Naser, S.S.: Email classification using artificial neural network. Int. J. Academic Dev. 2 (2018)
13. Kumar, N., Sonowal, S., Nishant: Email spam detection using machine learning algorithms. In: Proceedings of the 2020 Second International Conference on Inventive Research in Computing Applications (ICIRCA), pp. 108–113. IEEE, Coimbatore, India (2020)
14. Jain, G., Sharma, M., Agarwal, B.: Optimizing semantic lstm for spam detection. Int. J. Inform. Technol. **11**(2), 239–250 (2019)
15. Saleh, A.J., Karim, A., Shanmugam, B., et al.: An intelligent spam detection model based on artificial immune system. Information **10**(6), 209 (2019)
16. Kotsiantis, S.B., Zaharakis, I., Pintelas, P.: Supervised machine learning: a review of classification techniques. Emerg. Artific. Intell. Appl. Comput. Eng. **160**, 3–24 (2007)
17. Blanzieri, E., Bryl, A.: E-mail Spam Filtering with Local SVM Classifiers. University of Trento, Trento, Italy (2008)
18. Chinnasamy, P., Albakri, A., Mudassir Khan, A., Raja, A., Kiran, A., Babu, J.C.: Smart contract-enabled secure sharing of health data for a mobile cloud-based e-health system. Appl. Sci. **13**(6), 3970 (2023). https://doi.org/10.3390/app13063970
19. Patil, V., Madgi, M., Kiran, A.: Early prediction of Alzheimer's disease using convolutional neural network: a review. Egypt. J. Neurol. Psychiatry Neurosurg. **58**, 130 (2022). https://doi.org/10.1186/s41983-022-00571-w
20. Bhuiyan, H., Ashiquzzaman, A., Islam Juthi, T., Biswas, S., Ara, J.: A survey of existing e-mail spam filtering methods considering machine learning techniques. Global J. Comput. Sci. Technol. **18** (2018)
21. Asuncion, A., Newman, D.: UCI machine learning repository (2007). https://archive.ics.uci.edu/ml/index.php. 22. Vyas, T., Prajapati, P. Gadhwal, S.: A survey and evaluation of supervised machine learning techniques for spam e-mail filtering. In: Proceedings of the 2015 IEEE international conference on electrical, computer and communication technologies (ICECCT). IEEE, Tamil Nadu, India (2015)
22. Lee, D., Lee, M.J., Kim, B.J.: Deviation-based spam-filtering method via stochastic approach. EPL (Europhys. Lett.) **121**(6), 68004 (2018). https://doi.org/10.1209/0295-5075/121/68004
23. Jain, A.K., Gupta, B.B.: Towards detection of phishing websites on client-side using machine learning based approach. Telecommun. Syst. **68**(4), 687–700 (2018)
24. Olatunji, S.O.: Extreme Learning machines and Support Vector Machines models for email spam detection. In: Proceedings of the 2017 IEEE 30th Canadian Conference on Electrical and Computer Engineering (CCECE), IEEE, Windsor, Canada (2017)
25. Khan, M., Malviya, A.: Big data approach for sentiment analysis of twitter data using Hadoop framework and deep learning. In: 2020 International Conference on Emerging Trends in Information Technology and Engineering (ic-ETITE) (2020). https://doi.org/10.1109/ic-ETITE47903.2020.201, 978-1-7281-4142-8/$31.00 ©2020 IEEE
26. Khan, M., et al.: Challenges and Uses of Big Data Analytics for Social Media. Springer Nature Singapore Pte Ltd. (2020). https://doi.org/10.1007/978-981-15-1420-3_118
27. Singh Surinder S.T., Pal Singh, M., Gabhane, D., Mahamuni, C.: Study of machine learning and deep learning algorithms for the detection of email spam based on python implementation. In: 2023 International Conference on Disruptive Technologies (ICDT), pp. 637–642. Greater Noida, India (2023). https://doi.org/10.1109/ICDT57929.2023.10150836.
28. Sesha Sai Krishna Vineeth, G.V., Leela Venkata Sai, M., Mahesh, M.U., Varun, M., Shanmugapriya, S.: Email spam: a new strategy of screening spam emails using natural language processing. In: 2023 Third International Conference on Artificial Intelligence and Smart Energy

(ICAIS), pp. 710–715. Coimbatore, India (2023). https://doi.org/10.1109/ICAIS56108.2023.
10073758.
29. Karn,, R.K., Jesi, V.E., Aslam, S.M.: Spam email detection using machine learning integrated
in cloud. In: 2023 International Conference on Networking and Communications (ICNWC),
pp. 1–8. Chennai, India (2023). https://doi.org/10.1109/ICNWC57852.2023.10127237.

A Research Survey on Optimal Crop Recommendation Systems Integrating Market and Climatic Conditions

Dipmala Salunke, Rutwik Shinde[✉], Tejas Chechar, Ajay Biradar, Kiran Patil, Sonali Rangadale, and Pallavi Tekade

Department of Information Technology, JSPM's Rajarshi Shahu College of Engineering, Pune, Maharashtra 411033, India
rshinde_it@jspm.edu.in

Abstract. Recent years have seen a surge in interest in crop recommendation systems that consider the market and the weather to help farmers choose the right crops. The best crop can be predicted using machine learning approaches and algorithms based on market data, demand, and supply, and this survey article gives an overview of the available research in this area. The survey emphasises how crop recommendation systems use well-known techniques like Random Forests, Support Vector Machines, and Artificial Neural Networks. It covers its use in emerging market analysis, supply-demand information, and climatic factors to give farmers precise advice. Also mentioned are the drawbacks of the research papers under review, such as the lack of data, the narrow geographic reach, and the development of new technologies. The survey's results highlight how machine-learning approaches can boost agricultural output and profitability while taking market dynamics and climatic conditions into account. Future research could address the survey's limitations to create more reliable and useful crop recommendation systems that are adapted to various agricultural environments.

Keywords: Crop recommendation system · integrating market · SVM · Random Forest

1 Introduction

Agriculture, a cornerstone of human civilization, stands as a crucial economic sector, ensuring global food security and fostering economic prosperity. The quest for optimised crop production involves navigating a complex web of factors, including intricate weather dynamics, soil attributes, irrigation techniques, and the dynamic landscape of market forces. Modern agricultural paradigms integrate machine learning-based models to provide a data-driven approach, assisting farmers in maximising yields and profitability by recommending the most suitable crop choices.

However, achieving precision in crop recommendations encounters complexities, leading to incongruities across various research findings. The delicate balance between

A. Gupta et al. (Eds.): ICAIA 2023, CCIS 2308, pp. 15–25, 2025.
https://doi.org/10.1007/978-3-031-84394-5_2

enhanced production under favourable climatic conditions and its interplay with supply-demand dynamics and resulting profitability has yielded inconclusive outcomes. A central contributor to these disparities lies in the selection of training regions for predictive models, often rooted in data from specific locales, limiting adaptability and generalizability.

The influence of farmers extends beyond immediate cultivation zones, necessitating a comprehensive approach that integrates market dynamics at a macro-regional scale, spanning states or countries. This broader perspective calls for an assessment of supply-demand dynamics on a larger territorial canvas. Additionally, the diverse tapestry of irrigation infrastructure and climatic intricacies across locales underscores the need for yield production models functioning at finer geographic granularity, possibly at the level of talukas or districts.

To navigate these complexities, this survey introduces an innovative methodology to refine the advisory process for optimal crop selections. This approach advocates for the fusion of broader regional supply-demand factors with micro-regional yield production evaluations. At the heart of this framework is the synergy of machine learning algorithms, strategically harnessed to synthesise these multifarious considerations. The overarching objective is to elevate crop yields and bolster profitability by orchestrating a harmonious interplay of diverse determinants.

In this context, the survey serves a dual purpose: firstly, conducting a comprehensive review and synthesis of existing research in crop recommendation systems; secondly, charting potential avenues for further exploration and advancement. At the introductory stage, the paper provides insights into the kind of data analysed for this application, exploring questions such as whether it involves image datasets, sensor data, or other sources, and the volume of data required for effective analysis. Through this systematic examination, the survey aims to offer a comprehensive overview of current crop recommendation methodologies, shed light on limitations, and unveil potential pathways for future technological enhancements.

2 Observation

The examination of multiple research papers on crop recommendation systems reveals a prevalent reliance on diverse machine learning algorithms. However, a recurring limitation is the lack of detailed insights into the utilized features, posing challenges related to feature selection, data quality, and model interpretability. Researchers advocate for advanced techniques like deep learning to enhance accuracy, emphasizing the consideration of regional factors for market analysis and the creation of scalable systems adaptable to diverse conditions.

An intrinsic challenge lies in the robustness of features integrated into machine learning models. Meticulous feature selection becomes imperative, addressing the task of identifying salient variables that significantly contribute to predictive accuracy. Concurrently, data quality emerges as a pivotal concern, requiring rigorous preprocessing techniques to ensure integrity. Model interpretability is another focal point, emphasizing the development of transparent models for confident decision-making.

Amidst these challenges, there is a discernible trend towards considering regional factors in market analyses. Researchers highlight the importance of tailoring models to account for regional intricacies, aiming to enhance the applicability of crop recommendations. Scalability is also crucial, necessitating systems that can adapt seamlessly to diverse environmental and agricultural conditions. Despite challenges, the collective research underscores the substantial potential of machine learning for transformative and sustainable agricultural practices, envisioning enhanced productivity and resilience (Fig. 1 and Table 1).

Table 1. Literature on different methodologies and Future scope and shortcomings

Title	Literature Survey Summary	Shortcomings and Future Scope
Supervised Machine Learning Approach for Crop Yield Prediction in Agriculture Sector (2020)	Employs Random Forest with SVR for crop yield prediction. Limited insight into features used	Shortcomings include the absence of feature details and an opportunity to advance by building a comprehensive agriculture production and distribution recommender system
Crop Yield Prediction Using Machine Learning Algorithm (2021)	Utilises CNN, LSTM, DNN algorithms. Identifies a need for improvement in CYP	Challenges involve improving feature selection concerning temperature variations. Future work emphasises refinement of CYP models
Crop Yield Analysis Using Machine Learning Algorithms (2020)	Employs a hybrid SVR model and contemplates the impact of deep neural networks	Acknowledges potential computational challenges with deep neural networks. No specific mention of the features used
Machine learning Methodologies for Paddy Yield Estimation in India: A Case Study (2019)	Applies SVM, RF, and NN algorithms with a focus on paddy yield in India. Highlights challenges with optical imagery during monsoon seasons	Challenges involve cloud occlusion and spatial aggregation of satellite imagery, leading to information loss
Prediction of Land Suitability for Crop Cultivation Based on Soil and Environmental Characteristics Using Modified Recursive Feature Elimination Technique With Various Classifiers (2021)	Implements feature selection, specifically MRFE. Indicates a need for performance-wise improvements	Highlights the necessity for enhancements in the MRFE technique for large feature datasets
A Machine Learning-Based Framework for Crop Yield Prediction	Deploys Gradient Boosting algorithm. Identifies areas for model accuracy and interpretability improvement	Shortcomings include the need for improved model accuracy, interpretability, and handling of limited ground-truth data
Using Remote Sensing Data and Machine Learning Techniques for Crop Yield Prediction	Utilises Gradient Boosting algorithm, emphasises the challenges of limited ground-truth data	Challenges include improving model accuracy and interpretability, addressing limited ground-truth data, and handling missing data
Crop Recommendation Model Based on Multi-Source Data Fusion	Employs Random Forest with a precision of 91.83%. Suggests incorporating more data sources for better accuracy	Recommends integrating additional data sources, particularly remote sensing, for enhanced accuracy
Crop Recommendation System for Precision Agriculture Using Data Mining Techniques	Utilises Decision Tree, Naive Bayes, K-Nearest Neighbour with 95% precision. Suggests evaluating system performance under different environmental conditions	Highlights the need to evaluate the system's performance under varying environmental conditions

(*continued*)

Table 1. (*continued*)

Title	Literature Survey Summary	Shortcomings and Future Scope
Crop Recommendation System for Smart Agriculture Using IoT and Big Data	Implements Multiple Linear Regression, Naive Bayes, Random Forest with 87.2% accuracy. Suggests scaling for larger farms and regions	Recommends scaling the system for use in larger farms and regions
Design and Implementation of an Intelligent Decision Support System for Precision Agriculture	Employs K-Nearest Neighbour, Random Forest with a precision of 94.67%. Acknowledges limitations of data quality and quantity	Addresses challenges related to data quality and quantity, highlighting a need for improvement
Crop Recommendation System for Indian Agriculture Using Big Data Analytics	Utilises Random Forest with a precision of 86.4%. Suggests incorporating more granular data for better accuracy	Recommends incorporating more granular data to enhance prediction accuracy
A Survey on Crop Recommendation System Based on Data Mining Techniques	Applies Decision Tree, K-Nearest Neighbour, Random Forest with precision up to 90%. Suggests investigating deep learning algorithms	Recommends exploring deep learning algorithms for more accurate predictions
Crop Recommendation Model for Indian Farmers Using Machine Learning Algorithm	Employs Random Forest with an accuracy of 70%. Identifies challenges related to data quality, quantity, and regional variations	Addresses challenges of data quality, quantity, and regional variations in future research
Support System for Crop Selection Based on Market Analysis	Utilises Analytic Hierarchy Process, Fuzzy Logic, Regression Analysis. Specifies precision as 94.67%. Recommends developing more advanced algorithms for market analysis and forecasting	Suggests developing more advanced algorithms for market analysis and forecasting

The method's capacity to give farmers informed decision-making abilities can aid in enhancing agricultural yields and profitability while also taking market and meteorological variables into account. Farmers may make the best selections while assuring low supply and high output by using the median approach to combine the results of both the yield production model and the demand and supply model.

In conclusion, the suggested strategy has the potential to revolutionise the agricultural industry by offering farmers precise and individualised crop suggestions while maintaining food security and financial security.

3 Proposed Methodology

We suggest a methodology that comprises training the supply and demand model on a larger scale, such as the state or country level, and the yield production model on smaller regions, such as the taluka or district level, to address the issue of inconsistent results in earlier studies. This is due to the fact that regional differences in climatic conditions and irrigation infrastructure can have a big impact on agricultural production. We can identify the median between high output (in smaller locations) and low supply (across all regions) by training the models on various levels. This enables us to suggest other crops that might be more lucrative for farmers.

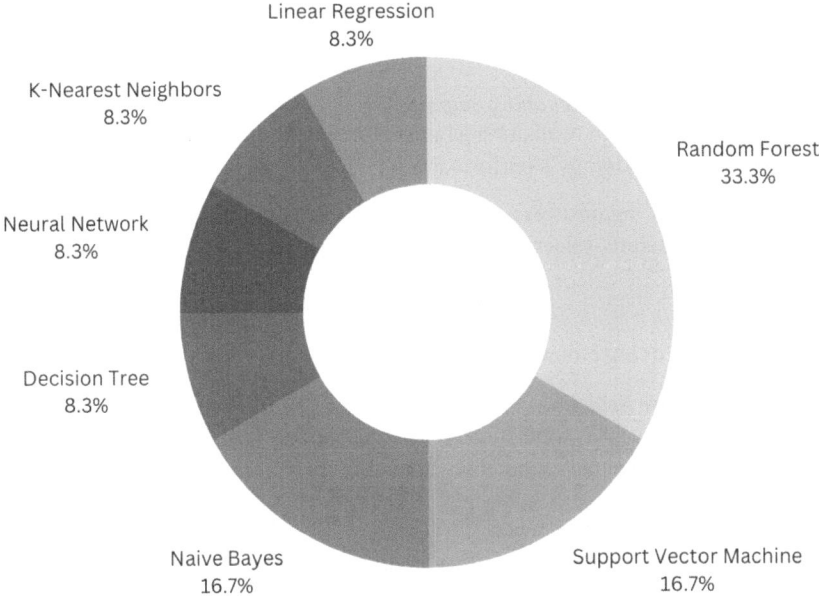

Fig. 1. Frequencies of the algorithms used in different research papers

A. *Data Acquisition and Preprocessing:*

Acquire extensive datasets encompassing historical market trends, climatic variables, soil quality metrics, and crop yields. Utilise satellite imagery and GIS data for geographical context. Conduct rigorous preprocessing, including standardisation, normalisation, and feature engineering to ensure data homogeneity.

B. *Novel Feature Selection - ERFE:*

Propose an Enhanced Recursive Feature Elimination (ERFE) method, integrating recursive elimination with Random Forest feature importance metrics. Validate ERFE against conventional methods (e.g., Recursive Feature Elimination, Principal Component Analysis) for efficacy.

C. *Hybrid Machine Learning Model:*

Implement an ensemble model leveraging Random Forest and a deep neural network. This hybrid approach aims to capture nuanced non-linear relationships within the data.

Train the model on regional data to account for geographical nuances. Evaluate performance metrics such as accuracy, precision, recall, and F1-score through cross-validation techniques.

D. *Market Demand Integration:*

Develop an adaptive market demand forecasting module utilising time-series analysis and external economic indicators.

Integrate real-time market demand predictions into the crop recommendation model, aligning recommendations with dynamic economic conditions.

E. *Scalability Enhancement:*
    Address scalability concerns with the incorporation of a distributed computing framework, specifically Apache Spark, optimising the model for handling substantial datasets efficiently.
F. *Comprehensive Evaluation and Comparison:*
    Conduct a thorough evaluation against benchmark datasets to objectively assess the proposed methodology's performance.

Compare results against existing state-of-the-art methodologies, emphasising advancements in feature selection, model complexity, and adaptability to regional variations.

3.1 Methodical Advancements

A. *Innovative Feature Selection:*
    ERFE introduces a sophisticated feature selection method, elevating interpretability and robustness, thereby enhancing the predictive capabilities of the model.
B. *Hybridized Learning Framework:*
    Integrating Random Forest with a deep neural network allows for a comprehensive exploration of both structured and unstructured data, outperforming models relying on singular algorithmic approaches.
C. *Real-time Market Dynamics:*
    The adaptive market demand forecasting module ensures that crop recommendations are dynamically attuned to prevailing economic trends, thus providing a more accurate reflection of real-world market conditions.
D. *Scalability and Distributed Computing:*
    The model's scalability is significantly improved through the strategic implementation of Apache Spark, facilitating efficient processing of large-scale datasets.
E. *Geographical Adaptability:*
    Regionalized training addresses the challenges posed by geographical variations, establishing a model that is both scalable and adaptable to diverse agricultural landscapes.
F. *Thorough Evaluation Framework:*
    A meticulous evaluation strategy against benchmark datasets, coupled with a comprehensive comparative analysis against existing methodologies, substantiates the efficacy and superiority of the proposed methodology.

This methodology represents a nuanced and comprehensive approach to crop recommendation systems, systematically overcoming key challenges prevalent in contemporary literature. The synthesis of advanced feature selection, hybrid machine learning models, real-time market integration, scalability enhancements, and thorough evaluations positions this research at the forefront of precision agriculture advancements (Fig. 2).

The illustrated decision tree structure is a fundamental element in a crop recommendation system, intricately managing the decision-making process by integrating market demand and climatic conditions. This hierarchical model systematically evaluates the

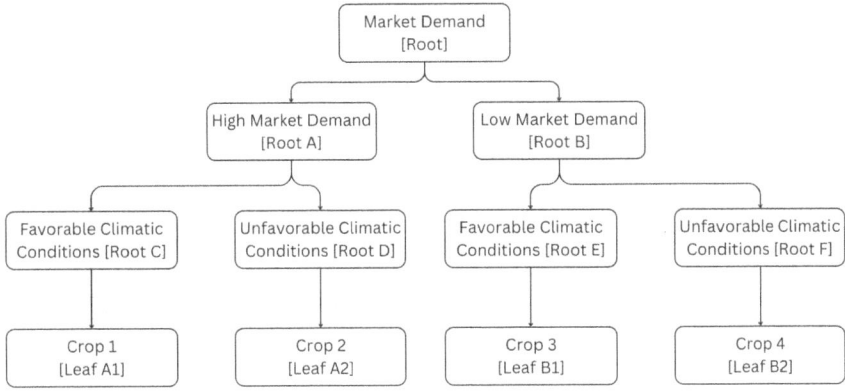

Fig. 2. Decision Tree for Crop Selection Based on Market Demand and Climatic Conditions

intricate interplay between these variables, culminating in precise crop recommendations at the leaf nodes. Its value lies in the transparency and interpretability it provides, offering farmers and stakeholders insights into the reasoning behind each recommendation. The decision tree's rule-based nature makes it particularly advantageous for implementation, ensuring practicality and ease of deployment.

The implementation of this decision tree involves leveraging machine learning algorithms, specifically those tailored for decision tree construction. The model is trained using historical data comprising market trends, climatic variables, and past successful crop selections. To ensure adaptability, real-time data can be continually integrated into the system, allowing for dynamic adjustments based on changing conditions. The modular and adaptable nature of the decision tree positions it as a powerful tool for farmers, providing personalised crop recommendations that align with both market dynamics and climatic variations (Fig. 3).

4 Datasets

4.1 Historical Climate Data

Source: The dataset is derived from authoritative sources such as the National Oceanic and Atmospheric Administration (NOAA), local meteorological departments, or reputable climate research organizations.

Description: This comprehensive dataset encompasses a spectrum of historical climate variables, including but not limited to temperature, precipitation, humidity, wind speed, solar radiation, and other pertinent climate parameters. This dataset serves as a fundamental resource for conducting nuanced analyses to discern the intricate relationships between diverse climate conditions and their impacts on crop growth and yield outcomes.

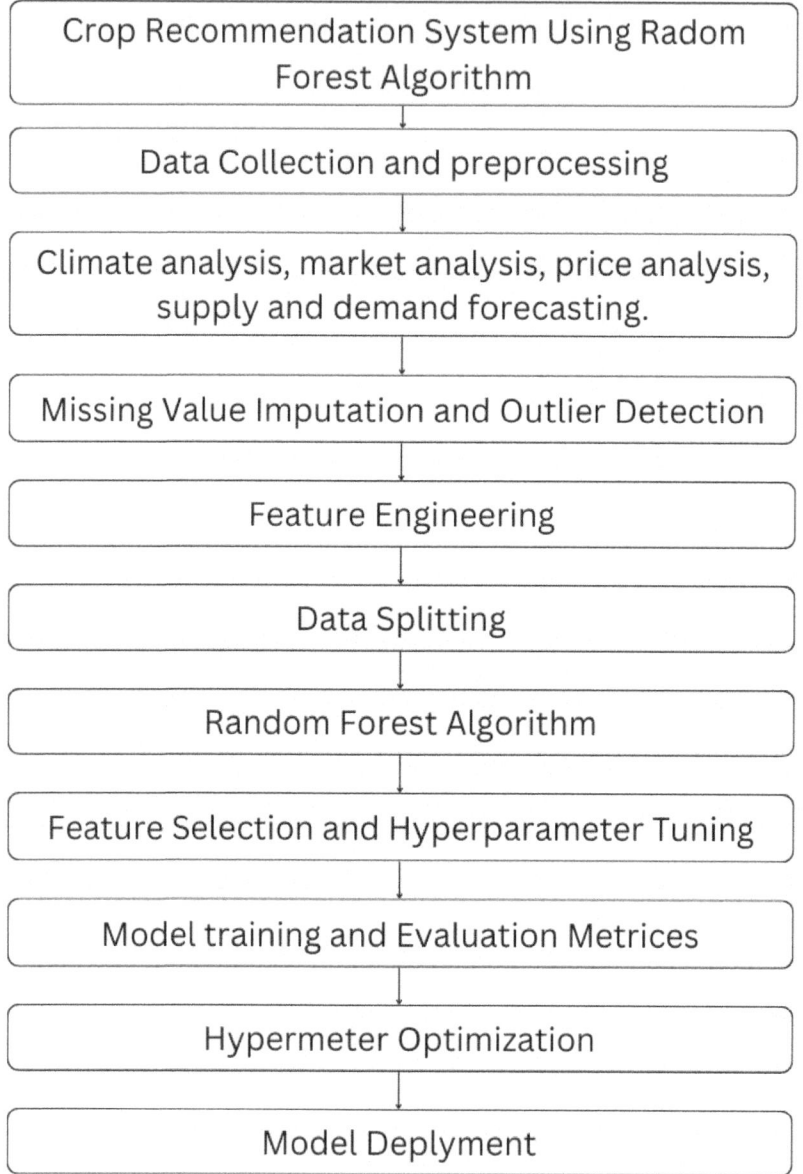

Fig. 3. Block Diagram of Crop Recommendation System using Random Forest Algorithm

4.2 Market Data

Source: Aggregated from reliable repositories such as the Agricultural Market Information System (AMIS), national agricultural departments, commodity exchanges, or esteemed market research firms.

Description: This dataset unfolds a wealth of information crucial for market-oriented analyses, featuring comprehensive data on crop prices, demand dynamics, supply variations, trade volumes, and other essential market indicators. By delving into this dataset, researchers can discern market trends, identify crops with favorable economic prospects, and assess the financial viability of specific crops within diverse regional contexts.

4.3 Soil Data

Source: Compiled from authoritative entities like the Soil Science Division, national soil databases, or esteemed agricultural research institutes.

Description: This dataset encapsulates a spectrum of soil-related parameters, encompassing soil type, pH levels, nutrient composition, organic matter content, and soil fertility indices. The dataset serves as a critical tool for evaluating soil suitability for diverse crops, as different crops exhibit distinct preferences for specific soil conditions. The utilization of this dataset facilitates precision in crop planning by aligning cultivation with optimal soil characteristics.

4.4 Crop Performance Data

Source: Aggregated from reputable sources such as agricultural research institutes, field trials, or sophisticated crop yield monitoring systems.

Description: This dataset comprises historical records detailing the performance of various crops, encompassing critical metrics such as crop yields, growth patterns, susceptibility to diseases, and responses to diverse climate conditions. Researchers leverage this dataset to conduct thorough evaluations of crop behavior under distinct market and climatic scenarios, enabling a nuanced understanding of the factors influencing crop productivity. The insights derived from this dataset inform strategic decision-making for crop selection and cultivation practices.

5 Limitations of Survey

This study's recommendations for crops based on market and meteorological conditions has a number of limitations to take into account:

1. *Data Availability*: The results and recommendations of the survey strongly depend on the availability and accuracy of data from multiple sources. The accuracy and generalizability of the results could be affected by a lack of access to complete and trustworthy datasets.
2. *Geographic Focus*: The study might have concentrated on particular geographic areas or datasets, which might not accurately reflect the varied agricultural landscapes and market dynamics present around the world. The results of the poll might not be as generalizable to other areas or nations as a result.

3. *Methodological Variations*: A variety of research publications that use various techniques, algorithms, and assessment criteria are included in the survey. Direct comparison and the development of firm conclusions among research are made difficult by the differences in these methodologies.
4. *Changing Technologies*: New technologies, algorithms, and methods are continuously emerging in the field of crop recommendation. Due to the study's constrained scope and period, it's possible that it didn't catch the most current developments in the field.
5. *Validation and Real*-World Implementation: Although the assessed research publications show encouraging results, there may be extra difficulties in the practical implementation and validation of these crop recommendation systems in real-world contexts. Further study should be done to determine the viability, scalability, and performance of these systems outside of research environments.
6. *Market Dynamics*: Although the survey's emphasis is on integrating market analysis into crop recommendation systems, market dynamics can be complicated and extremely volatile. Additional difficulties could arise from including real-time market data and modifying the models to account for changing market conditions.
7. *Human Factors*: Although the majority of the survey's attention is given to the technical components of crop recommendation systems, there is also potential for a large impact from human factors, such as farmer preferences, knowledge, and local information. The study publications that were surveyed might not have adequately addressed or captured these aspects.

6 Conclusion

In conclusion, recent research highlights the potential of machine learning algorithms, specifically Random Forest, in crop recommendation systems based on market and climatic conditions. These systems offer valuable insights for optimising crop selection by considering market demand and meteorological conditions. However, challenges related to geographical variations, scalability, and market integration need to be addressed for wider applicability and improved decision-making for farmers. Continued research and development in this field will refine and enhance crop recommendation systems, empowering farmers to achieve optimal yields and contribute to sustainable agriculture.

Overall, machine learning algorithms have the potential to transform the crop recommendation landscape by bridging the gap between supply, demand, and meteorological conditions. By incorporating advanced technologies and refining methodologies, more accurate and efficient crop recommendation systems can be developed, benefiting farmers and contributing to global food security.

References

1. Bannerjee, G., Sarkar, U., Das, S., Ghosh, I.: Artificial intelligence in agriculture: a literature survey. Int. J. Sci. Res. Comput. Sci. Appl. Manag. Stud. **7**(3), 1 (2018)
2. Haque, F.F., Abdelgawad, A., Yanambaka, V.P., Yelamarthi, K.: Crop yield analysis using machine learning algorithms. Proceedings of the IEEE (2020)
3. Guruprasad, R.B., Saurav, K., Randhawa, S.: Machine learning methodologies for paddy yield estimation in India: a case study. Proceedings of the IEEE (2019)

4. Nagendra Kumar, Y.J., Spandana, V., Vaishnavi, V.S., Neha, K.: Supervised machine learning approach for crop yield prediction in the agriculture sector. Proceedings of the IEEE (2020)
5. Mariammal, G., Suruliandi, A., Raja, S.P., Poongothai, E.: Prediction of land suitability for crop cultivation based on soil and environmental characteristics using modified recursive feature elimination technique with various classifiers. Proc. IEEE **8**, 1132 (2021)
6. Reddy, J., Kumar, M.R.: Crop yield prediction using machine learning algorithm. In: Proceedings of the ICICCS (2021)
7. Rakhra, M., Bhargava, A., Bhargava, D., Singh, R., Bhanot, A., Rahmani, A.W.: Implementing machine learning for supply-demand shifts and price impacts in the farmer market for tool and equipment sharing. Hindawi J. Food Qual. **2022**, 1 (2022)
8. Bhardwaj, M. R., Pawar, J., Bhat, A.: An innovative deep learning based approach for accurate agriculture crop price prediction. Int. J. Sci. Res. Comput. Sci. Appl. Manag. Stud. **6**(2)
9. Bharadiya, J.P., Tzenios, N., Reddy, M.: Forecasting of crop yield using remote sensing data, agrarian factors, and machine learning approaches. Journal of Engineering Research and Reports **24**, 29 (2023)
10. Dharani, M.K., Thamilselvan, R., Natesan, P., Kalaivani, P.C.D., Santhoshkumar, S.: Review on crop prediction using deep learning techniques. J. Phys. Conf. Ser. (2021). https://doi.org/10.1088/1742-6596/1767/1/012026
11. Rao, M.S., Singh, A., Reddy, N.V.S., Acharya, D.U.: Crop prediction using machine learning. J. Phys. Conf. Ser. (2022). https://doi.org/10.1088/1742-6596/2161/1/012033
12. Author(s). (Year). Integration of remote sensing data and market analysis for crop recommendation. New Zealand J. Crop Horticult. Sci.
13. Oikonomidis, A., Catal, C., Kassahun, A.: Deep learning for crop yield prediction: a systematic literature review. J. Agric. Inform. https://doi.org/10.1080/01140671.2022
14. Salunke, D., Tekade, P., Ranjan, N., Ujalambkar, D., Sangve, S., Mane, D.: Real-time dimension detection using customized canny edge detection algorithm. Int. J. Eng. Trends Technol. **71**(9), 375–384 (2023)
15. Sabne, P., Saini, H., Shivanagi, V., Jadhav, P.: Handwritten Devanagari word recognition using customized convolution neural network. In: 2021 International conference on computing, communication and green engineering (CCGE), pp. 1–5 (2021). https://doi.org/10.1109/Ccg e50943.2021.9776351
16. Joshi, R., Ranjan, N., Tekade, P., Panchal, G.: Sign language recognition system using customized convolution neural network. In: Saraswat, M., Chowdhury, C., Kumar Mandal, C., Gandomi, A.H. (eds.) Proceedings of international conference on data science and applications. Lecture notes in networks and systems, vol. 552 (2023)
17. Chavan, P., Ingale, H., Sakhare, S., Kadam, C., Tekade, P.: Enabling remote healthcare platform for rural & urban areas. In: 2023 3rd international conference on pervasive computing and social networking (ICPCSN), Salem, India, pp. 1702–1707 (2023)
18. Salunke, D., Peddi, P., Joshi, R.: The significance of image augmentation in deep learning: a review. Int. J. Adv. Res. Comput. Commun. Eng. **11**(3), 2319 (2022)
19. Salunke, D., Joshi, R., Peddi, P., Mane, D.T.: Deep learning techniques for dental image diagnostics: a survey. In: 2022 international conference on augmented intelligence and sustainable systems (ICAISS), Trichy, India, pp. 244–257 (2022)
20. Kalbhor, S., Gaikwad, A., Bhise, K., Salunke, D., Bangar, V.: A survey on digital signature. Inf. Technol. **5**(1), 279 (2015)

Trust Based Mechanism for the Isolation of Sink Hole Attack in Wireless Sensor Networks

Swedika Sharma[1(✉)] and Vishal Bharti[2]

[1] Chandigarh University, Mohali, India
swedika.sharma@gmail.com
[2] Maharishi Markandeshwar University, Mullana, India

Abstract. Wireless Sensor Network (WSN) technology operates without a centralized controller for data transmission. Due to the network's dynamic nature, it's possible for malicious nodes to join or leave, potentially launching attacks like sink hole assaults. These attacks involve impostor nodes pretending to be sinks in order to intercept data packets. This research proposes a two-stage approach to identify these malicious nodes. The first stage focuses on identifying potential attacker nodes in the network, while the subsequent stage aims to detect the actual malicious nodes. The proposed technique is evaluated using NS2 (Network Simulator 2), and the results are analyzed based on metrics such as throughput, packet loss, and delay. The findings demonstrate that this approach outperforms traditional methods in detecting malicious nodes.

Keywords: WSN · Sink Hole · Promiscuous mode · Two Step Scheme · Node Localization

1 Introduction

The widespread usage of wireless sensor networks in multiple applications, including as the monitoring of habitations, environmental pollutants, smart grids, and battle-grounds, has been facilitated by the WSNs' steady maturation. Numerous tiny sensor nodes make up a WSN. These nodes are set up in an uncontrolled, public setting. These sensor nodes have insufficient power sources. Because of the constrained communication range, inadequate computing power, and constrained space for storage at each sensor node (SN), WSN eventually becomes vulnerable to malicious intrusions [1]. These hostile invasions result in the waste of numerous electricity resources. To balance energy use and increase the network's lifespan, it is imperative to create a reliable system for detecting harmful attacks. Attackers use a variety of attack methods against WSN, including denial-of-service, sinkhole, and sybil attacks. WSN assaults are often split into two categories: internal attacks and external assaults. An assailant's external assaults are directed towards an external entity that has accessed the network. The objective of this assault is to compromise the overall performance of network. In an internal assault, the domain is attacked or another attack is started by breaching a sensor node [2].

Considering the autonomous nature of Wireless Sensor Networks (WSNs), the role of the sink in communication holds significant importance. Within WSNs, nodes often

A. Gupta et al. (Eds.): ICAIA 2023, CCIS 2308, pp. 26–36, 2025.
https://doi.org/10.1007/978-3-031-84394-5_3

utilize multi-hop routing to forward data to the Base Station (BS). Consequently, any attack capable of obstructing communication paths between sensors and the sink could seriously disrupt the functionality of the WSN. A sinkhole attack specifically aims to disrupt the exchange of messages between sensors and the sink.

In the context of WSNs, data is typically relayed from sensors to sinks using a hop-by-hop routing strategy. Each node maintains information about the smallest number of hops to reach sinks, both for itself and its neighboring nodes [3]. When transmitting data to the BS, a sensor node selects the closest mote with the least hops to convey the data. However, a sinkhole attack can be orchestrated as follows: The attacker mote, masquerading as a sink, initially presents a false hop count to the actual sink, indicating fewer hops than its true distance to the BS. For instance, an attacking node claiming zero hops to the sink is essentially implying its role as the sink. This prompts neighboring nodes to route their data through the malicious node, wrongly perceiving a direct link to the sink [4]. Utilizing this strategy, the malevolent node gains access to information about neighboring motes and can impede their communication with the actual sink, causing substantial disruption.

In more severe cases, even without actively launching attacks on other nodes, the sinkhole disrupts the surrounding region to the extent that it resembles a "black hole," absorbing all incoming information and connections, akin to the behavior of an actual black hole in space. Consequently, nodes within this affected area are drawn into the "black hole" region, unable to establish connections with nodes beyond this perimeter. As a result, the area encompassed by the sinkhole attack's coverage is often referred to as the "black hole region." The attacked mote sends the information which won't reach its intended destination because the sinkhole has collected it. As a result, it is highly challenging for the infected motes to recognise the attack occurrence. Additionally, the sinkhole intrusion disrupts coverage and connection by undermining the basic routing architecture [5]. The current routing system affects the attack's detection and isolation. As a result, the sinkhole attack is particularly difficult to detect using the traditional routing mechanism-based security approach. A sinkhole assault is depicted in a WSN in Fig. 1.

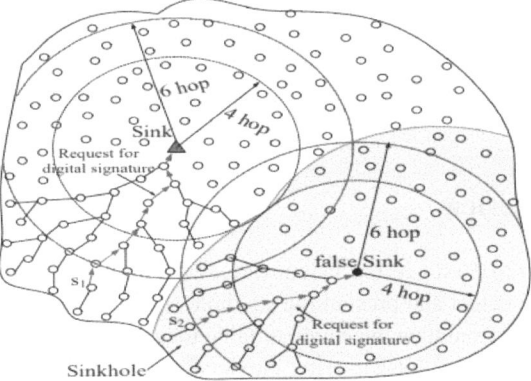

Fig. 1. An illustration of the sinkhole attack's scope

In the depicted diagram, the adversary node claims that the fake sink can be reached with 0 hops. The hop count increases once data convergence areas are established in the system, revolving around both a legitimate and a deceptive Base Station (BS). Nodes closer to the genuine sink transmit data to the false one, as opposed to the dishonest base station broadcasting data to the authentic one [6]. This phenomenon is represented in Fig. 1, where the Sinkhole attack range is delineated as a shadowy zone. The dividing line within the image, depicted in red, separates the system into an attack region and a safeguarded area. This line is linked by nodes sharing the same hop count for both the true and counterfeit BS.

Mitigating a sinkhole attack is profoundly challenging for two primary reasons [7]. Firstly, it's intricate to determine if a sinkhole breach has occurred within the system. Although each mote contributes to data transmission, discerning whether the data has ultimately been diverted to a fake sink is considerably complex. Secondly, even if the attack is detected using the attack range, locating the actual sink remains problematic, hindering the ability to notify the legitimate sink. Similar distinct characteristics are found in the hypothetical sink that initiates the sinkhole attack scenario.

(a) The attack's energy consumption is disregarded as the sinkhole node possesses unlimited energy. Additionally, the adversary can observe the internal state of the compromised SN [8].
(b) Node positions and identities are randomly determined.
(c) The node manipulates its hop count to be lower than its actual count to the BS while disseminating the fictitious value to neighboring motes. As these neighboring nodes receive the hop count from the sink, they transform the adversary node into a compromised one with a reduced hop count. Subsequently, they relay their adjusted hop count to create an attack region centered around the compromised node.

Current research struggles to effectively counter these attacks due to the gravity of sinkhole assaults and the challenges in detection and prevention. Consequently, a few existing issues must be addressed. Firstly, there are limited strategies for preventing sinkhole attacks [9]. Some studies can detect sinkhole assaults but fail to pinpoint their location. Furthermore, numerous sinkhole prevention methods lack strategies to circumvent the attacks. Data cannot reach the legitimate BS before the sinkhole is saturated, causing prolonged network damage. Thirdly, the absence of a bypassing mechanism hinders nodes within the attack radius from informing the sink about the attack occurrence. Hence, additional hardware or alternative methods are necessary to notify the system of a sinkhole's presence.

2 Literature Review

Karthigadevi, et al. (2019) suggested an innovative decentralized framework that utilized NDET algorithm to detect and prevent the sinkhole assault [10]. Every node deployed this method in order to maintain the neighbor table for storing the neighborhood details. Every node was responsible for gathering the details regarding neighbor. This method assisted in estimating the network density and recognizing the malevolent node in the area. The distribution of information related to recognized malevolent nodes was done to the neighboring nodes for avoiding the malicious node at subsequent transmissions. The suggested framework was applicable for mitigating the overhead and maximizing the throughput when the best effort traffic was maximized.

S. Padmanabhan, et al. (2022) investigated RSR (Reliable Self Reconfiguration) technique for removing the malicious sinkhole assault from the network [11]. The primary stage was executed for detecting the malevolent node. The next stage employed RM (reconfiguration mechanism) to correct it without considering resource loss. C++ based simulator applied to simulate RM to fix the sinkhole assault. This technique was computed on the basis of diverse components namely PDR (Packet Delivery Ratio) and energy utilization. The experimental outcomes revealed the superiority of the investigated technique over the traditional methods to discover and remove the sinkhole attack.

N. Al-Maslamani, et al. (2020) focused on constructing and implementing method to detect sinkhole intrusion with the deployment of SIO model [12]. In this method, a WE (weight estimation) method was integrated with ABCO framework with the objective of improving the accuracy to detect the sinkhole assault. MATLAB was executed in quantifying the constructed method with the regard to accuracy, detection time, convergence speed, overhead, and power usage. The experiments indicated the effectiveness and robust of the constructed technique against the sinkhole assault and offered superior precision.

A. A. Jasim, et al. (2019) designed an algorithm called Secure and Energy-Efficient Data Aggregation with the purpose of detecting sinkhole attack [13]. This protocol produced a random value and random timestamp via a secret key for making the network more authentic. The sink aimed to determine the false aggregated data after receiving the packets based on the produced key in advance. A SNA algorithm, DFA, FHE, and AC algorithms were deployed for detecting and preventing the attacks. The initial algorithm helped in preventing the assaults from achieving the access of network. The simulation results reported that this approach offered an accuracy 98.84% for detecting malevolent motes, energy usage of 3.04 J, the delay up to 0.038 secs, and resistance time up to 0.054.

N. D. L, et al. (2019) introduced a novel robust protocol on the basis of trust system for WSNs [14]. First of all, this algorithm was utilized to generate the SNs (sensor nodes) as clusters. After that, a secure path was developed using a TE (trust evaluation) technique for every SN at CH (Cluster Head) in order to transmit the data from SN to sink. The trust was computed at CH according to the social trust and data trust. Diverse parameters namely duration of network, MDR (Malicious Detection Rate) employed to evaluate the introduced algorithm in experimentation. The outcomes validated the applicability of the presented algorithm in contrast to the existing methods during the maximization of malevolent behavior of network.

A. K. Sangaiah, et al. (2022) discussed that the routing assaults such as sinkholes resulted in directing the network data to malevolent user and disrupting the network device [15]. Thus, a novel protocol was established based on CL-MLSP with AODV. The data was encrypted and decrypted using AES (Advanced Encryption Standard) algorithm. A clustering technique was deployed on the basis of power, mobility, and distribution for every mote to acquire the shorter route. The established protocol was computed in NS2 (Network Simulator 2) concerning duration of network, latency, PLL (packet loss), and security. The outcomes revealed that established protocol mitigated the energy consumption up to 6.54%, drop rate of 12.87%, delay, and maximized the throughput by 8.12%, and security up to 9.46%.

D. Kumar, et al. (2022) described that the military areas made the deployment of WSN (Wireless Sensor Network) for monitoring the activities of inconsistent sides [16]. The malicious nodes had potential for joining the system and activating the security intrusions. Then, SNs (sensor nodes) were employed to initialize the process to transmit the information to the malevolent mote rather than BS. A new algorithm was suggested for discovering and segregating the malicious nodes from the network. The suggested approach was computed on NS2 with respect to diverse parameters. The results demonstrated the adaptability of the suggested algorithm in comparison with the traditional methods for detecting the sinkhole attack.

K. E. Nwankwo, et al. (2019) discussed that the sinkhole assault was launched in WSN when the malicious mote pretended as the authentic node nearer to sink for transmitting the data, and modifying, dropping or delay the data [17]. Thus, a sinkhole detection system called ACO (Ant Colony Optimization) was presented and employed for detecting the sinkhole more effectively concerning packet drop, PDR, energy exchange and throughput in WSN. An analysis was conducted on the presented approach. According to simulation, the presented approach worked effectively to maximize the accuracy to detect the sinkhole attack and mitigating FAR (False Alarm Rate) in WSN.

B. M. Devaraju, et al. (2018) analyzed that the sensor nodes had susceptibility to failure due to their implementation in open regions, and tampering by intruders [18]. Malicious assaults such as DoS, Sinkhole assault, etc. aimed to modify the complex information which led to degrade the efficacy. Therefore, a CLMPI technique was projected for detecting and preventing the malevolent activities in WSN. The results depicted that the projected method was effectual to alleviate the processing delay and communicating delay subsequent to avoid the malevolent activities in networks.

3 Research Methodology

This research work is conducted to detect and isolate the sink hole attack in WSNs (wireless sensor networks). This attack focuses on sinking all the packets in the network via attacker nodes. These nodes masquerade themselves as a BS (base station). Consequently, this BS misleads all the nodes and they transmit the information to it rather than to base station. This work presents a technique to detect the malicious nodes in 2 stages. Initially, an analysis is conducted to determine the presence of attack in the network. When the attack is detected, the next stage is to discover the attacker mote. These stages are discussed as:

Step 1: This stage emphasizes on deploying the projected method for detecting the presence of assault in the network. Thus, the implementation of extract beacon nodes is done in WSN (wireless sensor network). The beacon nodes result in flooding the beacon frames within the network. The motes available in network are responsible to give response to the beacon node. These modes are executed for localizing the state of each mote on the basis of given response. The malevolent node aims to give reply very late and it leads to localize the improper position of the node. After localizing the incorrect position, exhibits the launching of attack in the network.

Step 2: It emphasizes on detecting and isolating the attack from the network. In case of localizing the improper location of SN, the beacon node results in flooding the control message. After receiving the control message, SN leads to change its mode to sleep mode. SN having more number of packets and transmitting small amount of packets is considered as the malevolent node.

3.1 Proposed Algorithm

1. Execute network with restricted number of senor nodes.
2. Sensor nodes in the network will transmit the ping messages in the network to localize the node.
3. Sensor nodes receive reply will define the position of sensor nodes. if (Position node is incorrect)
   3.1 Sensor nodes will monitor and maintain list of the network traffic.
   3.2 The senor node which receives maximum information and transmit least packets will be marked as malicious.
4. Else
   4.1 Communication continues in the network
5. Repeat Step 3 if position is incorrect in the network.

4 Result and Discussion

The purpose of this work is to detect suspicious motes in the network. The findings are analyzed using a variety of performance indicators, such as energy usage, throughput, and packet loss. Table 1 defines these simulation parameters as follows:

Table 1. Simulation Parameters

Parameter	Value
Number of Nodes	100
Area	800*800 meters
Standard	802.11
Queue Type	Priority Queue
Queue Size	50
Antenna Type	Omi-directional
Range	18 meter

Figure 2 illustrated the comparison between attach scenario, leach scenario and proposed scenario. When attack is triggered in the network energy consumption is increased at steady rate and when attack is detected from the network energy consumption is reduced. The leach scenario has less energy consumption as compared to proposed scenario because some energy is consumed during attack prevention.

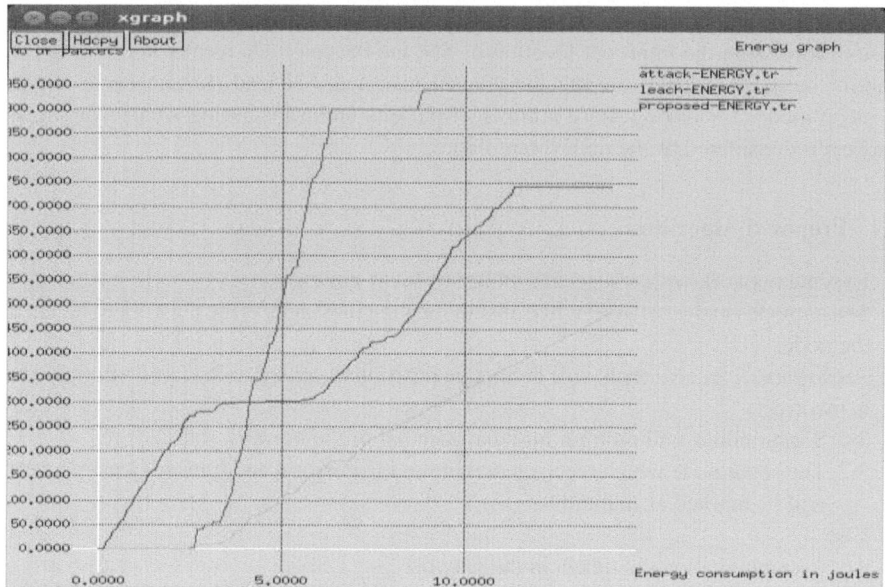

Fig. 2. Energy Comparison

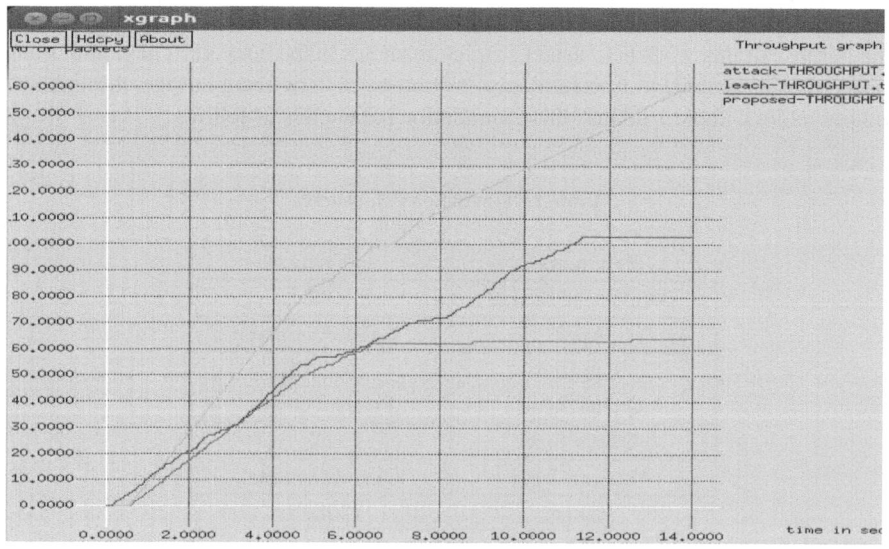

Fig. 3. Throughput Comparison

Figure 3 illustrates that the proposed technique has high throughput as compared to attack scenario. When the attack is detected and prevented from the network throughput is increased but it is less than leach protocol due to packet loss occurrence during attack.

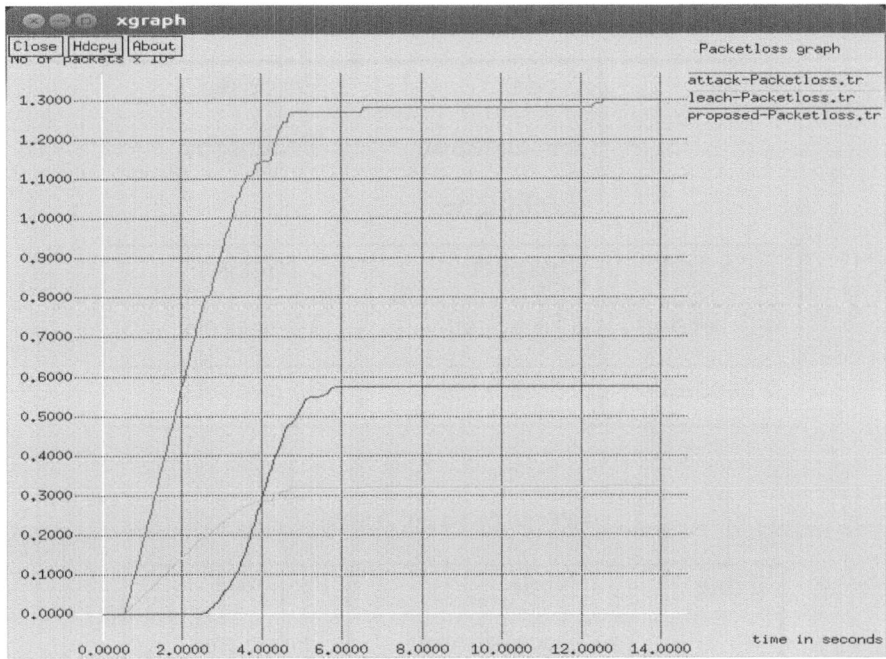

Fig. 4. Packet loss Comparison

As shown in Fig. 4, the packet loss of attack scenario, proposed scenario and leach protocol is compared for the performance analysis. It is analyzed that when attack is detected from the network packet loss is reduced as steady rate (Tables 2, 3, 4).

Table 2. Evaluation of energy usage

Time	Node Localization Technique	Proposed Technique
1 second	7 joule	3 joule
3.5 second	8 joule	4 joule
6 second	13 joule	7 joule

Table 3. Throughput Analysis

Time	Node Localization Technique	Proposed Technique
3 second	4 packets	5 packets
4.5 second	15 packets	10 packets
6 second	15 packets	55 packets

Table 4. Packet loss Analysis

Time	Node Localization Technique	Proposed Technique
2 second	2 packets	1 packets
4.5 second	22 packets	3 packets
6 second	5 packets	24 packets

5 Conclusion

WSN (Wireless Sensor Network) contains various tiny sized SNs (sensor nodes). An open environment makes the deployment of these nodes for which no supervision is considered. These sensor nodes contain a restricted amount of energy. Due to it, the issue of limited communication range, lower processing capacity and constricted storage space is occurred at every SN. Hence, WSN attains vulnerability against the malicious attacks. This network is more prone to the security attack. Therefore, attacker node launches the sinkhole attack on network to degrade the efficacy of network.

   This research work projects a technique in which 2 stages are executed to detect the malicious node. This technique is initially used to ascertain whether the compromised node is there or not. It then concentrates on finding the rogue node within the network. The projected technique is simulated on NS2 (Network simulator 2) with regard to throughput, packet loss and energy utilization. The results revealed that the projected technique leads to enhance the results up to 5–6% while detecting the malicious node.

References

1. Karthigadevi, K., Balamurali, S., Venkatesulu, M.: Based on neighbor density estimation technique to improve the quality of service and to detect and prevent the sinkhole attack in wireless sensor network. In: International conference on intelligent techniques in control, optimization and signal processing (INCOS), Tamilnadu, India, pp. 1–4 (2019)
2. S. Ali, S., Khan, M. A., Ahmad, J., Malik, A.W., Rehman, A.: Detection and prevention of black hole attacks in IOT & WSN. In: Third international conference on fog and mobile edge computing (FMEC), Barcelona, Spain, pp. 217–226 (2018)
3. Li, J., Yang, Z., Yi, X., Hong, T., Wang, X.: A secure routing mechanism for industrial wireless networks based on SDN. In: 14th international conference on mobile ad-hoc and sensor networks (2019)
4. Gebremariam, G.G., Panda, J., Indu, S.: Secure intrusion detection system for hierarchically distributed wireless sensor networks. In: International conference on industrial electronics research and applications (ICIERA), New Delhi, India, pp. 1–6 (2021)
5. Rizvi, S., Gualdoni, J., Razaque, A.: Securing wireless networks from sinkhole and sybil attacks using secure data aggregation protocol. In: 17th IEEE international conference on trust, security and privacy in computing and communications (2021)
6. Zheng, T.X., Chen, X., Wang, C., Wong, K.K., Yuan, J.: Physical layer security in large-scale random multiple access wireless sensor networks: a stochastic geometry approach. IEEE Trans. Commun. **70**(6), 4038–4051 (2022)
7. Kumar, M., Mukherjee, P., Verma, K., Verma, S., Rawat, D.B.: Improved deep convolutional neural network based malicious node detection and energy-efficient data transmission in wireless sensor networks. IEEE Trans. Netw. Sci. Eng. **9**(5), 3272–3281 (2022)
8. Ramasamy, L.K., Khan, F., Imoize, A.L., Ogbebor, J.O., Kadry, S., Rho, S.: Blockchain-based wireless sensor networks for malicious node detection: a survey. IEEE Access **9**, 128765–128785 (2021)
9. Xie, N., Chen, Y., Li, Z., Wu, D.O.: Lightweight secure localization approach in wireless sensor networks. IEEE Trans. Commun. **69**(10), 6879–6893 (2021)
10. Padmanabhan, S., Anitha, R.: An experimental study to recognize and mitigate the malevolent attack in wireless sensors networks. Glob. Trans. Proc. **3**, 55 (2022)
11. Al-Maslamani, N., Abdallah, M.: Malicious node detection in wireless sensor network using swarm intelligence optimization. In: IEEE international conference on informatics, IoT, and enabling technologies (ICIoT), Doha, Qatar, pp. 219–224 (2020)
12. Jasim, A.A.: Secure and energy-efficientdata aggregation method based on an access control model. IEEE Access **7**, 164327–164343 (2019)
13. Nirmala, D.L., Venkata, S.K.: Secure and composite routing strategy through clustering in WSN. In: 2nd international conference on innovations in electronics, signal processing and communication (IESC), Shillong, India, pp. 119–123 (2019)
14. Sangaiah, A.K., Javadpour, A., Zhang, W.: CL-MLSP: the design of a detection mechanism for sinkhole attacks in smart cities. Microprocess. Microsyst. **25**(9), 3416–3427 (2022)
15. Kumar, D., Kapoor, E.N.: Novel scheme for mutual authentication to isolate sinkhole attack in wireless sensor networks. In: International conference on engineering and emerging technologies (ICEET), Kuala Lumpur, Malaysia, pp. 1–5 (2022)
16. Nwankwo, K.E., Abdulhamid, S.M.: Sinkhole attack detection in a wireless sensor networks using enhanced ant colony optimizationto improve detection rate. In: 2019 2nd International conference of the IEEE Nigeria computer chapter (NigeriaComputConf), Zaria, Nigeria, pp. 1–6 (2019)
17. Devaraju, B.M., Raju, G.T.: Cross layer and management plane integration approach for detection and prevention of malicious activities in WSN. In: International conference on

electrical, electronics, communication, computer, and optimization techniques (ICEECCOT), Msyuru, India, pp. 1831–1838 (2018)

18. Kala, P.C., Agrawal, A.P., Sharma, R.R.: A novel approach for isolation of sinkhole attack in wireless sensor networks. Int. Conf. Cloud Comput. Data Sci. Eng. (2020)

Autonomous Driving System Based on Deep Q-Learning: A Survey of Attacks and Defenses

Ritu Gupta[✉], Nishtha Arora, Kangan, Reeya Ottalwar, Kriti Upadhyay, and Lithiga Jayaprakash

Chandigarh University, Gharuan, Punjab, India
erritugupta02@gmail.com

Abstract. Numerous learning-based motion planning techniques have been put forth in the literature for autonomous driving. These techniques can directly predict motion commands from the sensory data of the environment, but they are unable to predict multiple motion commands, such as steering angle, accelerator, and brake, or balance errors between various motion commands. This study examines the simulation outcomes of an autonomous vehicle learning to operate in a streamlined environment with just static impediments and lane lines.

This review offers a thorough examination of several threats that could endanger autonomous driving systems (ADS), as well as the associated cutting-edge defense techniques. The research begins by providing a comprehensive review of each stage of the ADS workflow, covering adversarial assaults for various deep learning models and attacks in both physical and virtual environments. These assaults inevitably pose a serious threat to the safety and security of deep learning-based autonomous driving, from which the remedies should be thoroughly researched and investigated to reduce any potential hazards. This review offers a detailed examination of several threats that might endanger ADSs, as well as the associated cutting-edge protection techniques. The adversarial assaults, which restrict the applications' performance, might target certain tensor perturbations in machine learning models. Implementing defensive models against adversarial assaults is thus a crucial research subject nowadays. For the purpose of enhancing the safety of deep learning-based autonomous driving, certain intriguing research avenues are also recommended. The various defense schemes have also been illustrated in this paper.

Keywords: Autonomous Driving System · Cyber Attacks · Defense · Cloud Services

1 Introduction

Independent driving has attracted a ton of consideration in both the scholarly world and industry because of the progression of computerized reasoning innovations. Quite possibly the earliest independent driving task, the Aha PROMETHEUS Venture (Program for a European Traffic of Most Noteworthy Effectiveness and Phenomenal Wellbeing) [1], was done by Daimler-Benz from 1987 to 1995. The Defense Advance Research Project

A. Gupta et al. (Eds.): ICAIA 2023, CCIS 2308, pp. 37–48, 2025.
https://doi.org/10.1007/978-3-031-84394-5_4

Agency (DARPA) [2] laid out prestigious independent driving contest, in 2005. Various turns of events and enhancements to cutting-edge independent driving frameworks (ADSs) have been proposed from that point forward.

Various methods, including AI-based picture handling strategies and sensor information combination procedures, have been utilized in the dynamic part. Handling the sensor information from an independent vehicle in a near ongoing utilizing any of these methods is generally troublesome [3]. Skillful human driving is instinctual and successful, as a matter of fact. Most recently, there are many opportunities to explore driving conduct thanks to the improvement of sensors and brilliant gadget innovation.

Most of organizations, including Tesla [4], focus on making level 3 ADSs, which are fit for restricted self-driving in specific situations (like on roadways). The main competitor Google Waymo [5] is given to explore and industrialize on Level 4 ADSs that, by and large, do not need human contact. However, self-driving vehicle research is still in its beginning phases. Prior to pushing ahead with industrialization at its fullest, a few significant issues, especially those relating to somewhere safe, should be successfully tended to. For instance, the latest Uber vehicles need of focusing on research on the security of independent driving as shown by a deadly mishap [6]. The most well-liked artificial intelligence method, deep learning, is frequently used in self-driving cars to carry out various perception jobs and make instantaneous selections. In a nutshell, raw data from various sensors and HD map data from the cloud are fed into deep learning models in the perception layer first to extract ambient information from the environment, and then various designated deep/reinforcement learning models in the decision layer start the real-time decision-making process. The ongoing development of deep learning-based ADSs using this pipeline topology is, however, hindered by several difficulties.

First, sensors are susceptible to a variety of physical attacks, which renders the majority of them incapable of collecting data as usual. They may get negative instructions to gather false data, which would severely harm the performance of all learning-based models in the subsequent layers. Deep neural networks are also susceptible to adversarial assaults, which are intended to cause learning-based models to make incorrect predictions, according to new study. The most typical adversarial assault involves creating so-called adversarial samples that differ only slightly from the original inputs in order to confuse the model's neurons. The degree of relevance of these threats to the security of deep learning-based ADSs is demonstrated by certain results from earlier research publications that focus on analyzing such adversarial assaults.

The present status of independent driving, in any case, is over the top expensive because of dependence on framework and innovation requires a lot of information, like Light Discovery and Running (LIDAR) for the route, Worldwide Situating Framework (GPS) for restriction, and Laser Reach Locater (LRF) for obstruction identification [1].

The Profound Q Organization (DQN) is a kind of support learning in which the Q-values (rewards) are given to activities because the info states, as opposed to classification, are the result of a CNN. Utilizing start-to-finish support learning, the DQN specialist gains successful arrangements straightforwardly from high-layered tactile data sources. DQN has as of late shown outcomes in the troublesome field of one-of-a-kind Atari 2600 games [7]. With enormous amounts of named data, any independent driving framework should be prepared to utilize genuine human driving. Figure 1 portrays the recommended framework. Information on driving ways of behaving is created by one or the other human or independent drivers. The data contains both ecological and driving-related information. The vehicle's current circumstance is portrayed by the climate information. Moreover, a future climate could be utilized to expect future activity.

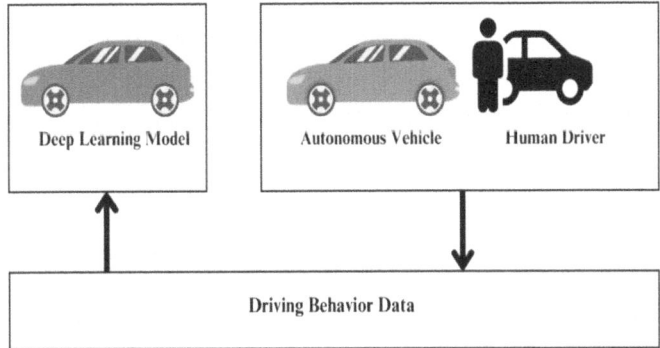

Fig. 1. A Recommended Framework

The industry requires this data and additional insights to further the development of ADSs' robustness and safety. Although several academics have researched the safety risks and defenses of autonomous cars and autonomous vehicular networks. However, none of these looked at security issues with ADSs based on deep learning. Contrarily, the majority of safety deep learning researchers concentrate on adversarial assaults on the picture classification job. This research did not, however, address relevant study on deep learning system vulnerabilities and countermeasures for more challenging autonomous driving tasks.

The process of deep learning-based ADSs, the most current attacks, and the associated defense tactics are all covered in-depth in this article as a result of the latest research efforts. The following is a list of this paper's contributions:

- A range of assaults on the deep learning-based ADS pipeline are examined and thoroughly studied.
- Deep learning-based ADSs' cutting-edge assaults and defense strategies are thoroughly explained.
- New attack applications as well as improvements to the security and robustness of deep learning-based ADSs are presented as future research objectives.

The rest of the paper is organized as follows: segment 2 contains important works, Segment 3 talks about issue definitions, Segment 4 examines the recommended system for learning the independent driving way of behaving, segment 5 finishes up.

2 Workflow of Deep Learning- Based ADSs

As shown in Fig. 2, a deep learning-based promotion typically consists of three beneficial layers: a detecting/sensing layer, a discernment/precipitation layer, and a choice/decision layer, as well as an additional cloud administration layer. Various heterogeneous sensors, including GPS, cameras, LiDAR, radar, and ultrasonic sensors are used in the detecting layer to collect ongoing environmental data, such as the current position and spatially-fleeting information (for example, time series picture outlines). The insight layer, on the other hand, includes deep learning models to examine the data obtained by the detecting layer and then separate important natural data from the raw data for further processing. The decision layer would act as a navigational aid to produce instructions regarding the speed differential as well as guiding points in light of the data that was isolated from the insight layer. The item that goes with this part will reveal how a profound learning-based promotion works.

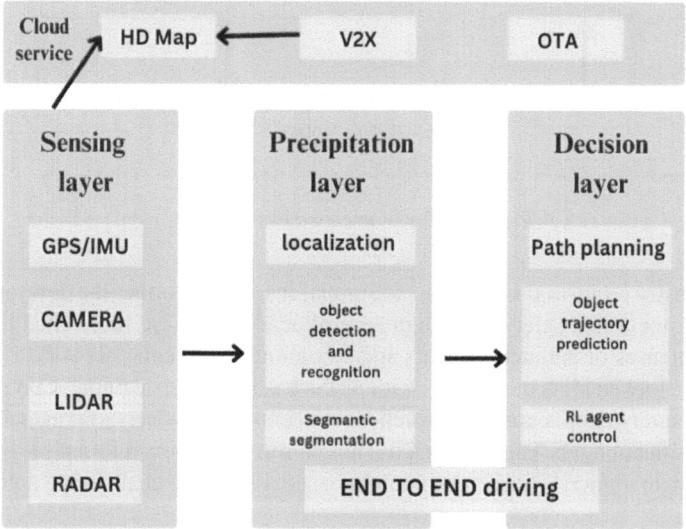

Fig. 2. Architecture of end-to-end driving

2.1 The Detecting/Sensing Layer

The detecting layer uses a variety of sensors to collect comprehensive information about a standalone vehicle. The cameras, light recognition and travelling (LiDAR), radio location and running (Radar), ultrasonic sensors, and Baidu GPS/Inertial Estimation Units (IMU) are the most commonly used sensors used by driving independent driving vehicle organizations. With the help of geostationary satellites, GPS may more explicitly provide precise position information, whilst IMU provides information on direction, speed, and speed increase.

2.2 The Discernment/Precipitation Layer

In the discernment layer, calculations such as optical stream [8] and profound learning models are used to separate semantic data from unstructured data. As of present, deep learning models in the discernment layer frequently use image data from cameras as well as cloud point data from LiDAR for various tasks such as confinement, object recognition, and semantic division.

Isolation. Limitation plays a fundamental role in the work of organizing the route in an advertisement. The independent car can understand the real-time weather and locate its precise location on the map by using confinement developments. At this time, the information from the combined GPS, IMU, and LiDAR point systems serves as the main source of constraint. The combined data is used to analyze odometry and direct recreation assignments. These tasks are designed to evaluate the progress of a standalone vehicle, replicate the environmental element's instruction manual, and ultimately determine the future location of the vehicle. In [9], CNN and RNN were also used to analyze the development and positions of a vehicle using regular images captured by a camera. A sophisticated autoencoder was used in [10] to encode observed images into a condensed configuration for map restoration and limitation.

Street Object Identification and Acknowledgment. The difficulty of accurately recognizing numerous items with different shapes, such as paths, traffic signs, and different vehicles, in addition to people on foot accurately in continuously and steadily changing general conditions, makes street object identification a major point of contention for independent vehicles. Quicker RCNN [11] is regarded as persuasive in the field of item identification to identify items in images.

Another well-known item identification formula, You Just Look Once (Consequences be damned) [12], transforms the discovery task into a relapse problem. The two experts and other business people are currently focusing mostly on LiDAR-based object location profound learning models. The main end-to-end model that predicts objects in view directly is Voxel Net [13]. Point A dominant presentation is achieved by RCNN [14], which modifies the design of RCNN to accept 3D point clouds as a contribution to object recognition.

Semantic Division. In independent driving, semantic division explicitly categorizes different elements of a scene into classes like automobiles, people on foot, and ground. It helps with vehicle restriction, object recognition, path verification, and guide revision. Completely Convolutional Organization (FCN) [15] is a foundational, highly effective deep-learning model for semantic segmentation. It essentially converts the fully associated layer of a conventional CNN to a convolutional layer. Another well-known semantic division organization, PSPNet, employs a Pyramid pooling engineering to concentrate data from images [16].

2.3 The Cloud Service

In the independent driving industry, the cloud server is typically used as a specialized organization for various asset-dependent services. First, independent driving organizations use LiDAR and other sensors to produce an early HD Guide that might be transmitted to the cloud. The HD Guide is filled with vital information on obstacles, signage, and street routes.

As a result, the vehicle may use this information to begin reroute planning and enhance its perception of the surrounding weather. In the interim, Vehicle to Everything

(V2X) administration could transfer ongoing raw data and perception information of other independent vehicles to the cloud to help keep HD Guides up to date, enabling HD Guides to provide more significant constant data, such as encompassing vehicles on a map the same street.

2.4 The Choice/Decision Layer

1) **Way Planning and Article Direction Forecasting.** Planning a route between a starting point and an endpoint is regarded as a vital task for independent vehicles. In order to achieve the goal and estimate the article's trajectory, it is necessary for independent vehicles to predict the locations of obstacles they have already encountered. Recently, several experts have tried to use converse support learning to achieve a dominating outcome in route planning. The vehicle is prepared to be equipped for generating a path that is more like a person by getting reward capabilities from human drivers [17]. A few variations of RNN and LSTM [18] are presented for direction expectation to achieve high expectation exactness and productivity. Additionally, a single CNN and 3D spatial-transient information is attempted to predict automobile trajectory using CNN [19].

2) **Vehicle Control Through In-depth Support Learning.** Conventional rule-based computations cannot simply account for all complex driving circumstances. Therefore, in situations involving independent driving, deep support learning that equips a specialist to determine the proper behavior under various circumstances is more promising. CNN-based Reverse Support Learning model has been suggested to construct a driving route using 2D and 3D data collected in a variety of realistic driving scenarios [20].

3) From Beginning to End Driving: An E2E driving model is a very profound learning model that combines insight and decision-making processes. In this case, based on the nearby detecting data, the model forecasts the current directing point and driving speed. Front-facing camera images are used as the input. For the CNN design E2E driving model DAVE-2 framework, which forecasts the ongoing controlling point [21].

3 Attacks in ADSs

We thoroughly describe several attacks on ADSs in this section. Figure 3 shows the general layout of the promotions' attacks on each component, which will be covered in-depth in this segment [22, 23].

3.1 Physical Attacks on Sensors

Enemies typically consider the detecting layer, which is typically viewed as the wilderness layer of an Advertisement, as an attack focus. Attackers anticipate degrading the sensor's nature. Sensors to obtain false information by faking information signals, adding noise signals, or both. Poor quality or even fake information would affect how profound learning models were presented in the discernment layer and the decision layer, as well as how an independent vehicle behaved.

In this threat model, adversaries are anticipated to have some knowledge of the tools and sensors used on an independent vehicle, but they are not required to understand the complexities of complex learning models at various tiers. Physical attacks on the detecting layer should have been prevented in this fashion Black-box attacks against the very educational-based Advertisements have been discernible. For genuine attacks on sensors, attackers may alter the data collected by the sensors or provide indications to fool the sensors with the use of external equipment. In this situation, sticking attacks and ridiculing assaults are the two most common real assaults.

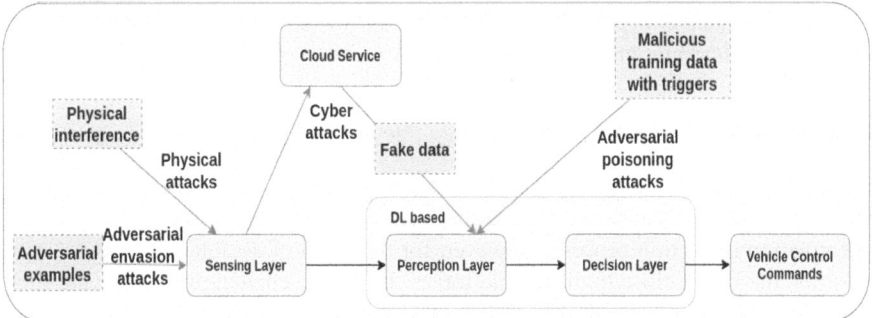

Fig. 3. The layout of the various attacks on each component

3.2 Cyberattacks on Cloud Services

Due to the continuous correspondence between the cloud and independent vehicles, the cloud may be the target of some attacks from the perspective of the enemy, making independent vehicles shaky.

Keep in mind that an HD Guide may be gradually updated with information from various cars using V2X. Attackers may be able to stop this cycle. Attacks on Sybil and attacks on message adulteration are two examples of how the planned path is meant to be interfered with. In fact, Sybil attacks target the continual HD map updating in V2X, creating many "counterfeit drivers" with phony GPS data in the target area framework. These attacks aim to trick the system out of the bottleneck while also interfering with the vehicle's limits and route allocations. They also modify the traffic data that is sent from the vehicle to the HD map server and attack several automobiles while the HD map data is refreshed through this server in order to carry out the message distortion assaults.

The V2X organization, in which autonomous vehicles are coupled with trade data, is being undermined by customary cloud attacks. Refusal of Administration (DoS) and Dispersed DoS (DDoS) both have the potential to wear out administrative resources, resulting in significant inactivity or even the inaccessibility of the V2X organization. Due to the current situation, autonomous vehicles' security is seriously jeopardized since they won't be able to connect with the HD map for precise route and discernment administration.

3.3   Adversarial Attacks on Deep Learning Models in Perception and Decision Layers

Late research demonstrates that maladaptive models that add nebulous commotions to specific information images are particularly helpless versus profound learning models. Even though negative models appear to humans to be like ordinary pictures, they may trick deep learning models into having unrealistic expectations. An attack with ill will is, by definition, an attack to create such antagonistic models. Due to the extensive use of deep learning models in both the discernment layer and the choice layer, antagonistic assaults pose noteworthy risks to ADSs. In this section, we first define hostile attacks and discuss a few related concepts. The written audit of the development of hostile attacks on various deep learning models in ADSs is then summarized.

3.4   Analysis of Attacks

1. Physical attacks are straightforward but limited in a certain range. Attacks on sensors themselves might disturb the foundational learning models by interfering with the information collection process. However, for this kind of assault, it is necessary to position the target close to the hostile forces. It may be challenging to execute such an assault, for example, if the laser light is pointed directly at the target vehicles. This may result in a camera-blinding attack.
2. Cyberattacks do harm while being tested. Cyberattacks on the cloud might seriously destroy many V2X-connected autonomous cars. For cyberattacks on the cloud, however, adversaries must fabricate data transfer between the cloud and the vehicle or else conduct DDoS attacks using a sizable Botnet. However, both attacks might be prevented by information transmission cycle encryption, and the cloud might send discovery frameworks like [24] to limit DDoS attacks.
3. Adversarial assaults are efficient and harmful in the actual world. Due to the existence of antagonistic bothers in the black-box setting, antagonistic assaults, in particular avoidance assaults, would pose serious risks to profound learning models in ADSs. In a simulation environment or in the real world, black-box evasion assaults have been used to target E2E driving models or object detectors in the perception layer of ADSs from a range of angles, distances, and lighting conditions. Adversaries might create harmful stickers for this type of assault in whatever way they choose and covertly place them wherever. Adversarial poisoning assaults might be hazardous and covert in a scenario where corporate espionage can taint training data. The outcome is, it is necessary to offer a review of the most recent studies on defenses against adversarial attacks. Additional research may be based on the possibility that more powerful attacks exist that can destroy autonomous vehicles.

4   Defense Methods

In this part, we focus on a few of the current defenses against adversarial and physical attacks. We also briefly consider cloud administration security measures.

A. Protection from Physical Sensor Assaults
Out of all the physical sensor attack countermeasures, redundancy offers the best chance of protecting against jamming attacks [22. 25, 26]. Redundancy is the employment of numerous identical sensors to gather a certain type of data and mix it with other data to create the final input for the perception layer.

For instance, even if an attacker blinds one camera, others may continue to gather regular photos in order to decipher the situation. This tactic unquestionably results in higher costs. Furthermore, sensor data fusion is frequently seen as a challenging research issue. Using a near-infrared-cut filter during the day to minimize near-infrared light and enhance the quality of photos taken is another method for making cameras more durable [27, 28]. Yet, this approach is ineffective after dark. Alternately, improving cameras may also entail using photochromic lenses to block out a specific wavelength of light. This is how sticking Attacks on these cameras could be reduced. Due to the rarity of disturbances in a typical working environment, ultrasonic sensors and radars can easily distinguish approaching sticking assaults by building an identification system.

Assailants may, for example, target LiDAR with precision attacks because LiDAR may receive signals during a respectable test window. As a consequence, by merging information from radars, ultrasonic sensors, LiDAR, cameras, and other sensors, the performance of the perception layer might be stabilized.

The existing sensor attacks are subject to a few obvious limitations. For example, many assaults call for the use of extra gear to produce sounds and false signals close to the vehicle being attacked. A human may see attacks like the camera-blinding attack from the front of the car and take action to stop an accident. Therefore, even if the development of autonomous vehicles reaches a highly computerized level, installing a security system remains essential for added assurance [29, 30].

B. Defence for Cloud Services
To combat the weaknesses of GPS, there are certain tactics. The system should recognize a spurious signal in the satellite as a possible assault. A signal strength detector, clock data, and time intervals can all be used to find the fake signal. In order to avoid spoofing signals, antennas might also be employed to identify the direction of arrival of the signal. Another tactic is to employ a cryptographic method that analyses an encrypted GPS code to detect the presence of spoofing signals. The received signal's authenticity might likewise be ascertained using an authentication approach [31, 32].

By using defense strategies like managing how it receives and emits signals, the lidar system can fend off threats. There are also in-vehicle vulnerabilities associated with access control systems, including keyless entry systems, voice-controlled systems, and vehicle immobilizers.

C. Defense Against Adversarial Evasion Attacks
Current ideas for defenses against adversarial evasion assaults are many. We examined and categorize the numerous existing defenses in use in this survey. Autonomous vehicles

(AVs) have advanced significantly in recent years thanks to the continuous development and advancement of deep learning technologies. Despite having enormous potential, deep learning-based AV currently faces significant security risks that prevent it from being used on a big basis.

On the other hand, the EAPSO (evasion attack with particle swarm optimization) focuses on the interference process of the deep learning algorithms, where the attacker inserts some barely noticeable disturbances to the intended test sample, causing a mis-classification on it. Numerous tests are performed in order to determine how effective the defense strategies are that explored by Jiang et al. [33].

The five primary categories of proactive defense techniques are adversarial training, network distillation, network regularization, model ensemble, and certified defense. The two main reactive defenses are adversarial transformation and adversarial detection.

D. Defence Against Adversarial Poisoning Attacks
Numerous defenses against poisoning attacks have been put forth in recent studies. The fundamental concept is to simply determine whether the image being entered at the moment is a hijacked image with triggers. Another fundamental principle is to identify the damaging attack in the model and then remove the Trojan or secondary channel. Both ideas have a place with adaptable hostile recognition defenses.

To be more precise, we first take advantage of the PAPSO (poisoning attack with particle swarm optimization), which targets the deep learning algorithms that train the traffic sign recognition system. In this attack, the attacker inserts crafted samples into the training dataset, lowering the classification accuracy of the system. In a poisoning attack, which takes place during the deep learning training process, the attacker injects harmful samples into the training dataset, lowering the predictive accuracy of the learnt model. For instance, certain well produced harmful samples may fool handwritten digit identification. PDF malware detection and recommendation algorithms. The PAPSO was also utilized by the author in [33].

This study examines First and foremost, different triggers were created to chase after each mark, and at that point, loads of neurons activated by the recognized trigger were removed to render the trigger useless. The experiment's findings revealed that the success rates of some poisoning assaults can even decline from above 90% to 0% with this technique.

5   Conclusion

The only way to get a more intelligent self-driving system is through ADS, which is based on deep learning. However, the system is vulnerable to numerous attacks. In this study, the workflow of the deep learning-based ADS is examined for safe threatening assaults, such as adversarial, physical, and online attacks. Although the physical assault is simple, it reveals some flaws that defense tactics could successfully address. Large-scale cyberattacks are regarded to be challenging to launch, but system defense techniques are straightforward to put into practice. Traditional defenses don't work well for self-driving cars, thus we need additional strategies to counter the aggressive onslaught. Future research should look into malicious attacks on Li-DAR and strong support models as well as identifying attacks as potential attacks. Model to improve the ADS's robustness, in-depth research should be done on adversarial threats detection in real-time, model testing and verification, and robustness training.

References

1. Eureka. Programme for a European traffic system with highest efficiency and unprecedented safety. https://www.eurekanetwork.org/. Accessed 1 Dec. 220
2. Buehler, M., Iagnemma, K., Singh, S.: The 2005 DARPA grand challenge: the great robot race. Springer (2007)
3. Hallac, D., Sharang, A., Stahlmann, R., Lamprecht, A., Huber, M., Roehder, M.: Driver Identification Using Automobile Sensor Data from a Single Turn. Volkswagen Electronics Research Laboratory
4. Tesla. Telsa autopilot, https://www.tesla.com/autopilot. Accessed 30 Sept 2019
5. Waymo. Waymo llc, https://waymo.com/. Accessed 30 Sept 2019
6. Berboucha, M.: Uber self-driving car crash: what really happened. https://bit.ly/2YKu9WN. Accessed 30 Sep. 2019
7. Tramer, A., Kurakin, N., Papernot, I.J., Goodfellow, D., Boneh, McDaniel, P.D.: Ensemble adversarial training: attacks and defences. In: Proceedings ICLR, Vancouver, BC, Canada (2018)
8. Agarwal, A., Gupta, S., Singh, D.K.: Review of optical flow technique for moving object detection. In: Procedings IC3I, Noida, India, pp. 409–413 (2016)
9. Wang, S., Clark, R., Wen, H., Trigoni, N.: DeepVO: towards end-to end visual odometry with deep recurrent convolutional neural networks (2017)
10. Bloesch, M., Czarnowski, J., Clark, R., Leutenegger, S., Davison, A.J.: CodeSLAM-learning a compact, optimisable representation for dense visual SLAM. In: Proceedings CVPR, Salt Lake City, UT, USA, pp. 2560–2568 (2018)
11. Girshick, R.B.: Fast R-CNN. In: Proceedings ICCV, Santiago, Chile, pp. 1440–1448 (2015).
12. Redmon, J., Divvala, S.K., Girshick, R.B., Farhadi, A.: You only look once: unified, real-time object detection. In: Proceedings of CVPR, Las Vegas, NV, USA. pp. 779–788 (2016)
13. Zhou, Y., Tuzel, O.: Voxelnet: end-to-end learning for point cloud based 3D object detection. In: Proceedingds of CVPR, Salt Lake City, UT, USA (2018)
14. Shi, S., Wang, X., Li, H.: Pointrcnn: 3D object proposal generation and detection from point cloud. In: IEEE conference on computer vision and pattern recognition (CVPR) (2019)
15. Zhao, Z.-Q., Zheng, P., Xu, S.-T., Wu, X.: Object detection with deep learning: a review. Accepted by IEEE Transactions on Neural Networks and Learning Systems (2019)
16. Chen, J., et al.: CSPP-IQA: a multi-scale spatial pyramid pooling-based approach for blind image quality assessment. Neural Comput. Appl. (2022)
17. Gupta, A., Anpalagan, A., Guan, L., Khwaja, A.S.: Deep learning for object detection and scene perception in self-driving cars: survey, challenges, and open issues. Array **10** (2021)
18. Siami-Namini, S., Tavakoli, N., Namin, A.S.: The Performance of LSTM and BiLSTM in Forecasting Time Series. IEEE Int. Conf. Big Data (2019)
19. Xie, G., Shangguan, A., Fei, R., Ji, W.: Motion trajectory prediction based on a CNN-LSTM sequential model. Sci. China Inf. Sci. **63**(11), 25 (2020). https://doi.org/10.1007/s11432-019-2761-y
20. Feng, L., Li, Q., Peng, Z., Tan, S., Zhou, B., Zurich, E.T.H.: Traffic gen: learning to generate diverse and realistic traffic scenarios (2023)
21. Islam, M., Chowdhury, M., Li, H., Hu, H.: Vision-based navigation of autonomous vehicles in roadway environments with unexpected hazards article (2019)

22. Xing, K., Sundhar, S., Srinivasan, R., Rivera, M., Li, J., Cheng, X.: Attacks and countermeasures in sensor networks: a survey. In: Network security, pp. 251–272 (2010)
23. Yussoff, Y.M., Hashim, H., Rosli, R., Baba, M.D.: A review of physical attacks and trusted platforms. Wirel. Sensor Netw. Procedia Eng. **41**, 580–587 (2012)
24. Chen, Y.-W., Sheu, J.-P., Kuo, Y.-C., Van Cuong, N.,: Design and implementation of IoT DDoS attacks detection system based on machine learning. In: European conference on networks and communications (EuCNC): vertical applications and Internet of Things (VAP), pp 122–127 (2020)
25. W. Xu, W. Trappe, Y. Zhang.: The Feasibility of Launching and Detecting jamming attacks in Wireless Networks. In proceedings of MobiHoc, (2005)
26. Jilani S.A., Koner C., Nandi S.: Security in wireless sensor networks: attacks and evasion. National conference on emerging trends on sustainable technology and engineering applications (NCETSTEA), pp. 1–5 (2020)
27. Süsstrunk, S., Fredembach, C.: Enhancing the visible with the invisible: exploiting near-infrared to advance computational photography and computer vision. In SID Symp. Dig. Tech. Pap. **41**, 90–93 (2010)
28. Park, Y., Jeon, B.: An acquisition method for visible and near infrared images from single CMYG color filter array-based sensor (2020)
29. Meshcheryakov, R., et al.: A probabilistic approach to estimating allowed SNR values for automotive LiDARs in "Smart Cities" under various external influences
30. Cao, Y., Xiao, C., Yang, D., Fang, J., Yang, R., Liu, M., Li, B.: Adversarial objects against LiDAR-based autonomous driving systems
31. https://www.avertium.com/blog/cloud-security-defense-in-depth (2020)
32. Giordani J.: Cyberattacks on vehicles pose a threat to drivers and manufacturers Forbes technology council (2021)
33. Jiang, W., Li, H., Liu, S., Luo, X., Lu, R.: Poisoning and evasion attacks against deep learning algorithms in autonomous vehicles. IEEE Trans. Veh. Technol.Veh. Technol. **69**(4), 4439–4449 (2020)

Machine Learning-Based Intelligent Approach for Energy Efficient Transmission in Wireless Communication: A Review

Jyoti Saini$^{(\boxtimes)}$ and Ramesh Kait

Department of Computer Science and Applications, Kurukshetra University, Kurukshetra, India
{jyoti.dcsa,rameshkait}@kuk.ac.in

Abstract. The most popular technologies nowadays are wireless sensor networks, which have advantages like low cost, small size, and mobility. But the networks like Wireless Sensor Network (WSNs) are resource constrained with respect to energy utilization that comprises tiny solitary sensor nodes with limited bandwidth. Furthermore, some of the other main issues to pay attention are: accumulating sensed information from the network, transferring the data to the base station while focusing on network coverage, lifetime, and power conservation. More researchers have recently become interested in using machine learning techniques in wireless networks to solve these problems that arise in the specified network. This paper reviews the existing approaches of machine learning that have already proposed. Furthermore, detailed analysis of exiting work also discussed in the form of table that serve as a reference for anyone interested in learning more about designing suitable Machine Learning solutions for the wireless sensor netwok's applications.

Keywords: Machine learning Energy efficient · Routing algorithms · Wireless sensor networks

1 Introduction

The unconventional discoveries in the disciplines of the creation and widespread adoption of reliable and multifunctional sensor nodes that are compact and affordable have been fueled by micro-electromechanical systems, digital electronics, and wireless communication. Sensor nodes have the capacity to sense, examine, and transmit unaltered data over short distances. This feature eventually results in widespread collaboration between these sensor nodes, resulting in wireless sensor networks (WSNs).

The networks similar to WSNs are resource constrained where the energy consumption is a significant issue. It is especially well-suited to applications on the battlefield, in hazardous chemical factories, and in high-temperature settings, where installing traditional network infrastructure is difficult or impossible. Sensor-based applications are used in most vital surveillance and security applications, which is not unique. Small and low-cost sensors can be utilized to save energy in various applications. Because of the batteries or electricity power of the sensor nodes, the energy-efficient multi-Heuristic routing model is created, including metaheuristic optimization methodologies to reduce WSN routing energy consumption. These methods make use of separate energy properties to accelerate convergence.

1.1 Wireless Sensor Network

A sensor node detects and responds to physical and environmental input such as pressure, heat, light, etc. The sensor's output is typically an electrical signal that is transferred to a controller for additional processing. Various smart sensor nodes create a network where these nodes collect the data to generate relevant decisions, and such a network is called a WSN. In these years, the practical intent of a WSN has become a prominent field of study. It is an interconnected set of specialized sensors that are dispersed in space and are used to track environmental variables and send data to a central point. Various sensors can be used in WSNs, depending on the application. WSNs can be structured or unstructured, and are frequently devoid of infrastructure. The sensed facts are communicated to the BS via radio and then broadcast to the user. Batteries are used to power sensor nodes. Sometimes, energy-collecting techniques are employed for additional reasons [1].

Monitoring for radiation and nuclear threats, ship-mounted weapon sensors, poison detection, and source tracing in public gathering places, earthquake detection, seismic monitoring, etc are the few applications where WSN is used.

Sensor Network Communication Architecture

A WSN is made up of sensor nodes that receive and communicate data about the monitored environment to a base station(BS) or sink via wireless networks. The data is sent to the BS/sink over a single or several hops. Sensor nodes might be mobile or fixed and can be heterogeneous or homogenous. The sensor node communicates with other sensor nodes as well as a BS. The work of the BS is to provide commands to the sensor nodes so they can work together to complete the task. After completing the task, these sensor nodes are linked to the other network via the internet, and they transmit data back to the BS. In the end, after receiving the information, the BS performs minimal data processing before sending the updated information to the end user via the internet [27].

The OSI model is most commonly used in WSN design. The physical, data connection, network, transport, and application layers are the five core layers of the WSN architecture. Aside from these layers, three cross-layer planes coordinate the complete sensor node network and monitor its overall efficiency [8]. Those are the power management plane, task management plane, and mobility management plane. The task of coordinating energy use falls under the purview of the power management plane. The sensing procedure is scheduled by the task management plane, while the mobility management plane keeps node mobility. These cross-layer optimizations aid in power and resource conservation. The WSN OSI architecture, also known as layered network architecture, uses less power than other sensor network architectures since each node only transmits over short distances and with low power to its neighbors' nodes. It is more fault-tolerant and scalable.

Energy Consumption

The sensor nodes come in a variety of shapes and sizes and have a variety of purposes. A

CPU, a radio transceiver, a power source, and a sensing device are the main components of a wireless sensor node. [20]. A communication unit has a radio system that transmits and receives the data. As all the components embedded in this network are low-powered, they are all controlled by a single battery CR-2032 [9]. A Sensor Node, despite its name, contains not only the sensing component but also performs other critical functions such as processing, communication, and storage. A Sensor Node collects all the data from the physical world, performs an analysis of the network, correlates the data, and fuses the data from other sensors with its data [28]. With all of these characteristics, components, and advancements, The WSN's energy usage is determined by how much energy each component uses for each of its numerous tasks. The sensor module needs energy for sensing the signals, sampling the signals and after that for modulation/ demodulation process. When the CPU module is running, sleeping, or idle, it uses energy. The primary duties of this module are to regulate the sensing process, communicate, and process data. Many factors influence the energy consumption of wireless communication modules, including the hardware, operation frequency, and so on. The node's manufacturer and model determine the capacity of the power supply module.

Routing Protocols
In wireless networks, the routing techniques could be either static or dynamic. They are categorized using a variety of factors, including topology used, time and position, communication model, routing structure, routing choice, and adaptive nature.

Based on routing information from source to destination, the routing protocols are divided into three categories: proactive, reactive, and hybrid. Routes are computed in advance but produced on demand in proactive procedures. Hybrid procedures use a combination of both proactive and reactive approaches. Another type of routing protocol is cooperative routing, whereby nodes transmit data to a central node. The processing and aggregation of data are done at the central node, which reduces the cost of the energy route.

Based on the temporal information, the routing protocols are categorized as path selection using past temporal information and path selection using future temporal information. When choosing a path for a protocol, past temporal information is used to make judgments regarding the links' past state, current status, or status at the time of routing. In contrast, the protocol path selection using future temporal information makes approximation routing decisions based on knowledge about the anticipated future status of the wireless networks.

Based on the routing topology, the protocols can be separated into flat topology routing protocol and hierarchical routing protocol. The flat routing protocols operate on the assumption that each node in a wireless network has a globally unique addressing scheme. Protocols that use hierarchical routing make advantage of the network's logical structure and a related addressing scheme. The hierarchy might be determined by geographic data or hop distance. Based on utilizing specific resources, various classifications are power-aware routing, routing using geographical information, and routing with efficient flooding.

The term "adaptive routing" refers to routing protocols whose many characteristics can be altered in response to environmental factors and energy availability. These protocols are further classified according to the operation using coherence, multipath, query,

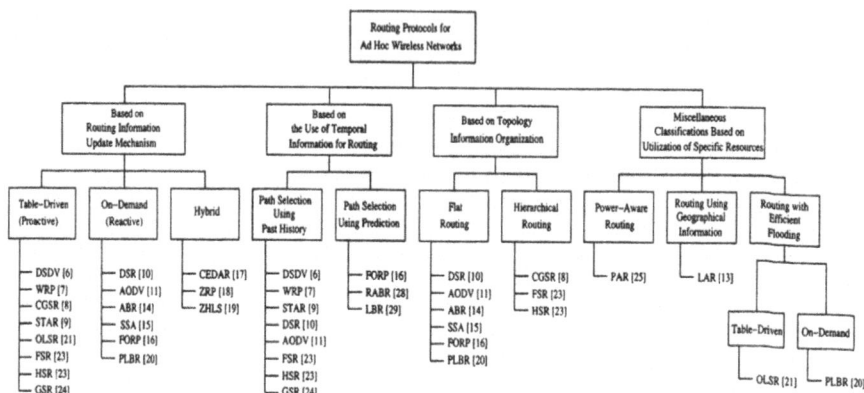

Fig. 1. Routing Protocols

negotiation, and QoS routing techniques. Figure 1 illustrates an outline of the several routing protocols utilized in WSNs.

The Communication Model indicates how packets should be routed in the network in a routing protocol. These protocols for broadcast and point-to-point paradigms are distinguished by their high data transfer rate for a given quantity of energy. The data delivery ratio, however, is quite low and cannot be guaranteed. There are several distinct kinds of protocols in this architecture, including query-based protocols, negotiation-based protocols, and coherent/non-coherent based protocols [4].

Need for Energy Efficiency Routing
Over many years, extensive research has explored the possibility of sensor collaboration in diverse tasks such as sensing, data gathering, and processing. Because WSNs have distinct characteristics, novel strategies for extending the network's lifetime are critical [18]. There are many limitations on strategy and administration due to the enormous number of sensor nodes and the environmental restrictions imposed by the deployed location dependent on the application.

The WSN protocol stack has multiple layers, each of which uses energy. One of them is the network layer since it performs routing, which is the activity that uses the greatest energy [6]. The communication process consumes more energy than sensing and processing combined. If any sensor node loses power, the network will lose connectivity, rendering the deployment useless. To keep and extend the network's life, it is necessary to use the available resources efficiently. Using an energy-efficient routing approach can save significant energy, extending the lifetime of WSNs [4].

Since these protocols perform excellently by developing load balancing mechanisms and satisfying specified QoS criteria, schemes using the reliable routing protocol are more resilient to rout catastrophes. The inability to maintain routing tables at sensor nodes, however, is a drawback. These protocols include the multipath-based protocol and the quality-of-service protocol. [14].

Energy Efficient Routing using AI techniques

WSNs and Artificial Intelligence (AI) are essential new computer science technologies. It is well known that the energy is a valuable resource in WSNs and that it should be used intelligently to extend their lives. Uneven load distribution across sensor devices is also a cause of energy depletion, which can cause network operations to be disrupted [7]. As a result, artificial intelligence plays a critical role in addressing the energy challenges of WSNs by reducing sensor node energy consumption. A smart wireless sensor is another name for the artificial intelligence system used by WSNs. It modifies its internal behavior to enhance its capacity to collect data from the physical world and swiftly transmit it to a BS or a host system [9]. Intelligent sensors have the capacities for self-validation, self-calibration, and compensation.

Machine learning is known for its self-experiencing nature and does not require reprogramming. ML is a successful strategy that delivers a computing process that is efficient, dependable, and cost-effective. The energy harvesting makes WSNs self-powered, long-lasting, and low-maintenance, even when deployed in severe environments [26]. In such cases, the ML approach increases network longevity, enhancing WSN's lifetime and making them acceptable for forecasting the amount of energy gathered over a specific period. Because there are fixed sensor nodes in the coverage region, ML approaches can also tackle coverage difficulties in the target area. There is a transmission overhead by transmitting all the data to the BS. By lowering the dimensionality of the data at the cluster head, ML methods can tackle this problem. ML can improve the efficiency of WSNs as well. They aid in separating active sensor nodes from dead or malfunctioning sensor nodes [24].

1.2   Machine Learning

Algorithms of machine learning(ML) are typically used to forecast or classify data. Supervised learning, unsupervised learning, and reinforcement learning are the three main categories.

Supervised Learning

Supervised learning entails developing a model for categorizing incoming data by generating predictions about predefined classes. As a result, the learning algorithm is provided with training data that contains data points and labels. In other words, the training system already knows the class or category labels associated with the data used for training, enabling the learning algorithm to estimate using the learned data. [20]. Generally, supervised learning is used to solve issues involving classification and regression. Data must be sorted into groups according to the stated categories to solve classification issues. These established categories may be used to teach the system, and as it learns, it will classify fresh incoming data in a similar manner. The challenge with supervised learning is that it relies on the accuracy of class predictions because the learning algorithm develops generalized rules to support the training dataset [2]. By providing a significant number of images, the accuracy of the trained model increases. However, over-fitting occurs as a result of focusing too much on the training dataset, which suggests that the trained model performs badly under real-world conditions.

Unsupervised Learning

Unsupervised learning does not need a labeled dataset, and the machine does not provide any training data at all, in contrast to supervised learning. Unstructured data is provided to the computer rather than a training dataset, and the job for the machine is to look for structures in the data. [24]. Frequently, unsupervised learning is utilized to discover patterns. It examines data similarities to cluster relevant data together. As a result, this technique is well-suited for clustering issues. Data should be sorted into groups based on their similarity to address clustering issues, with each group representing a cluster [18]. To decrease cluster similarities, data with values that differ from those in other groups should not be included in the same group and generate a new one. In terms of data processing, the machine examines new data and assigns it to a related data cluster. Unsupervised learning is used to recognize faces in a large dataset of pictures; the robots group the photos based on facial similarity. The machine aims to determine which photographs belong to the same person while also determining the number of distinct persons in the dataset.

Reinforcement Learning

In reinforcement learning, the computer determines the best suitable course of action in certain instances. The system must investigate cause-and-effect relationships and determine if these acts have immediate or long-term implications. One example of reinforcement learning is when an agent in the same environment gives a mouse a stimulating sound to motivate it to go and grab the food. Consequently, the machine learns via its actions, resulting in a gift or punishment based on prior attempts, resulting in favorable or bad outcomes.

2   Machine Learning Methods Applied in WSN Routing

Computer networks have developed rapidly thanks to the rapid expansion of machine learning applications, with a variety of formation methodologies and flexible, localised communication channels. The three categories of machine learning techniques are supervised, unsupervised, and reinforcement learning. In the literature that has already been written, all three of them have been used to create secure energy-efficient routing for WSNs.

To reduce energy consumption and extend the lifespan of the WSN, an energy-efficient routing based on fuzzy logic and neural networks has been proposed [10]. According to the author, to intelligently choose cluster heads (CH) that will optimally consume equal amounts of energy from the sensors, the program used fuzzy logic and neural network techniques. The DeepSense IoT-MANET technology was developed for efficient routing of packets from IoT nodes through mobile sensor nodes in MANETs[11]. According to the authors, DeepSense uses DeepSense's deep neural network (DNN) learning techniques to organize MANET routing.

Elephant Herding Optimization (EHO[12]., a bio-inspired technique, and the well-known Ad Hoc On-Demand Multipath Distance Vector (AOMDV) routing protocol were used to create a new, energy-efficient routing system. Some Artificial Neural Network (ANN) based techniques were also proposed for detecting [13] [18]. The findings are compared between the black hole AODV and the Secure AODV security mechanism (SAODV). To overcome issues, an ACO algorithm with two strategies for reducing network overhead by anticipating node motions was developed [14].

The suggested method, according to the authors, dynamically generated network heuristic parameters for choosing the best node for each cluster, enabling quicker cluster generation and head selection. On the basis of ACO, the Multi-Objective Constraints Applied Energy-Efficient Routing (MCER-ACO) protocol was created. [15]. Based on the quantity of packets on the path, node energy, and dynamic topology movement, the proposed protocol selected the next hop. IoT's significance in WSN was demonstrated by [16]. According to the authors, IoT devices are considered dynamic and are not performance-effective in an IoT-enabled system. As a result, the algorithms were improved, and fuzzy-based strategies were highlighted using an adaptable neural network that could adapt to a dynamic network.

Analysis of existing literatures based on ML methods is done in Table 1. Various machine learning techniques are discussed which are used to achieve the objective of the research. The merits/demerits of the research is discussed in the review column. From the above table we can analyze the great work has been done in the area of energy efficiency of sensor nodes preserving the energy utilized by the sensor network, too high packet delivery ratio for testing the speed of data transmission and high system latency are still big issues for researchers.

Table 1. The existing literatures based on ML methods

Author's Name	Objectives	Dataset	Techniques	Outcome	Remarks
Preeth et al. [19]	To create a low-energy cluster and an immune-inspired routing mechanism	Data collected from sensor nodes	Adaptive fuzzy multi criteria decision making	Packet delivery ratio = 99%	There is no optimal way for improving sensor lifetime that has a high residual energy
Mittal et al. [20]	To improve energy efficiency in wireless sensor network	NSLKDD dataset	Levenberg–Marquardt neural network (LMNN), Support vector machine	Accuracy = 96.15%	To improve accuracy, more advanced machine learning algorithms must be included
Kirubasri et al. [21]	To establish link quality prediction technique for predicting the link accuracy	Packet traces had been taken from Motelab	LQEuML	Accuracy = 95.8%	The results showed that using the data traffic approach, the link quality estimate was more trustworthy
Anandh et al [22]	To construct energy efficient path towards BS using optimization techniques	Sensor node data	RACO	Energy consumption 79.7%	The routing system could be expanded to accommodate uncontrolled mobility situations where regular disconnections can stifle network throughput
Townsend et al. [23]	To make it possible for nodes to cluster in a near-optimal arrangement for energy efficiency	Data collected from sensor nodes	Genetic algorithm	Compared to a typical genetic algorithm, it requires less time and energy	There was no consideration of transmission range variation. and only static nodes were evaluated

(continued)

Table 1. (*continued*)

Author's Name	Objectives	Dataset	Techniques	Outcome	Remarks
Gao et al. [24]	To analyze the performance of different learning models for different shapes of building	Data had been collected from centre for machine learning and intelligent systems	Random forest, Lazy k star, AMT	RAE = 4.76 RRSE = 6.52 RMSE = 0.62	High computational complexity
Kumar et al. [25]	To ensure the validity of the network by using adequate amount of energy	Data collected from sensors	Reinforcement learning	Packet delivery ratio = 89.87%	To boost the value of the packet delivery ratio, further work must be done
Yuldashev et al. [26]	To reduce the consumed energy in WSNs	Data collected from sensor	Decision Tree	Accuracy = 80%	The approach regulates all available external environment characteristics and only selects the bare minimum
Sridhar et al. [27]	To enable effective routing in wsn for optimizing data transmission	Number of route paths	Softmax-Regressed-Tanimoto-Reweight-Boost-Classification (SRTRBC)	Data delivery rate = 98%	A higher level of energy efficiency is necessary, as well as a lower rate of delay
Kumar et al. [28]	To curtail the energy efficiency issues in WSN	Data taken from 100 nodes	divide-and-rule sectorization (DRS)	Packet reception ratio = 100%	In comparison to existing procedures, the proposed scheme's energy consumption is significantly lowered
Thangaramyaet al.[17]	To form cluster for efficient energy routing of packets in wsn	Data collected from IoT based sensor nodes	Neuro fuzzy logics	Average network lifetime = 1096.27	Every node is assumed to be trustworthy

3   Conclusion and Future Scope

An overview of numerous machine learning approaches to design secured energy efficient routing strategies for WSNs is presented. To summarize, a great deal of effort has been done in the subject of energy efficiency by establishing various strategies, but there are still certain difficulties to be addressed in future research. Researchers had faced issue in preserving the energy utilized by the sensor network. Hence new methodologies should be incorporated that can improve energy efficiency and also minimize system latency during data transmission. From the above study it is concluded that some of the study had high packet delivery ratio, thus efforts should be done to improve the packet delivery ratio so the speed of data transmission can be tested, and static nodes should not be targeted throughout the experiment. The work had been conducted only by using few machine learning based techniques hence it is suggestible to explore using more machine learning models. From review, it is found that machine learning technique's properties are more suitable for optimizing the WSN. Researcher can have further survey to apply numerous machine learning approaches as well as the optimal path for improving the system's performance.

References

1. Zhao, F., Guibas, L.: Wireless Sensor Networks: An Information Processing Approach. Morgan Kaufman Publishers, Elsevier (2014)
2. Usha Kumari, C., Padma, T.: Energy-efficient routing protocols for wireless sensor networks. In: Wang, J., Mohana Reddy, G.R., Prasad, V.K., Sivakumar Reddy, V. (eds.) Soft Computing and Signal Processing: Proceedings of ICSCSP 2018, Volume 2, pp. 377–384. Springer Singapore, Singapore (2019). https://doi.org/10.1007/978-981-13-3393-4_39
3. Lepagnot, I., Siarry, P.: A survey on optimization metaheuristics. Inform. Sci. **237**(1), 82–117 (2013)
4. Kait, R., Chauhan, R.K., Kherwal, K.: Route mechanisms for wireless adhocnetworks:- classifications and comparison analysis. Int. J. Sci. Environ. Technol. **1**(2) (2012)
5. Norden, W.V., Jong, J., D., Bolderheij, F., Rothkrantz, L.: Intelligent task scheduling in sensor networks. In: International Conference on Information Fusion. IEEE (2005)
6. Jan, B., Farman, H., Javed, H., Montrucchio, B., Khan, M., Ali, S.: Energy efficient hierarchical clustering approaches in wireless sensor networks: a survey. Wireless Commun. Mobile Comput. **2017**, 1–14 (2017)
7. Chen, J.I.Z., Lai, K.L.: Machine learning based energy management at internet of things network nodes. J. Trends Comput. Sci. Smart Technol. **2020**(3), 127–133 (2020)
8. Ghosh, A., Ho, C.C., Bestak, R.: Secured energy-efficient routing in wireless sensor networks using machine learning algorithm: fundamentals and applications. In: Martin Sagayam, K., Bharat Bhushan, A., Andrushia, Diana, Victor, Hugo C., de Albuquerque, (eds.) Deep Learning Strategies for Security Enhancement in Wireless Sensor Networks:, pp. 23–41. IGI Global (2020). https://doi.org/10.4018/978-1-7998-5068-7.ch002
9. Khamayseh, Y.M., Mardini, W., Aldwairi, M., Mouftah, H.T.: On the optimality of route selection in grid wireless sensor networks: theory and applications. J. Wirel. Mob. Networks Ubiquitous Comput. Dependable Appl. **11**(2), 87–105 (2020)
10. Varun, R. K., Gangwar, R. C., Kaiwartya, O., Aggarwal, G.: Energy-efficient routing using fuzzy neural network in wireless sensor networks. Wireless Commun. Mobile Comput. 2021 (2021)

11. Chandrasekaran, S., Kannan, S., Subburathinam, K.: DeepSense—Deep neural network framework to improve the network lifetime of IoT-MANETs. Int. J. Commun Syst **34**(3), e4650 (2021)
12. Sarhan, S., Sarhan, S.: Elephant herding optimization Ad Hoc on-demand multipath distance vector routing protocol for MANET. IEEE Access **9**, 39489–39499 (2021)
13. Rani, P., Verma, S., Nguyen, G.N.: Mitigation of black hole and grayhole attack using swarm inspired algorithm with artificial neural network. IEEE Access **8**, 121755–121764 (2020)
14. Sathiamoorthy, J., Ramakrishnan, B.: Energy and delay efficient dynamic cluster formation using hybrid AGA with FACO in EAACK MANETs. Wireless Netw. **23**(2), 371–385 (2017)
15. Malar, A., Kowsigan, M., Krishnamoorthy, N., Karthick, S., Prabhu, E., Venkatachalam, K.: Multi constraints applied energy efficient routing technique based on ant colony optimization used for disaster resilient location detection in mobile ad-hoc network. J. Ambient. Intell. Humaniz. Comput. **12**(3), 4007–4017 (2021)
16. Nayak, P., Anurag, D., Bhargavi, V.V.N.A.: FM-SCHM: fuzzy method based super CH electio election for wireless sensor network with mobile base station (FM-SCHM). In: Proceeding 2nd International Conference Advance Computational Methodology, pp. 422–427. Hyderabad, India (2013)
17. Thangaramya, K., Kulothungan, K., Logambigai, R., Selvi, M., Ganapathy, S., Kannan, A.: Energy aware cluster and neuro-fuzzy based routing algorithm for wireless sensor networks in IoT. Comput. Netw. **151**, 211–223 (2019)
18. Pandey, S., Singh, V.: Blackhole attack detection using machine learning approach on MANET. In: 2020 International Conference on Electronics and Sustainable Communication Systems (ICESC), pp. 797–802. IEEE (2020)
19. Preeth, S. K., Dhanalakshmi, R., Kumar, R., Shakeel, P.M.: An adaptive fuzzy rule based energy efficient clustering and immune-inspired routing protocol for WSN-assisted IoT system. J. Ambient Intell. Humanized Comput. 1–13 (2018)
20. Mittal, M., de Prado, R.P., Kawai, Y., Nakajima, S., Muñoz-Expósito, J.E.: Machine learning techniques for energy efficiency and anomaly detection in hybrid wireless sensor networks. Energies **14**(11), 3125 (2021)
21. Kirubasri, G.: Energy efficient routing using machine learning based link quality estimation for WMSNs. Turkish J. Comput. Math. Educ. **12**(11), 3767–3775 (2021)
22. Anandh, S.J., Baburaj, E.: Energy efficient routing technique for wireless sensor networks using ant-colony optimization. Wireless Pers. Commun. **114**(4), 3419–3433 (2020)
23. Townsend, L.: Wireless sensor network clustering with machine learning, Doctoral dissertation, Nova Southeastern University (2018)
24. Gao, W., Alsarraf, J., Moayedi, H., Shahsavar, A., Nguyen, H.: Comprehensive preference learning and feature validity for designing energy-efficient residential buildings using machine learning paradigms. Appl. Soft Comput. **84**, 105748 (2019)
25. Saravana Kumar, N.M., Suryaprabha, E., Hariprasath, K.: Machine learning based hybrid model for energy efficient secured transmission in wireless sensor networks. J. Ambient Intell. Humanized Comput. **13**(2), 887–902 (2022). https://doi.org/10.1007/s12652-021-029 46-y
26. Yuldashev, M.N., Vlasov, A.I., Novikov, A.N.: Energy-efficient algorithm for classification of states of wireless sensor network using machine learning methods. J. Phys.: Conf. Ser. **1015**, 032153 (2018)
27. Sridhar, V., Ranga Rao, K.V., Vinay Kumar, V., Mukred, M., Ullah, S.S., AlSalman, H.: A machine learning-based intelligence approach for multiple-input/multiple-output routing in wireless sensor networks. Math. Problem. Eng. **2022**, 1–13 (2022)
28. Kumar, S., Gautam, P.R., Rashid, T., Verma, A., Kumar, A.: Division algorithm based energy-efficient routing in wireless sensor networks. Wireless Pers. Commun. **122**(3), 2335–2354 (2022)

Exploring the Use of Educational Data Mining and Learning Analytics Through AI to Improve Instructional Practices and Student Performance

Nidhi Agarwal[1]([envelope]) [ORCID], Yogendra Babu[2], Ram Awadh[2], and Vikas Mishra[3]

[1] Faculty of Social Science and Humanities, Lincoln University College, Kota Bharu, Malaysia
dr.nidhi@lincoln.edu.my
[2] Department of Education, Mahatma Gandhi Antarrashtriya Hindi Vishwavidyalaya, Wardha, MH, India
[3] Department of Education, Akbarpur Degree College, Akbarpur, Kanpur Dehat, U.P., India

Abstract. AI learning analytics (AI LA) and educational data mining (EDM) are two new fields that use the power of data analysis to better educational practises and student performance. These approaches give teachers access to information about the behaviour, learning patterns, and performance of their students by analysing big datasets gathered from diverse educational resources. Personalising learning experiences, offering early interventions and assistance, informing curriculum and instructional design, utilising predictive analytics for preventative measures, improving assessment and feedback systems, and guiding institutional decision-making are all possible uses for this information. To achieve proper and efficient implementation, ethical issues like prejudice, consent, and data protection must be properly considered. Overall, educational data mining and AI Learning analytics have enormous potential to alter education and give teachers the tools they need to optimise teaching learning Process.

Keywords: AI Learning analytics · Educational Data Mining · Performance · Instructional Practices

1 Introduction

In order to improve educational results, AI Learning analytics refers to the gathering, examination, and interpretation of data produced throughout the learning process. To collect and analyse student data, including their interactions with learning management systems, online platforms, digital materials, and exams, a variety of methodologies and tools are used. The knowledge acquired from AI Learning analytics can improve student success and engagement, personalise learning experiences, and inform instructional decision-making.

The landscape of education is undergoing a significant transition in the fast-paced digital era, driven by developments in technology and data analytics. Globally, educational institutions are using cutting-edge strategies to improve teaching methods and raise student achievement. Two cutting-edge technologies, Educational Data Mining (EDM) and AI Learning Analytics (AI LA), stand out among these revolutionary techniques. Together, they have the power to transform education in the future by empowering both teachers and students with data-driven insights and individualised learning experiences [1].

A more individualised and adaptive learning environment is gradually replacing the conventional "one-size-fits-all" approach to education. This paradigm change is significantly accelerated by educational data mining. EDM reveals hidden patterns and important information by methodically gathering and examining enormous amounts of educational data, such as student assessments, interactions with learning resources, and performance indicators. In order to make data-informed decisions to optimise instructional tactics, educators can use this plethora of data to better understand the learning behaviours, strengths, and weaknesses of students [2].

The incorporation of AI Learning Analytics works in tandem with the strength of Educational Data Mining. Data processing powers are greatly enhanced by artificial intelligence and machine learning algorithms, allowing instructors to quickly and effectively sort through massive datasets. With its predictive and prescriptive insights, AI LA goes beyond just descriptive analytics. Based on past performance, it can predict students' future performance and suggest customised learning routes to meet particular learning needs. In addition to maximising individual learning outcomes, this proactive strategy helps educators spot possible problems and offer prompt responses.

In this study project, we examine the significant effects that these revolutionary technologies will have on education in the future. We seek to emphasise the contributions that Educational Data Mining and AI Learning Analytics make to improving instructional strategies and raising student achievement by exploring their applications. We'll look at how data-driven decision-making may revolutionise the educational process for students of all backgrounds and abilities, resulting in a more inclusive, interesting, and successful educational experience.

But like with any technological development, moral issues must come first [3]. We will examine issues about student privacy, data security, and the ethical use of educational data during the course of our examination. To make sure that these developments serve as tools of progress rather than unintentional causes of harm, it is essential to strike a balance between data usage and protecting individual rights.

It is important to be aware of the opportunities and potential difficulties that may lie ahead as we set out on this trip into the future of education [4, 5]. We can usher in a new era of education that not only meets the individual needs of each learner but also equips teachers to become better knowledge facilitators by utilising the synergistic potential of Educational Data Mining and AI Learning Analytics. The ultimate goal of this study is to add to the continuing discussion about how to use technology to create a better and more inclusive future for education. We will look at case studies and real-world examples to highlight how these technologies have been successfully used to enhance teaching methods and student outcomes. We can better appreciate how EDM and AI LA can change the learning process if we are aware of their practical effects.

The role of personalised learning through data-driven insights will be one important area of investigation. We will look at how EDM and AI LA might support teachers in designing individualised learning routes that take into account each student's learning preferences, shortcomings, and strengths. By adjusting the content and teaching strategies to meet the individual requirements of each learner, this strategy has the potential to increase student engagement, motivation, and overall academic success.

This study will look at how collaborative data-driven education is. In order to promote a culture of continuous improvement, educational institutions should encourage teachers to exchange best practises, pool their data analysis skills, and work together to improve their lesson plans. We will look into how collaboration between educators, administrators, and stakeholders is facilitated by data analytics, creating a more dynamic and interconnected learning ecosystem [6–8].

AI Learning Analytics and Educational Data Mining have an impact beyond of traditional classroom instruction. We'll look at how these tools can help institutions make decisions based on solid facts. Administrators can pinpoint systemic problems, improve resource allocation, and create focused interventions to boost student performance on a larger scale by analysing large-scale educational data.

It is critical to be aware of potential obstacles to using these technologies, such as data privacy concerns, security concerns, and the possibility of bias perpetuation. In order to ensure that EDM and AI LA are used responsibly and with an emphasis on equity and inclusivity, as part of our research, we will examine ethical aspects and best practises in data gathering and analysis.

We will also look into the changing role of educators to get a full picture of the future of education. As more administrative and analytical work are handled by technology, instructors can move their attention to mentoring, promoting critical thinking, and encouraging innovation. We will examine the best ways for teachers to incorporate EDM and AI LA into their lesson plans, enabling them to become flexible and adaptable knowledge facilitators.

Finally, our study will evaluate the long-term effects of implementing EDM and AI LA in schooling. We will examine potential patterns and new developments as these technologies grow further in order to better understand how they might affect the future of education. By taking a proactive stance, we can foresee obstacles and opportunities, ensuring that the use of technology in education is progressive and advantageous for all parties involved.

This study's goal is to illuminate the revolutionary potential of AI Learning Analytics and Educational Data Mining in reshaping the course of education. We aim to contribute to the continuing discussion on utilising technology to create a more personalised, inclusive, and effective educational experience for learners throughout the world by looking at their real-world applications, ethical issues, and wider consequences. The key to realising every student's potential lies in the union of data-driven insights and the art of teaching, and this study aims to determine how to get there [9].

AI Learning analytics (AI LA) and educational data mining (EDM) are two closely connected topics that use data analysis approaches to gain understanding and make decisions regarding teaching strategies and student performance. These strategies seek to enhance teaching and learning outcomes by making use of data. An investigation into the application of AI Learning analytics and educational data mining to these objectives follows:

Finding Learning Patterns. The analysis of huge datasets in educational data mining is used to find patterns and trends in student performance, behaviour, and learning processes. Teachers can learn a lot about how students learn, their strengths, limitations, and areas where they might need more help by looking at data gathered through learning management systems, online platforms, tests, and other educational tools.

AI learning. Learning analytics can be used to provide personalised learning experiences that are catered to specific students. In order to increase student engagement and achievement, educators can create personalised learning pathways, suggest pertinent resources, and give timely feedback by analysing data on students' learning preferences, progress, and performance.

Early Intervention and Support. Educational data mining and AI Learning analytics can assist in identifying pupils who are in danger of falling behind or having academic difficulties. Educators can spot early warning signs and take necessary action by keeping an eye on indicators including attendance, participation, assessment results, and engagement levels. This proactive strategy can assist in preventing student disengagement and can offer focused treatments to meet each student's individual requirements [10–12].

Curriculum Creation and Instructional Design. This can be influenced by data-driven insights from AI Learning analytics and educational data mining. By reviewing student performance data, educators can pinpoint areas in which students frequently struggle and decide on appropriate curriculum changes, instructional approaches to better align with student needs [16].

Using Predictive Modelling Techniques. AI Learning analytics can be used to forecast student outcomes, such as test performance or the likelihood of dropping out. Teachers can foresee problems and take proactive steps to support struggling pupils by analysing previous data and identifying key predictors. To increase overall student success rates, predictive analytics can also aid to optimise resource allocation and lesson planning.

Evaluation and Feedback. Educational data mining and AI Learning analytics might make it easier to create efficient evaluation procedures. Educators can learn more about the validity and reliability of assessment tools, spot areas for improvement, and modify assessments to be more in line with learning objectives by analysing assessment data. Additionally, AI Learning analytics can provide students with timely and relevant feedback, encouraging self-reflection and directing their learning process.

Decision-Making at the Institutional Level. This can be influenced by educational data mining and AI Learning analytics. Educational institutions can get a comprehensive understanding of student performance, instructional efficacy, and resource allocation by collecting and analysing data from various courses, departments, or schools. To improve educational outcomes, these insights can direct the creation of policies, strategic planning, and resource allocation [13, 14].

With the incorporation of cutting-edge technologies like Educational Data Mining (EDM) and AI Learning Analytics (AI LA), the future of education offers enormous promise. These cutting-edge tools have the potential to completely alter how students are taught and perform in the classroom.

Massive volumes of educational data must be gathered, analysed, and interpreted in order to find patterns, trends, and correlations. This process is known as "educational data mining." Utilising this plethora of data, educators can learn a lot about the learning styles, aptitudes, and weaknesses of their pupils. This gives them the ability to adapt curriculum, provide targeted interventions, and personalise education to meet the specific needs of each student.

AI learning analytics uses machine learning and artificial intelligence to handle data quickly and effectively, complementing EDM. It can identify subtle learning trends, forecast student performance, and provide individualised learning pathways. With this proactive strategy, educators are better equipped to address possible learning gaps before they prevent students from progressing [15].

Educational institutions can promote data-driven decision-making, optimising instructional tactics and resources, by integrating EDM and AI LA. Teachers have access to real-time performance metrics, can track student growth, and can modify their pedagogical approaches as necessary. Learning experiences that are interesting and flexible help students become more motivated and succeed in school.

The seamless integration of these technology goes outside the classroom, as well. The partnership between home and school can be improved when administrators can make well-informed policy decisions based on complete data and parents can actively engage in their children's educational journeys.

While adopting these technology improvements is essential for the future of education, it is necessary to strike a balance between data usage and protecting student privacy. Leading this revolutionary movement should continue to be ensuring the moral and proper use of educational data.

The combination of Educational Data Mining and AI Learning Analytics has huge potential to improve teaching and raise student achievement in the constantly changing educational context. We can create a more inclusive, personalised, and effective learning environment and create a better future for future generations by utilising the power of data and artificial intelligence.

1.1 Research Objectives

1. Introduce the concepts of EDM and AI LA and their relevance in collecting and analyzing educational data.
2. Examine different approaches and tools utilized in EDM and AI LA to gain insights into student learning behaviors and identify learning difficulties.
3. Investigate the ethical considerations and challenges associated with the use of educational data, ensuring student privacy and data security.
4. Explore the benefits of implementing EDM and AI LA techniques, such as personalized learning, early identification of at-risk students, adaptive instruction, and curriculum development.

2 Theoretical Framework

By employing this framework, the research aims to provide a systematic and comprehensive exploration of the use of EDM and AI LA in improving instructional practices and student performance. It will facilitate a structured approach to data collection, analysis, implementation, evaluation, and ethical considerations, contributing to the understanding and application of these techniques in educational contexts.

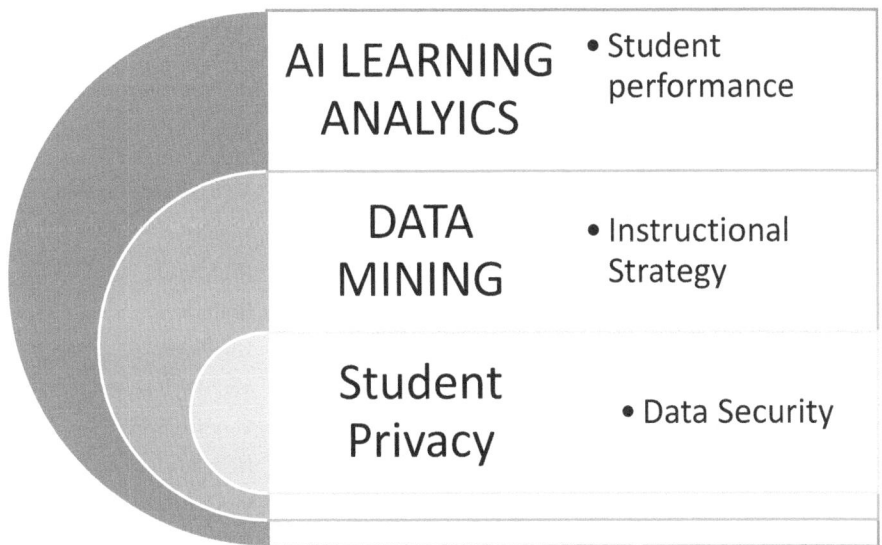

3   Methodology

3.1   Data Collection

Respondents of this study were the collegiate students enrolled in higher educational institutions. They were selected using a random sampling technique. Group of students from each higher educational institution. The 5-point Likert scale was used to determine the level of respondents' digital and information literacy skills as a result of 2-year utilization of educational technology such as computer assisted education through Learning Management System, online and electronic educational resources. Two separate Google forms were made for each of the higher educational institutions. Data were generated and analyzed using descriptive statistics.

3.2   Data Analysis

Analyse the collected data using the proper data analysis methods. To find patterns, trends, and correlations in the educational data, quantitative analysis, such as statistical analysis, data mining methods, and machine learning approaches, may be used. Insights from qualitative data sources like interviews or open-ended survey responses can also be gleaned using qualitative analysis techniques like thematic analysis and content analysis.

AI Learning analytics enables personalized learning experiences by utilizing student data to tailor instruction and resources to individual needs. By analyzing data on students' strengths, weaknesses, learning preferences, and progress, educators can provide targeted interventions, adaptive learning pathways, and customized feedback to optimize learning outcomes.

4 Results and Discussion

The results and discussion of the study based on the survey conducted for the AI Learning analytics on the two variables instructional practices and student performance with collegiate students as respondents are hereby presented (Table 1).

Table 1. AI learning analytics of collegiate students for instructional practices based

Items	Mean	Verbal Interpretation
Time To Proficiency	4.01	High level
Knowledge	3.97	High level
Transfer Of Training	4.08	High level
Impact On Organizational Performance Metrics	4.2	High level
Employee Engagement	4.26	Very High level
Net Promoter Score	4.36	Very High level
Stakeholder Satisfaction	4.18	High level
Skill Retention	4.47	Very High level
Grand Mean	4.19	High level

The table presents the AI Learning analytics of collegiate students based for instructional practices. The highest recorded mean of college students is 4.47 or Very High Level for Respect for Skill retention. It means that college students value skill retention so much. Having a very high level on skill retention also signifies that college students are more knowledgeable on the skills and they can recall their skills to transfer learning for long term memory. However, the lowest mean of the respondents is in the area of Content Knowledge with 3.97 (Table 2).

The table shows the AI Learning analytics of collegiate students based for performance. The highest mean is 4.45 or Very High Level for the item skill retention.

The highest recorded mean of college students is 4.47 or Very High Level for Respect for Skill retention. It means that college students value skill retention so much. Having a very high level on skill retention also signifies that college students are more knowledgeable on the skills and they can recall their skills to transfer learning for long term memory. However, the lowest mean of the respondents is in the area of Content Knowledge with 3.97. The grand mean of all items is 4.15 or High Level (Tables 3, 4, 5 and 6).

The results of this study offer a number of significant insights into how instructional practises affect student performance. These instructional tactics may improve student accomplishment, according to the positive associations between active learning strategies, project-based learning, personalised learning, and formative assessment. The study also emphasises how crucial it is to use a variety of teaching strategies in order to meet different learners' requirements and preferences. In order to ascertain the frequency and variety of instructional practises among the sample, the survey responses were examined. This investigation highlighted both conventional and cutting-edge techniques used

Table 2. AI learning analytics of collegiate students based for performance

Items	Mean	Verbal Interpretation
Time To Proficiency	4	High Level
Knowledge	3.97	High Level
Transfer Of Training	4.07	High Level
Impact On Organizational Performance Metrics	4.07	High Level
Employee Engagement	4.22	Very High Level
Net Promoter Score	4.36	Very High Level
Stakeholder Satisfaction	4.25	Very High Level
Skill Retention	4.45	Very High Level
Grand Mean	4.15	High Level

Table 3. Recorded variable

N	Valid	1
	Missing	0

Table 4. Statistics value of recorded variable

	Frequency	Percentage	Valid Percentage	Cumulative Percentage
Validity of Mean	1	100.00	100.00	100.00

Table 5. Statistical value of variable

Variable	Count	Mean	S.D	Coefficient of Variation
	6	4.4950	0.46198	10.3%
Knowledge	1	3.9700	0.00214	0.01%
Skill Retention	1	4.9800	0.01064	0.02%
Total	8	4.4900	0.47476	10.6%

by educators in a wide spectrum of educational methods. To determine the connection between instructional practices and student's performance, a correlation analysis was carried out. The findings showed that active learning techniques and project-based learning, for example, have a substantial positive link with better student performance While controlling for other pertinent characteristics, multiple regression models were built to investigate the predictive potential of instructional practises on student performance. The data showed how particular teaching strategies, such individualised instruction and

Table 6. Modality of dependent variable

	Number of Levels	Number of Parameters
Fixed Effects	1	1
Residual		1
Total	1	2

formative evaluation, had a big impact on how well students performed and skill retention played an important role for college students in performance based and instructional practices.

5 Conclusion

AI Learning analytics can identify students who may be at risk of falling behind or struggling academically. By analyzing data on student engagement, attendance, assessment scores, and other indicators, educators can detect early warning signs and provide timely interventions and support to prevent further difficulties. AI Learning analytics can inform instructional design and delivery by providing insights into the effectiveness of teaching strategies, learning resources, and assessment methods. Educators can analyze data on student interactions, engagement levels, and performance to refine their instructional practices and make data-informed decisions. AI Learning analytics enhances the assessment and feedback process by analyzing student performance data. Educators can gain insights into the validity and reliability of assessments, identify areas of improvement, and provide timely and targeted feedback to students to enhance their learning progress. AI Learning analytics can inform institutional-level decision making by analyzing data across multiple courses, departments, or schools. Educational institutions can gain insights into student success rates, resource allocation, and instructional effectiveness, enabling informed policy development, strategic planning, and allocation of resources. AI Learning analytics has the potential to revolutionize education by leveraging data to understand and support student learning. However, ethical considerations, such as data privacy, consent, and transparency, must be prioritized to ensure the responsible and ethical use of student data in the AI Learning analytics process.

With the merger of Educational Data Mining (EDM) and AI Learning Analytics (AI LA) at its core, the future of education is poised to see a dramatic revolution. This study has illuminated the enormous potential of these cutting-edge technologies to improve teaching methods and boost student achievement, opening the door to a more individualised, inclusive, and efficient learning environment.

EDM's capacity to gather and analyse huge amounts of educational data gives teachers insightful knowledge on the learning preferences and specific requirements of pupils. Students may achieve their full potential when instructors use data-driven decision-making to customise instruction, spot learning gaps, and put targeted interventions into place.

In addition to EDM, AI Learning Analytics uses machine learning and artificial intelligence algorithms to process data quickly and accurately while providing predictive and prescriptive insights. With the use of this proactive method, teachers can predict student results and provide personalised learning pathways that will increase engagement and academic performance.

We have emphasised the significance of maintaining an ethical approach to data utilisation, guaranteeing student privacy, and guarding against biases throughout this study. To establish an egalitarian and just learning environment, EDM and AI LA must be implemented properly.

Case studies in real life have shown how these tools have already improved education. EDM and AI LA have shown to be effective instruments in raising the bar for educational practises, from individualised learning experiences to evidence-based decision-making at the institutional level [16, 17].

We have also looked at how educators' roles are changing as they use these tools. Teachers should concentrate on encouraging critical thinking, creativity, and mentoring pupils as administrative and analytical responsibilities are increasingly handled by EDM and AI LA. The improvement of teaching methods as a result of this fosters the development of knowledgeable and independent students.

This research has looked at anticipated future trends and advancements in the field of educational technology. Educational institutions can proactively innovate and adapt in order to ensure that the incorporation of technology remains progressive and significant by anticipating difficulties and opportunities.

In conclusion, the merger of educational data mining and artificial intelligence learning analytics holds the key to realising every student's full potential by making education a dynamic and adaptive process that meets the various needs of students. It is crucial to keep in mind that, while we embark on this transforming path, the ultimate objective is still the holistic development of people, creating a generation of lifelong learners who are capable of thriving in a constantly shifting environment. We can create a future of education that empowers both students and teachers by responsibly and ethically utilising these technologies, creating a better and more promising tomorrow.

5.1 Implications

For numerous stakeholders and the educational environment as a whole, the integration of Educational Data Mining (EDM) and AI Learning Analytics (AI LA) into education has major ramifications. To fully utilise these technologies and maximise their beneficial effects, it is essential to comprehend these ramifications. Here are some significant ramifications:

Learning experiences that are specifically catered to each student's strengths, weaknesses, and learning preferences are made possible through the use of EDM and AI LA. An increase in student involvement, motivation, and academic accomplishment may result from this individualised strategy.

Data-Driven Decision-Making. Based on data insights, instructional strategy optimisation, and resource allocation, educators and administrators can make wise choices. Making decisions based on data can increase the efficacy of instructional strategies and promote ongoing development.

Early Intervention and Assistance. Thanks to AI LA's predictive powers, educators may see difficult children early on and offer timely interventions and assistance to stop learning gaps from growing.

Better Teacher Professional Development. By giving educators information on the efficacy of their instructional strategies, EDM and AI LA can support teacher professional development. This feedback loop can promote a culture of continuous learning and enable focused improvement.

Ensuring that all students have an equal chance to succeed, educators can use data analysis to pinpoint obstacles to learning and solve concerns with inclusion and accessibility.

EDM and AI LA can be used by educational institutions and policymakers to gather data for establishing educational policies and initiatives. Education improvements may become more successful and efficient as a result of this data-driven methodology.

Ethical Issues. Concerns about student privacy, data security, and the possibility for bias perpetuation are raised by the collecting and use of educational data. To ensure responsible data usage, best practises and protections must be implemented.

A more holistic approach to education can be promoted by educators focusing on strengthening students' critical thinking, creativity, and emotional intelligence when some administrative responsibilities are taken over by technology [18].

Personalised learning experiences can provide students with the knowledge and abilities needed for the quickly changing work market, encouraging lifelong learning and flexibility.

Research and Innovation Advances. The adoption of EDM and AI LA in the classroom creates new opportunities for the study of educational technology. Researchers can investigate new uses and improve current processes.

Educational Data Mining and AI Learning Analytics have a wide range of effects on education. The way we approach teaching and learning might be completely changed by these technologies, which would encourage individualised instruction, evidence-based decision-making, and continual improvement. To guarantee that these technologies empower learners while respecting privacy and inclusion, thorough examination of ethical issues and the proper use of data is necessary. Responsibly embracing these innovations can help to design a future of education that equips students for today's problems and inspires a lifelong love of learning.

References

1. Deswal, A., Kumar, S.: Technological aspects in cloud computing. Cosmos J. Eng. Techno. **12**(2), 05–07 (2022)
2. Kasan, F.D., Buenavides, E.C.: Development of web-mobile based teachers' performance evaluation system (W-M BTPES). Globus Int. J. Manag. IT **14**(1), 32–39 (2022)
3. Kapri, T., Kumar, P.: Web Content Management System. Information and Communication Technology: Challenges and Business Opportunities. Excel Publishers, pp. 56–62. ISBN: 978-93-81361-00-9 (2010)

4. Dwivedi, A.: Development of educational model for e-learning. Globus J. Progressive Educ. **12**(2), 95–98 (2022)

5. Jaramillo, R.N., et al.: Projected action plan and timeline in pilot testing of simulation-based learning in the college of nursing at Phinma-University of Pangasinan. Cosmos An Int. J. Art High. Educ. **11**(2), 04–15 (2022)

6. Agarwal, N., Pundir, N.: A comparative study of personality traits and thinking styles of ICT users and non users. Int. J. Dyn. Educ. Res. Soc. **1**(1), 74–83 (2019)

7. Salinda, M.T., Tuzaon, A., lachica, P.: Integrity of third year nursing students to online related learning experiences: a concept analysis. Globus Int. J. Med. Sci. Eng. Technol. **10**(2), 87-98 (2021).

8. Deswal, A., Kumar, S.: A study on enterprise architecture in cloud computing. Cosmos Int. J. Manage. **12**(1), 04–07 (2022)

9. Agarwal, N., Verma, M.: A study on taxonomy of innovations. Globus Int. J. Manag. & IT **11**(1), 57–64 (2019). https://doi.org/10.5281/zenodo.3872090

10. Punla, N.F.D., Regalado, D., Rick, B., Navarro, R.M.: Perceived social support as moderator between self-stigma and help-seeking attitudes. Globus Int. J. Med. Sci. Eng. Technol. **11**(1), 87–98 (2022)

11. Singh, M.K., Tripathi, S., Singh, P.: Cybersecurity and data privacy. Cosmos Int. J. Manag. **11**(2), 01–04 (2022)

12. Agarwal, P.K., Verma, R.K.: Mathematical solution for analysing of ordinary differential equations using data assimilation approach. J. Crit. Rev. **7**(15), 4844–4848 (2020)

13. Cabaguing, J.M., Lacaba, T.V.G.: Predictors of faculty readiness to flexible learning management system. Globus Int. J. Manag. IT **14**(1), 40–49 (2022)

14. Mendoza, R., Marie, N., Dayao, E.F.: LCUP student use of Gmetrix system: an evaluation of usefulness, ease of use, and satisfaction. Cosmos J. Eng. Technol. **11**(2), 63–67 (2021)

15. Agarwal, N., Shiju, P.S.: A study on content generation for internet usage. Int. J. Adv. Res. Dev. **3**(2), 1380–1382 (2018). https://doi.org/10.5281/zenodo.3764806

16. Reyla, J.S.: Translanguaging in the classroom: impact on the academic performance of the learners. Cosmos An Int. J. Art High. Educ. **11**(1), 72–87 (2022)

17. Cahapin, E.L., Malabag, B.A., Samson, B.D., Santiago, C.S.: Stakeholders' awareness and acceptance of the institution's vision, mission and goals and information technology program objectives in a State University in the Philippines. Globus J. Progressive Educ. **12**(1), 51–60 (2022)

18. Galvez, R.S.: Graduate studies students' scholastic abilities and their effects on comprehensive examination. Cosmos Int. J. Manage. **11**(1), 112–119 (2021)

Empowering Education with Artificial Intelligence: Advancing Instructional Settings and Personalization

Yogendra Babu[1], Nidhi Agarwal[2]([✉]) [iD], and Ram Awadh[1]

[1] Department of Education, Mahatma Gandhi Antarrashtriya Hindi Vishwavidyalaya, Wardha, M.H., India
[2] Faculty of Social Science and Humanities, Lincoln University College, Kota Bharu, Malaysia
dr.nidhi@lincoln.edu.my

Abstract. Artificial Intelligence (AI) in education is revolutionising the way that learning experiences are personalised and instructional environments are defined in contemporary pedagogy. This study examines the various ways that artificial intelligence (AI) is changing education, with a focus on how these innovations have the potential to transform conventional teaching strategies and improve learning experiences for students of all ages. (Mendoza, Rose Marie N. and Dayao, Edna F. 2021 [20]).

AI-driven learning environments provide adaptable and customised learning experiences catered to individual needs as traditional classroom structures change. AI algorithms may determine students' learning preferences, skills, and weaknesses through the examination of data-driven insights. This information enables teachers to tailor their pedagogical strategies. AI-powered educational systems can also give prompt feedback, which encourages self-evaluation and progress.

This study looks at several AI-related applications in education, such as virtual mentors, intelligent tutoring systems, and intelligent content recommendations. With the use of these technologies, students may participate in realistic simulations, gain access to a multitude of materials, and get help instantly - all of which enhance the quality and inclusivity of the learning environment (Dabbagh, N. & Bannan-Ritland, B. 2005 [7]).

AI also makes it easier to automate administrative duties, freeing up teachers' time for instructional design and one-on-one student interactions. AI-powered analytics also support curriculum creation and programme assessment, helping educational institutions stay flexible and adaptable to the ever-changing demands of their students.

Even if AI has a lot to offer education, privacy and ethical issues need to be taken care of. This essay examines algorithmic biases, the ethical ramifications of data collecting, and the possibility that artificial intelligence will worsen educational disparities. It highlights how important it is to implement AI in educational settings in a transparent and responsible manner.

This study concludes by highlighting the ways that AI is improving educational settings and personalisation, which in turn is empowering education. It emphasises how AI has the ability to raise teaching standards, boost student performance, and create a more inclusive and egalitarian learning environment. To optimise the beneficial effects on education, it also emphasises how crucial ethical issues and responsible AI application are.

A. Gupta et al. (Eds.): ICAIA 2023, CCIS 2308, pp. 72–89, 2025.
https://doi.org/10.1007/978-3-031-84394-5_7

Keywords: Artificial Intelligence · Instructional Settings · Personalisation

1   Introduction

The field of education is not an exception to how artificial intelligence (AI) has emerged as a transformational force in a variety of fields. Researchers and educators have been exploring the use of AI technology to improve instructional environments and personalise learning for students more and more in recent years (Koehler, M. J., Mishra, P. & Cain, W. 2013 [13]). A cutting-edge application of AI technology, artificial intelligence (AI) in education aims to transform and improve the current educational landscape (Feng, S. & Law, N. 2021 [1]). AI in education strives to enhance students' overall learning experiences, simplify administrative duties for teachers, and deliver individualised instruction based on each learner's particular needs by utilising cutting-edge algorithms and data analysis (Jaber, Rowena D. and Reyes, Herbert Glenn P., 2023 [15]). Artificial Intelligence in Education (AIED) is an interdisciplinary research area that combines techniques and instruments from various fields, including computer science and information technology, to address educational issues. To enhance teaching and learning activities, AIED is particularly interested in creating intelligent problem-solving systems (Graesser et al., 2005). The use of digital technologies into learning activities has aided in the collection of a massive amount of data on user-generated material and traceable learning behaviours.

The capacity to provide personalised learning experiences is one of the main advantages of AI in education. With the help of AI-powered technologies, instructors may adapt the pace and substance of the learning process to each student's strengths and limitations by analysing large volumes of data about each student's performance, learning preferences, and development. In addition to increasing student engagement, this personalised strategy maximises learning outcomes by filling in knowledge gaps and promoting a better comprehension of the subject. In the past few decades, the widespread use of information technology in teaching and learning activities has given rise to a number of new research issues and foci in the subject of education. Researchers have looked at the adoption of information technologies by different stakeholders in educational settings and learning activities (Buchanan et al., 2013 [4]; Buabeng-Andoh, 2012 [2]; Hu et al., 2003 [12]), the pedagogical adjustments needed for the immersive use of information technology in teaching and learning (Ertmer & Ottenbreit-Leftwich, 2013 [8]; Koehler et al., 2013 [13]; Graham 2006), and the psychological effects of these changes. Along with the rapid advancement of digital technologies, we have seen the emergence of new learning paradigms, such as online learning (Dabbagh & Bannan-Ritland, 2005 [7]), distance learning (Valentine, 2002), e-learning (Moore et al., 2011 [14]), and game-based learning (Prensky, 2003).

Intelligent tutoring systems (ITS) are one example of how AI can be used to deliver individualised teaching. These systems engage with students using AI algorithms, providing real-time feedback and content that adapts to each student's replies. Through focused guidance and the identification of problem areas, ITS can help students advance towards mastery of difficult ideas. As a result, students can advance at their own rate, resulting in a more fruitful and fulfilling learning experience.

AI also automates administrative work, giving instructors more time to concentrate on teaching and coaching. Automated grading and assessment systems review homework, examinations, and quizzes using machine learning (ML) and natural language processing (NLP), giving students immediate feedback. This helps educators save time while also guaranteeing the evaluation process's consistency and neutrality.

The effects of AI on education go beyond conventional classroom settings. AI is being used more and more by online learning systems to provide adaptive learning paths. In order to design personalised learning journeys that deliver pertinent material and exercises tailored to each learner's abilities and interests, these platforms analyse unique learning patterns and preferences. This strategy encourages a self-paced and interesting learning environment, enabling students to better understand subjects.

The incorporation of AI in education does present obstacles despite its many benefits. It is important to carefully evaluate ethical issues such data privacy, security, and the possibility of algorithmic prejudice. It is imperative for educators and legislators to protect student data and provide fair and equal access to AI-driven services.

The use of artificial intelligence in education has enormous potential to improve both student learning and teacher effectiveness. AI makes education more efficient and inclusive by offering personalised instruction, automating administrative processes, and encouraging adaptive learning environments (Buga-ay, Cynthia U. and Causing, Ruby D. 2023 [3]). To ensure that AI actually improves education for all students, however, its deployment necessitates thorough consideration of ethical issues and responsible use. A future of educational innovation and excellence will be shaped by the continual interaction between human expertise and AI capabilities as technology develops. AI has the potential to revolutionise the conventional classroom model by analysing massive amounts of data, spotting patterns, and adapting its answers in real-time. In order to usher in a new era of educational innovation, this article aims to study the application of AI in education, with a particular focus on how it may enhance instructional environments and encourage personalised learning.

1.1 Improving Instructional Environments

Intelligent Tutoring Systems (ITS): ITS uses AI algorithms to provide students with specialised support and direction. ITS can identify knowledge gaps and adapt instructional content to meet the unique needs of each learner by analysing student performance data. The system may modify the pace, level of difficulty, and content delivery to ensure that each student receives individualised training and mastery of the material.

Automated Grading and Assessment: Time-consuming assignment, test, and exam evaluation is streamlined by AI-driven grading and assessment technologies (Moore, J. L., Dickson-Deane, C. & Galyen, K. 2011 [14]). These systems can accurately grade written responses, evaluate written responses, and provide constructive comments thanks to Natural Language Processing (NLP) and Machine Learning (ML). Due to their increased efficiency, instructors may concentrate more on creating engaging lessons and customised interventions for each student.

Curation and Smart Content Creation: AI-powered content creation technologies help teachers create interactive and interesting learning materials. AI is able to provide relevant content, multimedia components, and forms that appeal to students by examining the available resources and their preferences. The comprehension and memory of the knowledge by the students are improved by this personalisation.

1.2   Encouragement of Personalised Learning

Adaptive Learning Paths: Data analytics are used by AI-driven adaptive learning platforms to pinpoint each student's strengths, limitations, and preferred learning style. Based on this research, the system creates personalised learning paths that take into account each user's interests and needs. Adaptive learning maximises learning results by delivering knowledge in a way that meets the student's pace and capacity for learning.

AI is capable of giving pupils personalised feedback on their performance and progress in real time. Additionally, it can make specific suggestions for extra readings, practise problems, and enrichment activities, letting students explore deeper into subjects that interest them or need more practise.

Emotional Intelligence and Learning: Artificial intelligence (AI) can be programmed with emotional intelligence algorithms to identify students' feelings and emotional reactions as they learn. AI can provide a helpful and sympathetic learning environment by adapting its interactions and answers in accordance with students' emotional states.

Artificial intelligence in education has the power to completely change how we teach and learn (Ertmer, P. A. & Ottenbreit-Leftwich, A. [8]). AI can enable teachers to successfully meet the needs of individual students by enhancing learning environments and fostering personalised learning experiences. The use of AI in education must, however, be approached from a balanced standpoint, taking into account both the advantages and difficulties presented by this technology. Mendoza, Rose Marie N. and Dayao, Edna F. (2021) [20]. To guarantee that every learner has the best possible educational experience as AI develops, it is crucial to create a balance between human expertise and AI's capabilities. AI has the potential to significantly improve education and pave the path for a more promising and open future of learning with careful and appropriate integration.

1.3   Addressing Special Needs and Learning Disabilities

Personalised Support for Students with Learning impairments: By providing personalised interventions and support, AI can be very helpful for students with learning impairments. These technologies can recognise the individual difficulties that disabled students have and modify the teaching materials to fit their particular learning preferences. This individualised approach can increase their self-assurance and support them in more successfully overcoming academic obstacles.

Assistive technology: Students with dyslexia, visual or auditory impairments, or other disabilities can benefit from AI-driven assistive technology like speech recognition software and text-to-speech applications (Ertmer, P. A. & Ottenbreit-Leftwich, A. 2013 [8]). These tools can promote equitable chances for all learners by facilitating information access and fostering inclusive learning environments.

1.4 Professional Development for Teachers

AI-Enabled Teacher Development: By analysing classroom data and providing insights on instructional approaches and student performance, AI can help with educators' professional development. These AI-driven suggestions can assist instructors in enhancing their techniques and customising their strategies to successfully address the various demands of their pupils.

Data-Driven Decision Making: AI is capable of processing enormous volumes of information related to education, such as student evaluations, attendance statistics, and engagement levels. In order to improve student results, educators can identify areas for improvement by analysing this data and developing focused interventions.

1.5 Improved Learning Analytics

Predictive analytics: Predictive analytics driven by AI can detect trends in student performance and spot possible problems before they materialise. Early detection of academic difficulties allows educators to step in quickly, offer additional support, and stop learning gaps from growing.

Big Data and Learning Patterns: The processing power of AI in handling big data enables a thorough comprehension of learning patterns across various racial and geographic groupings. Dwivedi, Archana (2023). The creation of more comprehensive, evidence-based educational policies and initiatives can benefit from this information.

1.6 Considerations of Ethics

Data Privacy and Security: Because AI in education significantly relies on student data, it is crucial to implement effective data privacy and security safeguards. To prevent unauthorised access to or misuse of kids' sensitive information, educators and policymakers must set stringent rules and regulations.

Fairness and Bias: AI algorithms may unintentionally reinforce prejudices found in the data they were educated on (Singh, Mukul Kumar, Tripathi, Sakshi and Singh, Preeti 2022 [9]). To ensure that AI systems do not support discrimination and that they offer equal chances to all students, regardless of their origins, it is crucial to continuously monitor and evaluate them.

The use of artificial intelligence in education opens up a wide range of opportunities for improving learning environments and customising student experiences. Educators may give pupils individualised help, improve their teaching methods, and make data-driven decisions by utilising AI-driven solutions (Punla, Nara Fe D., Regalado, Danniel Rick B. and Navarro, Rica M. 2022 [10]). Stakeholders must work together as AI develops to address ethical questions and guarantee that it continues to be a tool that empowers both instructors and students. AI can open the door for a more inclusive and successful educational system, educating students for the opportunities and challenges of the future, with a responsible and balanced approach.

2   Need for the Research

It should make clear in this section the particular issues, gaps, or queries that your study seeks to resolve. Consider upon these things:

- Finding a Gap in Information: Describe what has been under- or under-researched in the current body of information, research, or practises. What unsolved problems or unanswered questions exist in your field?
- Relevance to Current difficulties: Emphasise how your findings relate to problems or difficulties that are facing society now. Give facts or instances to support the topic's significance.
- Potential Benefits: Summarise the possible advantages or results of your study. What issues could your research help with, or what advancements could be made?
- Alignment with Previous Research: Show how your study advances or enhances earlier research in the same or relevant fields. Describe how it adds to the corpus of knowledge already in existence.

3   Drawbacks of Conventional Techniques

Although conventional teaching methods have been in use for a long time, they do have certain disadvantages. The following are some typical disadvantages of using traditional teaching methods:

3.1   Lesson-Based Instruction

- Learning Passively: In lecture-based instruction, students frequently adopt a passive posture, simply listening to the teacher without contributing in any way. Reduced comprehension and retention may result from this.
- Absence of Personalization: Lectures usually use a one-size-fits-all strategy, ignoring the differences in each student's learning style and speed.

3.2   Textbook Education

- Limited Interactivity: In comparison to digital resources and multimedia content, textbooks offer very little in the way of interaction and engagement.
- Outdated Information: With the rapid evolution of subjects like science and technology, printed textbooks may soon become out of date.

3.3   Standardised Examinations

- Narrow Assessment: Memorization and information recitation are frequently the main focus of standardised assessments, as opposed to critical thinking, creativity, and problem-solving skills.
- Test Anxiety: Students' performance and general well-being may be hampered by test anxiety, which is a result of high stakes standardised testing.

3.4 Work Assignments at Home

- Massive Assignments: Students who have too much homework may experience stress, sleep deprivation, and a shortage of time for extracurricular activities and other crucial life skills.
- Absence of Immediate Feedback: Students may find it difficult to address misconceptions in a timely manner when they do not receive feedback on homework assignments.

3.5 Teacher-Focused Methodology

- Restricted Student Autonomy: In traditional education, teachers are frequently at the centre of the learning process, which leaves students with fewer options for independent study and inquiry.
- Lack of flexibility: Different learning needs and preferences might not be accommodated in traditional classroom settings.

3.6 Repetitive Memorization

- Superficial Understanding: Learners who are encouraged to memorise data without necessarily comprehending the underlying concepts may struggle to solve problems as a result of rote memory.
- Short-Term Memory Loss: After an exam, information that was memorised may be easily lost.

3.7 Lack of Practical Application

- Restricted Practical Skills: Students who get instruction via traditional methods may not necessarily be equipped with practical skills, leaving them with theoretical information that may not be easily applied in the workplace.

3.8 Teacher Fatigue

- High Workload: Grading papers, planning lectures, and maintaining classroom order are just a few of the tasks that traditional teaching methods can add to a teacher's workload, which can eventually lead to burnout.

3.9 Accessibility and Inequity

- Limited Accessibility: Students with impairments or those in need of flexible learning arrangements may not be able to attend classes in traditional classroom settings.
- Economic Disparities: Since not all students have equal access to resources like textbooks or extra materials, conventional methods can worsen educational inequities.

3.10 Insufficient Real-Time Data

- Limited Data for the Assessment: It may be challenging for teachers to make timely interventions when using traditional approaches because they may not provide real-time data on student performance.

It's crucial to remember that, despite these disadvantages, traditional teaching techniques offer benefits of their own and are frequently combined with cutting-edge strategies to produce a well-rounded educational experience. Furthermore, by focusing on student-centered learning, integrating technology into the classroom, and using deliberate pedagogical practises, the aforementioned downsides can be lessened.

4 Limitations of the Study

It is imperative to recognise and openly address the research's limits in any given study. Understanding these limits aids readers and researchers in comprehending the circumstances surrounding the study's execution as well as the restrictions that might have an impact on the reliability or generalizability of the results. The following typical restrictions may be applied to research projects:

4.1 The Size of the Sample and Reliability

- A small sample size may lower the study's statistical power and generalizability. Furthermore, there could be biases because the sample could not accurately reflect the larger population of interest.

4.2 Limitations on Data Collection

- Certain data collection techniques, such self-reporting biases, recall mistakes, or dependence on secondary data sources, may have drawbacks. These problems may have an impact on the data's accuracy.

4.3 Restraints on Resources

- The study's breadth and depth may be limited by a lack of funds, time, or access to specialised tools or software.

4.4 Ethics-Related Matters

- The study's design may be impacted by ethical limits, such as prohibitions on data collecting from populations that are considered vulnerable or on intrusive research techniques.

4.5 Undertakings for Data Analysis

- For data analysis, it may be necessary to make certain statistical or analytical assumptions; breaking these assumptions may have an impact on the validity of the findings.

4.6 Investigator Bias

- The study design, data interpretation, and results might be influenced by the biases, preconceptions, or perspectives held by researchers themselves.

4.7 Outside Events

- Unexpected outside events or alterations in the research period's surroundings could have an impact on the study's applicability or findings.

4.8 Cross-Linguistic and Linguistic Restrictions

- Challenges in data collecting and interpretation may arise from language hurdles and cultural variations while conducting a study involving different cultures or languages.

4.9 Bias in Publications

- Publication bias, in which only studies with noteworthy or encouraging outcomes are published and made available to the public, may be present in the literature evaluated for this study.

4.10 Availability of Data

- Constraints imposed by data providers may restrict the availability of data, particularly in investigations involving secondary data analysis.

    In order to show transparency and a clear awareness of the limitations of the study, it is crucial that researchers clearly state these limitations in their research articles. Talking about the study's limits can also point up areas that need more investigation and emphasise the study's positive aspects in relation to its shortcomings.

5 Research Objectives

The following are the study's goals in relation to improving instructional settings and personalization with artificial intelligence in education:

- to evaluate how well Intelligent Tutoring Systems (ITS) improve educational environments by giving students individualised learning experiences in comparison to conventional classroom instruction.
- to compare automated grading and assessment with manual grading techniques and analyse the effects on student performance and teacher burden.
- to look into how different subject areas and age groups' levels of student engagement and comprehension are affected by AI-driven content creation and curation tools.

6 Theoretical Framework

The integration and influence of AI in educational contexts are supported by a number of important theoretical views that serve as the foundation for the study on Artificial Intelligence in Education: Enhancing Instructional Settings and Personalization. These theoretical frameworks offer a conceptual framework for comprehending the use of AI in education and its ramifications. The following theoretical viewpoints are pertinent to the investigation:

Theories of Cognitive Learning. Theories of cognitive learning, including constructivism and schema theory, place an emphasis on the part that learners' mental processes play in picking up information and comprehending concepts. By customising learning experiences to individual students' cognitive ability, past knowledge, and learning preferences, the integration of AI in education, notably through adaptive learning paths and personalised instruction, corresponds with cognitive theories.

Data-Driven Decision Making and Learning Analytics. Predictive modelling and educational data mining ideas are included into learning analytics. AI is able to spot patterns and trends in student engagement and performance by analysing massive datasets. With learning analytics, educators can choose effective instructional strategies and interventions based on evidence from AI-generated insights, which is in line with the concept of data-driven decision making.

Human-Computer Interaction (HCI) and Artificial Intelligence. These are the theoretical underpinnings of adaptive learning and personalisation in education. AI-driven solutions follow the concepts of personalised learning by adapting the pace and substance of the lesson to the needs and progress of each learner. This theoretical framework emphasises the significance of personalising learning opportunities for every student to maximise learning outcomes.

Ethical Frameworks. Utilitarianism, deontology, and virtue ethics are just a few examples of the ethical theories that can help us responsibly integrate AI into education. When implementing AI technology in educational contexts, theoretical considerations of data privacy, algorithmic fairness, and equal access are crucial. The ethical framework acts as a compass to help people navigate the advantages and disadvantages of using AI in education while also making sure that it is consistent with broader social values.

Social learning theories, such as social constructivism and contextual learning, emphasise the value of interpersonal communication and group work in the classroom. Even in online learning contexts, AI technology can support social learning by encouraging student participation and offering opportunities for peer contact and feedback.

The research on AI in Education can get a thorough knowledge of the underlying principles governing the effectiveness and impact of AI technologies in boosting educational settings and personalization by incorporating these theoretical frameworks. These theoretical frameworks offer a strong theoretical foundation for analysing the study results and coming to relevant conclusions that can guide practises and policies in education.

7   Data Analysis

Processing and interpreting the collected data will be part of the data analysis for the study Artificial Intelligence in Education: Enhancing Instructional Settings and Personalization. As was already indicated, the research may include both quantitative and qualitative data, necessitating the employment of various analytical techniques. An outline of the data analysis procedure is provided below:

7.1   Analysing Quantitative Data

To summarise the quantitative data gleaned from surveys and questionnaires, compute measures like mean, median, standard deviation, and frequency distributions.

Inferential Statistics. Examine relationships and draw conclusions about the population using statistical tests like t-tests, ANOVA, correlation, and regression analysis. For instance, inferential analysis can be used to determine how student performance and AI integration are related.

Create diagrams, graphs, and plots to visually convey quantitative data using data visualisation. In doing so, it may be possible to spot patterns and trends in the data.

7.2   Analysing Qualitative Data

Thematic Analysis. Examine the qualitative information gained from focus groups, open-ended survey responses, and interviews to find themes, patterns, and insights regarding the influence of AI on education. The data are coded and put into useful categories throughout this phase.

Analyse student interactions in online learning environments and AI-generated content, when appropriate, to determine how effective and engaging AI-driven content is.

7.3   Data Integration

Utilise triangulation to combine and validate the results of quantitative and qualitative research. The results of the research are more credible and reliable thanks to triangulation.

Comparisons. To understand the differences in perceptions and experiences of AI in education, compare data from various participant groups (such as students, educators, and administrators).

Ethical Assessment. Examine data pertaining to ethical issues, such as algorithmic bias and data privacy, to gauge the ethical application of AI in education.

8   Result and Discussion

This query highlights the possibility for providing educators with useful insights resulting from data-driven decision making through the integration of AI in education. AI can give teachers fact-based information on student performance, engagement, and learning habits by analysing enormous amounts of educational data. These insights help teachers make wise choices, pinpoint areas for development, and successfully adapt their lesson plans to fit the different requirements of their pupils. Ultimately, AI's analytical powers can help teachers maximise their pupils' learning opportunities, which will improve education's overall effectiveness and quality. The overall response to this question is summarized below in the table and chart (Table 1 and Fig. 1).

Table 1. Response for the question number 1

Question	Options	Data Collected Responses 100	Percentage	Mean
Is the use of AI in education can empower educators with valuable insights for data-driven decision making?	Strongly Agreed	50	50	**4.21**
	Agreed	34	34	
	Neutral	7	7	
	Disagreed	5	5	
	Strongly Disagreed	4	4	

Number of Respondents

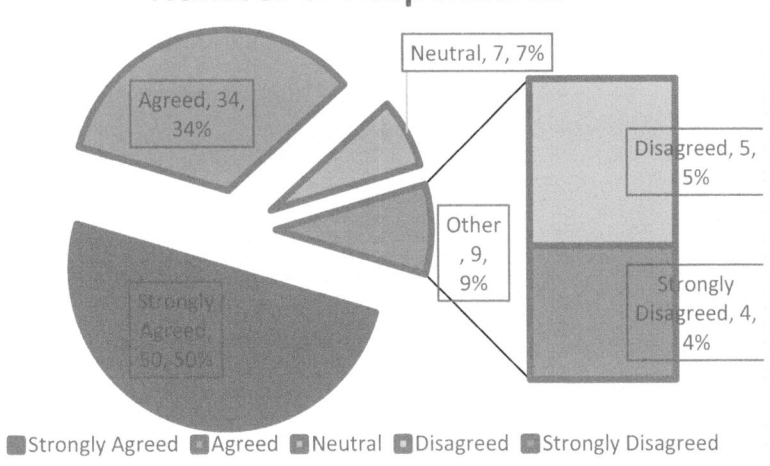

Strongly Agreed ■ Agreed ■ Neutral ■ Disagreed ■ Strongly Disagreed

Fig. 1. Chart representing response for the question number 1

This question demonstrates the potential of AI technologies to effectively serve students with unique needs and learning difficulties, fostering a more inclusive learning environment. Students with various learning needs can get specialised help, helping them to overcome obstacles and take part more completely in their educational journey, by utilising AI's adaptive capabilities, personalised learning routes, and assistive tools. By guaranteeing that all students have access to resources and support that are tailored to their specific learning requirements, this encourages inclusivity and equity. Respondents strongly agreed. The overall response to this question is summarized below in the table and chart (Table 2 and Fig. 2).

Table 2. Response for the question number 2

Question	Options	Data Collected Responses 100	Percentage	Mean
Do you agree that AI technologies can effectively support students with learning disabilities and special needs, promoting inclusivity in education?	Strongly Agreed	69	69	**4.60**
	Agreed	27	27	
	Neutral	1	1	
	Disagreed	1	1	
	Strongly Disagreed	2	2	

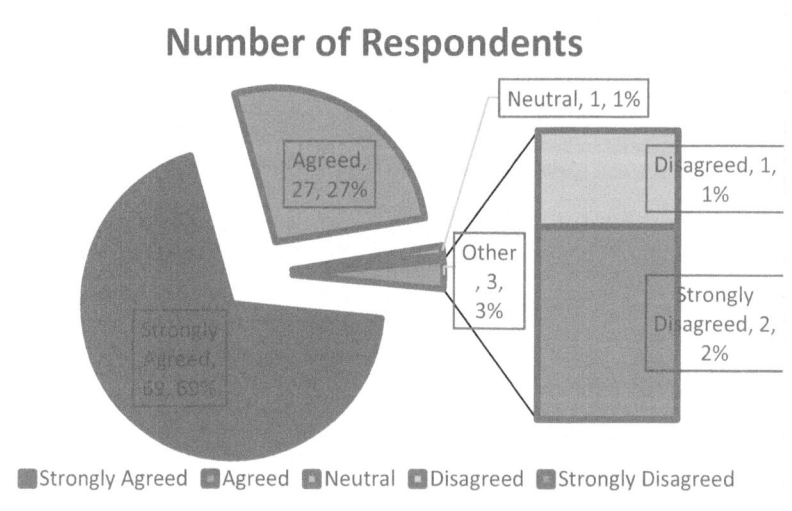

Fig. 2. Chart representing Response for the question number 2

The question highlights the significance of striking a balance when incorporating AI in education between human expertise and the capabilities of AI. It implies that judicious blending of human insights with AI-driven technologies is necessary for the ethical usage of AI in educational contexts. By striking this balance, it is made possible for AI to improve instructional settings and personalization while also taking ethical issues, data protection, and the equal distribution of AI-driven resources for all learners into account. Whether they were satisfied with their work relationships with the people around them, respondents agreed with this question. Responses are summarized below in the table and chart (Table 3 and Fig. 3).

Table 3. Response for the question number 3

Question	Options	Data Collected Responses 100	Percentage	Mean
Does AI integration in education should be approached with a balance between human expertise and AI capabilities to ensure responsible use?	Strongly Agreed	56	56	**4.34**
	Agreed	31	31	
	Neutral	7	7	
	Disagreed	3	3	
	Strongly Disagreed	3	3	

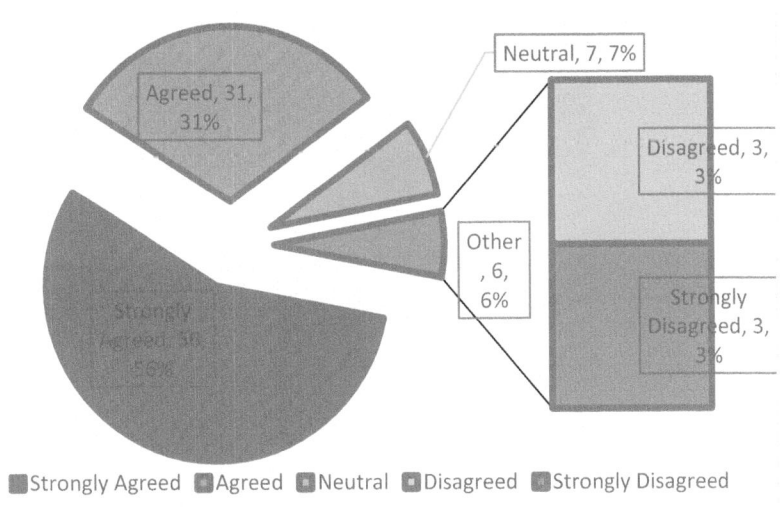

Fig. 3. Chart representing Response for the question number 3

The potential influence of AI-generated personalised feedback and recommendations on student learning results is highlighted by this question. AI can provide individualised feedback and recommendations for supplemental resources that are in line with each student's needs and learning style by analysing the data for each individual student. This individualised approach has the potential to increase student motivation, engagement, and understanding, which will likely result in better academic performance and overall learning outcomes. The learning experience is improved by AI's capacity to deliver timely and pertinent feedback, resulting in a more efficient and student-centered learning environment. When we asked this question to the respondents, all of them were similarly responsive. Few respondents strongly agreed. The overall response to this question is summarized below in the table and chart (Table 4 and Fig. 4).

Table 4. Response for the question number 4

Question	Options	Data Collected Responses 100	Percentage	Mean
Do you agree that AI-generated personalized feedback and recommendations can positively influence student learning outcomes?	Strongly Agreed	59	59	**4.32**
	Agreed	28	28	
	Neutral	4	4	
	Disagreed	4	4	
	Strongly Disagreed	5	5	

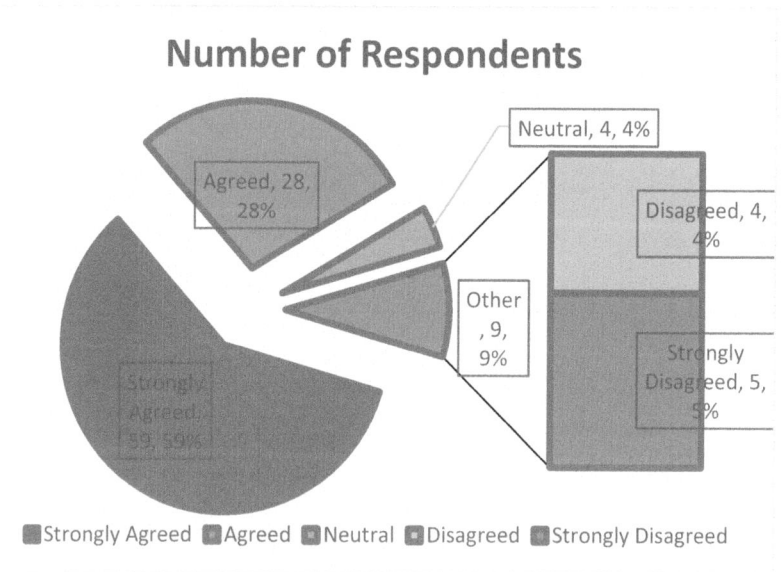

Fig. 4. Chart representing Response for the question number 4

The ethical issues surrounding the use of AI in education are highlighted by this query, notably those relating to algorithmic bias and data privacy. As AI systems rely on enormous volumes of student data, it is vital to ensure the security and privacy of this data. It's also important to address the possibility of algorithmic bias, in which AI programmes unintentionally support unfair or discriminating practises. To ensure the ethical and responsible integration of AI in education and to promote a safe and welcoming learning environment for all students, it is crucial to address these ethical problems. Respondents strongly agreed. The overall response to this question is summarized below in the table and chart (Table 5 and Fig. 5).

Table 5. Response for the question number 5

Question	Options	Data Collected Responses 100	Percentage	Mean
Does AI in education pose potential ethical concerns related to data privacy and algorithmic bias that need to be addressed?	Strongly Agreed	34	34	**3.90**
	Agreed	43	43	
	Neutral	10	10	
	Disagreed	5	5	
	Strongly Disagreed	8	8	

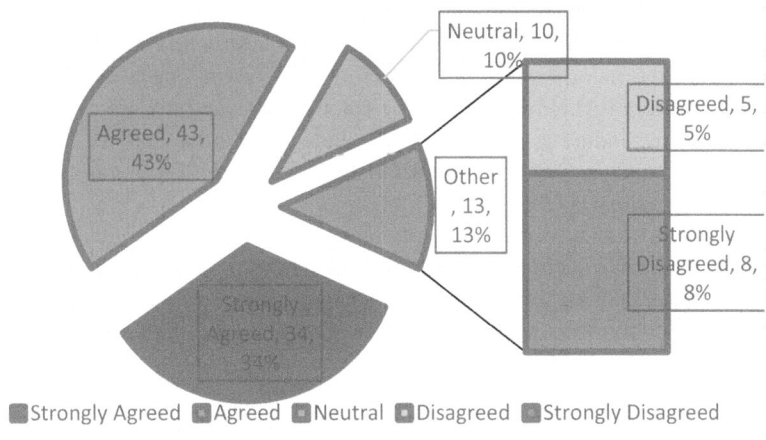

Fig. 5. Chart representing Response for the question number 5

9   Conclusion

Understanding the potential effects of AI technologies on the educational environment is now possible thanks to the research on Artificial Intelligence in Education: Improving Instructional Settings and Personalization. The study has shown the usefulness of AI in enriching instructional environments and fostering personalised learning experiences through a thorough investigation of its integration in education.

The results show that AI has a lot of potential for education, providing customised learning routes and adaptable content that are tailored to the needs and learning preferences of specific students. Intelligent tutoring systems (ITS) have demonstrated considerable promise for offering students individualised guidance, assisting them in overcoming obstacles and enhancing their academic achievement. Tools for content curation and development that are AI-driven have proved crucial in attracting students' attention and promoting a greater comprehension of the subject matter.

The study also emphasises how personalised feedback and recommendations created by AI have a favourable impact on students' learning outcomes. Students' engagement, motivation, and academic success have increased as a result of the timely and pertinent feedback provided by AI technologies. A more encouraging and sympathetic learning environment has been made possible by AI's emotional intelligence skills, which has benefited students' general wellbeing.

The study also emphasises the necessity to address any potential ethical issues with algorithmic bias and data protection when integrating AI. It is crucial to ensure that AI is used responsibly in education, with an emphasis on protecting student information and encouraging fair access to AI-driven resources for all students.

The study has shown that AI can give teachers useful information for making data-driven decisions. AI systems analyse enormous amounts of educational data and provide instructors with evidence-based knowledge to assist them in modifying their instructional practises and providing tailored interventions to enhance student outcomes.

As a result, there are several chances to improve instructional settings and personalization with the incorporation of artificial intelligence in education. Education can harness the power of AI to build more efficient, inclusive, and student-centered learning environments by finding a balance between human expertise and AI capabilities. To make sure AI in education is a responsible and fair tool, however, careful attention to ethical considerations is essential.

The research's conclusions and suggestions can act as a guide for educators, decision-makers, and other stakeholders as they decide how to responsibly implement and use AI in education. As long as human instructors and AI-driven technologies work together, education will continue to innovate and succeed in the future, preparing students for the opportunities and challenges of the digital age.

References

1. Feng, S., Law, N.: Mapping artificial intelligence in education research: a network-based keyword analysis. Int. J. Artif. Intell. Educ. **31**, 277–303 (2021)
2. Buabeng-Andoh, C.: Factors influencing teachers' adoption and integration of information and communication technology into teaching: a review of the literature. Int. J. Educ. Dev. Using Inf. Commun. Technol. **8**, 136–155 (2012)
3. Buga-ay, C.U., Causing, R.D.: Funneling teaching strategies to enhance students' learning, retention, and academic performance. Globus J. Progressive Educ. **13**(1), 129–137 (2023)
4. Buchanan, T., Sainter, P., Saunders, G.: Factors affecting faculty use of learning technologies: Implications for models of technology adoption. J. Comput. High. Educ. **25**(1), 1–11 (2013)
5. Deswal, A., Kumar, S.: Technological aspects in cloud computing. Cosmos J. Eng. Technol. **12**(2), 5–7 (2022)

6. Dunlosky, J., Rawson, K.A., Marsh, E.J., Nathan, M.J., Willingham, D.T.: Improving students' learning with effective learning techniques: promising directions from cognitive and educational psychology. Psychol. Sci. Public Interest **14**(1), 4–58 (2013)
7. Dabbagh, N., Bannan-Ritland, B.: Online learning: concepts, strategies, and application, pp. 68–107. Pearson/Merrill/Prentice Hall (2005)
8. Ertmer, P.A., Ottenbreit-Leftwich, A.: Removing obstacles to the pedagogical changes required by Jonassen's vision of authentic technology-enabled learning. Comput. Educ. **64**, 175–182 (2013)
9. Singh, M.K., Tripathi, S., Singh, P.: Cybersecurity and Data Privacy. Cosmos Int. J. Manag. **11**(2), 01–04 (2022)
10. Punla, N.F.D., Regalado, D.R.B., Navarro, R.M.: Perceived social support as moderator between self-stigma and help-seeking attitudes. Globus Int. J. Med. Sci. Eng. Technol. **11**(1), 87–98 (2022)
11. Jaramillo, R.N., et al.: Projected action plan and timeline in pilot testing of simulation-based learning in the college of nursing at Phinma-University of Pangasinan. Cosmos Int. J. Art High. Educ. **11**(2), 4–15 (2022)
12. Hu, H., Wang, X.: Evolution of a large online social network. Phys. Lett. A **373**(12–13), 1105–1110 (2009)
13. Koehler, M.J., Mishra, P., Cain, W.: What is technological pedagogical content knowledge (TPACK)? J. Educ. **193**(3), 13–19 (2013)
14. Moore, J.L., Dickson-Deane, C., Galyen, K.: E-Learning, online learning, and distance learning environments: are they the same? Internet High. Educ. **14**(2), 129–135 (2011)
15. Jaber, R.D., Reyes, H.G.P.: Assessing english language skills of higher education students through DYNED software: an input to policy formulation. Globus J. Progressive Educ. **13**(1), 112–119 (2023)
16. Galvez, R.S.: Graduate studies students' scholastic abilities and their effects on comprehensive examination. Cosmos Int. J. Manag. **11**(1), 112–119 (2021)
17. Tolentino, R.B.: Supervised industrial training online monitoring system with short message service (SMS) notification. Globus Int. J. Manag. IT **14**(2), 71–75 (2023)
18. Reyla, J.S.: Translanguaging in the classroom: impact on the academic performance of the learners. Cosmos Int. J. Art High. Educ. **11**(1), 72–87 (2022)
19. Dwivedi, A.: Psychological and technical barrier for teachers to shift face to face to online education during pandemic. Globus Int. J. Manag. IT **14**(2), 76–80 (2023)
20. Mendoza, R.M.N., Dayao, E.F.: LCUP student use of gmetrix system: an evaluation of usefulness, ease of use, and satisfaction. Cosmos J. Eng. Technol. **11**(2), 63–67 (2021)
21. Salinda, M.T., Tuzaon, A., Lachica, P.: Integrity of third year nursing students to online related learning experiences: a concept analysis. Globus Int. J. Med. Sci. Eng. Technol. **10**(2), 87–98 (2021)

Arduino-Based Wearable Health Monitoring System for Elderly Care Using Internet of Things

Danish Ather[1], Anupam Singh[2(✉)], Rubina Liyakat Khan[3], Richa Vijay[4], Ramveer Singh[5], and Raj Kumar[6]

[1] Amity University in Tashkent, Tashkent, Uzbekistan
[2] Department of Computer Science and Engineering, Graphic Era Hill University, Dehradun, India
anupam2007@gmail.com
[3] The Applied College, Imam Abdulrahman Bin Faisal University, Dammam, Saudi Arabia
[4] Amity University, Noida, Noida, Uttar Pradesh, India
[5] Department of CSE, Galgotias College of Engineering and Technology, Greater Noida, India
[6] Department of Computer Applications, Manav Rachna International Institute of Research and Studies (MRIIRS), Faridabad, India

Abstract. Concerns regarding the health and well-being of the old have been raised globally due to the elderly population's rapid rise. In order to overcome these difficulties, this study introduces an Internet of Things (IoT)-based wearable health monitoring system for senior care. The suggested system combines a variety of sensors with wireless communication capabilities to continuously monitor the health and physical activities of elderly people. The central element for data collection, processing, and transmission is the Arduino platform, which is renowned for its simplicity and adaptability. In addition to an accelerometer for tracking movement and fall detection, the wearable gadget has sensors for assessing vital signs like heart rate, blood pressure, body temperature, and oxygen saturation. The Arduino board processes the data locally, and it is then wirelessly transferred to a centralized monitoring system using Wi-Fi or Bluetooth Low Energy (BLE) communication. Caretakers or medical experts can access the central monitoring system, which offers real-time visualization and analysis of the gathered data. It makes it possible to discover aberrant health situations early, to act quickly, and to perform remote monitoring. Additionally, the system may produce warnings and notifications in the event of crises, ensuring quick access to medical care. The suggested wearable health monitoring system built on Arduino has a number of benefits, including affordability, mobility, and simplicity of use. It gives elderly people the ability to preserve their independence while offering a safety net through ongoing health monitoring. Additionally, the system enables healthcare professionals to give individualized and timely interventions based on real-time health data and simplifies remote caregiving.

Keywords: Arduino · Wearable · Health Monitoring · Elderly Care · Internet of Things

1 Introduction

A global phenomena, the aging population raises the need for efficient healthcare solutions catered to the unique requirements of senior citizens. The use of wearable health monitoring systems, which offer continuous and remote monitoring of vital signs and physical activity, has emerged as a viable strategy to meet this challenge. Early diagnosis of health issues, prompt therapies, and enhanced healthcare management are all made possible by these systems. Wearable technologies that are coupled with Arduino platforms provide an affordable and adaptable option for senior care within the Internet of Things (IoT) framework. This article explores wearable health monitoring systems for aged care in the IoT that are designed and implemented on the Arduino platform, highlighting the system's potential advantages and uses [1].

Numerous health issues, such as chronic illnesses, mobility issues, and cognitive impairment, are faced by the aging population. Traditional healthcare procedures frequently need to be improved in order to address these problems because they rely on routine medical facility visits that might not fully reflect a person's health status. In order to enable early diagnosis of health anomalies and customized therapies, wearable health monitoring devices offer continuous and real-time data collecting.

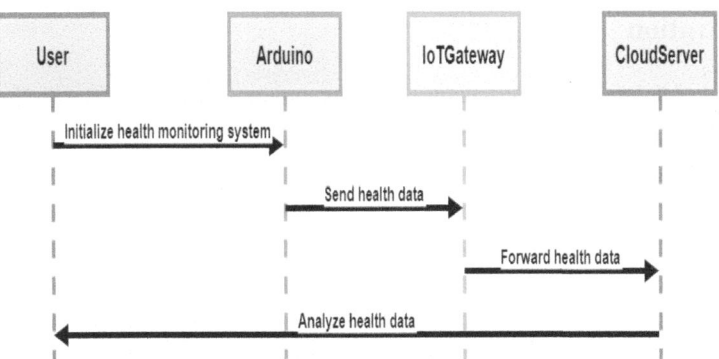

Fig. 1. Process Flow Representation

For wearable health monitoring systems, the Arduino platform, known for its ease of use, low cost, and versatility, is a great starting point. Arduino boards offer various input and output options, making them suitable for integrating sensors and communication modules. By leveraging Arduino, researchers and developers can create wearable devices capable of collecting and processing vital signs and activity data, facilitating seamless integration into an IoT ecosystem [2].

Accurate measurement of vital signs forms a crucial aspect of wearable health monitoring systems as represented in Fig. 1. Various sensors can be integrated into Arduino-based wearable devices to capture vital signs such as heart rate, blood pressure, body temperature, and oxygen saturation. For example, photoplethysmography (PPG) sensors can measure heart rate and blood oxygen saturation, while temperature sensors and blood pressure cuffs can provide additional health insights.

Physical activity monitoring is essential for assessing an individual's health and well-being. Accelerometers and gyroscopes can be incorporated into wearable devices to track movement patterns, step counts, and gait analysis. These sensors enable the detection of sedentary behavior, assessment of mobility, and identification of fall events, which are critical for the safety of elderly individuals. The IoT paradigm enables seamless connectivity and data exchange between wearable devices and centralized monitoring systems. Arduino boards equipped with wireless communication modules, such as Wi-Fi or Bluetooth Low Energy (BLE), facilitate real-time transmission of collected data to a centralized server or caregiver's smartphone. It enables remote monitoring, data analysis, and timely interventions based on the individual's health status [3].

Integrating Arduino with wearable health monitoring systems opens many benefits and applications for elderly care. These systems empower elderly individuals to actively participate in healthcare management, promoting independence and control over their well-being. Caregivers and healthcare professionals can remotely monitor vital signs, detect health anomalies, and intervene promptly, reducing the risk of adverse events and hospitalizations. Furthermore, the collected data can be analyzed to derive meaningful insights, enabling personalized healthcare approaches and improved disease management.

2   Motivation

The global population is experiencing a significant demographic shift, with the elderly population growing at an unprecedented rate. With this demographic change comes an increased need for healthcare solutions that cater to the unique needs of older adults. Wearable health monitoring systems, which enable continuous and remote monitoring of vital signs and physical activity, have emerged as a viable solution to address this difficulty, as was covered in the preceding section.

Integrating wearable Arduino-based devices with medical monitoring systems in an Internet of Things (IoT) setting has a number of benefits, including affordability, adaptability, and real-time data transmission. The justification for adopting an Arduino-based wearable health monitoring system for elderly care in the IoT is described in more detail in the following section, with an emphasis on the possible advantages and the demand for such solutions. This motive is mentioned in Fig. 2.

2.1   Growing Elderly Population and Healthcare Challenges

The aging of the world's population is causing an increase in healthcare issues associated to aging. Elderly people frequently deal with chronic illnesses, decreased mobility, and a higher risk of falling. The provision of prompt interventions and monitoring in traditional healthcare techniques, which rely on recurrent visits to medical institutions, frequently needs to be enhanced. By providing continuous and remote monitoring, wearable health monitoring systems have the potential to close this gap by facilitating the early detection of health issues and proactive interventions [4].

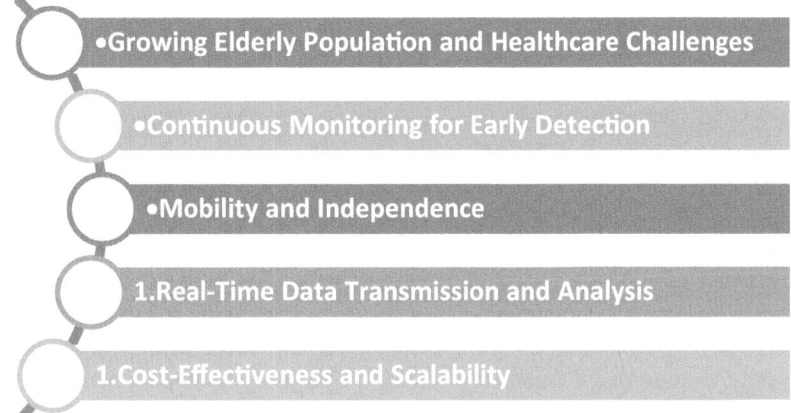

Fig. 2. Motivation for the Purposed Work

2.2 Continuous Monitoring for Early Detection

Elderly individuals require regular monitoring of their vital signs, including heart rate, blood pressure, body temperature, and oxygen saturation. Wearable devices integrated with Arduino platforms enable continuous monitoring, capturing real-time data on these vital signs. By continuously monitoring these parameters, potential health issues can be identified early, leading to prompt interventions and improved health outcomes [4].

2.3 Mobility and Independence

Age-related mobility issues frequently prevent elderly people from accessing healthcare treatments. The benefit of mobility offered by wearable health monitoring systems enables people to go about their regular lives while being continuously monitored. It promotes independence and enhances the quality of life for older adults, who can receive healthcare services remotely without frequent visits to healthcare facilities.

2.4 Real-Time Data Transmission and Analysis

Integrating Arduino with IoT technologies enables real-time data transmission from wearable devices to centralized monitoring systems. It allows healthcare providers and caregivers to access and analyze the collected data remotely. Real-time data analysis facilitates early detection of health anomalies, enables personalized healthcare interventions, and enhances disease management. Timely interventions can help prevent adverse events, reduce hospitalizations, and improve healthcare outcomes for elderly individuals [5].

2.5 Cost-Effectiveness and Scalability

Arduino platforms are known for their affordability and versatility, making them suitable for widespread adoption in healthcare applications. Compared to traditional healthcare monitoring systems, Arduino-based wearable devices offer cost-effective solutions for continuous monitoring. The scalability of Arduino-based systems allows for easy replication and deployment, making them accessible to a larger population of elderly individuals [6].

3 Literature Survey

An attempt by [7] proposes an Arduino-based wearable ECG monitoring system for remote healthcare. The developed system in the study integrates Arduino with IoT technologies to enable real-time monitoring and transmission of ECG signals, which facilitates early detection of cardiovascular abnormalities. In [8] authors discussed the potential of IoT-based wearable sensor devices in preventive medicine. The study highlights using Arduino-based wearable devices to continuously monitor vital signs and physical activity, facilitating personalized healthcare interventions. The authors in [9] propose a real-time healthcare monitoring system for the elderly using wireless sensor networks and Arduino. The proposed system enables continuous monitoring of vital signs, fall detection, and remote data transmission for prompt interventions.

Another study by [10] explores the benefits of IoT in healthcare, emphasizing the use of Arduino-based wearable devices for elderly care. The integration of Arduino with IoT technologies for ongoing health monitoring and individualized healthcare management is covered in the article. The authors of [11] offer a wearable health monitoring system that is IoT-based. The system includes sensors for vital sign monitoring and fall detection, allowing for remote data transmission and continuous monitoring for proactive healthcare management. The authors of [4] suggested a wearable sensor for geriatric fall detection that is Arduino-based. Accelerometers and gyroscopes are used by the system to detect falls and send out prompt signals for quick help.

An IoT-based intelligent healthcare system using Arduino is designed and implemented in a recent development in [12]. Different vital sign and activity tracking sensors are incorporated into the system, enabling remote data processing and ongoing health monitoring. Authors of [4] display a wearable sensor system for Arduino-based senior health monitoring. Real-time monitoring and alarms are provided by the system, which includes sensors for detecting vital signs, physical activity, and fall detection. The creation of an IoT-based wearable healthcare monitoring device utilizing Arduino is encouraged in a study by [13]. The system incorporates numerous sensors for measuring bodily activity and vital signs, enabling ongoing remote monitoring and medical interventions. The authors of [14] mentioned the development of an Arduino-based Internet of Things-based wearable sensor system for healthcare monitoring. The system includes sensors for vital sign monitoring, fall detection, and activity tracking, allowing for remote data transmission and continuous monitoring.

4 Methodology

The mathematical equations used for DS18B20 (temperature sensor), MAX30102 (pulse oximeter sensor), and AD8232 (ECG sensor), as shown in Fig. 3 and Fig. 4, are as follows:

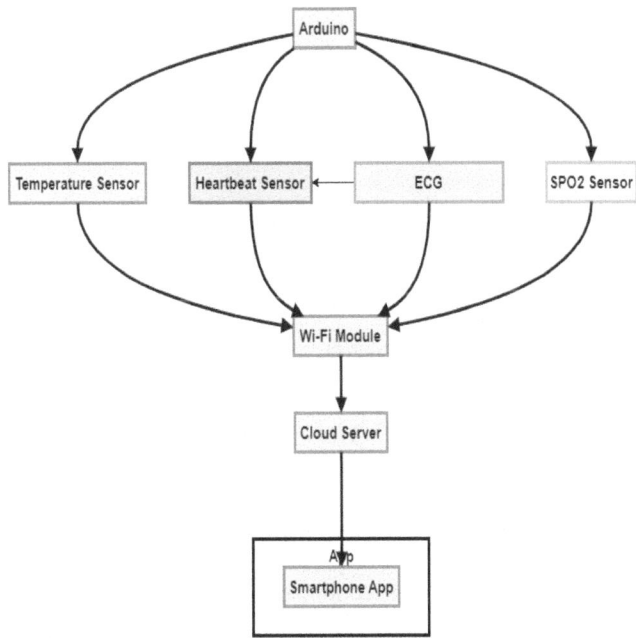

Fig. 3. Arduino-Based Wearable Health Monitoring System

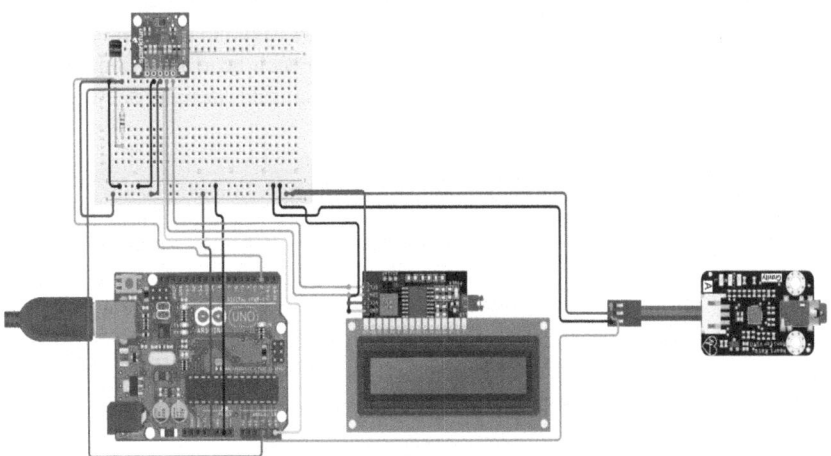

Fig. 4. Implementation Diagram

4.1 DS18B20 (Temperature Sensor)

The DS18B20 provides temperature measurements in digital format. The sensor communicates over a 1-Wire bus and provides the temperature reading as a 12-bit value. The mathematical equation for converting the raw sensor data into temperature depends on the specific library or code implementation. However, a commonly used formula is the following as shown in Eq. 1:

$$\text{Temperature} = (\text{RawData}/16) - \text{Offset} \tag{1}$$

Here, RawData is the 12-bit value read from the DS18B20 sensor, and Offset is an optional offset value to account for any calibration adjustments or temperature compensation required.

4.2 MAX30102 (Pulse Oximeter Sensor)

The MAX30102 is a sensor that combines a photoplethysmography (PPG) sensor and an IR LED to measure heart rate and SpO2 (blood oxygen saturation). The sensor provides raw PPG data and calculates the heart rate and SpO2 using built-in algorithms. The mathematical equations for heart rate and SpO2 calculation are typically implemented within the library or code provided by the sensor manufacturer. The specific equations may vary based on the implementation, but they usually involve complex signal-processing techniques, including filtering, peak detection, and calibration algorithms.

4.3 AD8232 (ECG Sensor)

The AD8232 is an ECG sensor that measures the heart's electrical activity. It provides analog ECG waveforms that must be processed and analyzed to extract heart rate information. The mathematical equation for calculating heart rate from the ECG signal typically involves detecting R-peaks (the peak corresponding to each heartbeat) and measuring the time intervals between consecutive R-peaks. The heart rate (in beats per minute) can be calculated using the formula as shown in Eq. 2:

$$\text{Heart Rate} = 60 / (\text{RR Interval}) \tag{2}$$

RR Interval is the duration between two consecutive R-peaks, typically measured in seconds. It is important to note that the actual implementation of these equations may vary based on the specific libraries, algorithms, and processing techniques used in the code or library associated with each sensor. It is recommended to refer to the documentation, datasheets, and example code provided by the sensor manufacturers or library developers for the most accurate and up-to-date equations and implementation details.

5   Implementation

To measure the parameters of the patient body temperature, heart rate, SpO2 (blood oxygen saturation), and ECG (electrocardiogram) using Arduino, one can follow the general steps as described below:

5.1 Hardware Setup

Connecting appropriate sensors to the Arduino board is crucial based on the parameters one wishes to measure. For the current system, a digital temperature based on sensor DS18B20 is used for body temperature measurement, and for heart rate and SpO2 measurement, a pulse oximeter based on sensor MAX30102 is used. Along with ECG measurement, an ECG module based on sensor AD8232 is used.

5.2 Sensor Calibration

Before taking measurements, it is to be ensured that the sensors are correctly calibrated according to their respective specifications (Refer to the sensor datasheets for calibration procedures).

5.3 Software Configuration

An Arduino IDE is set up on the computer with the necessary libraries for each sensor being installed. After that, a code that initializes the sensors and defines the necessary variables to store the measured values is developed.

5.4 Body Temperature Measurement

We Use the library file to communicate with the temperature sensor, which reads the temperature data from the sensor using the Arduino's analogue pins. Then we convert the raw sensor data into a temperature value using the conversion formula provided by the sensor manufacturer.

5.5 Heart Rate and SpO2 Measurement

Use the library specific to the pulse oximeter sensor. Next, we initialize the pulse oximeter sensor and configure it to measure heart rate and SpO2, which helps in reading the sensor's heart rate and SpO2 values. The system then stores and processes the data as needed.

5.6 ECG Measurement

The library specific to the installed ECG module or sensor is used to measure ECG readings. The system can read the ECG data from the sensor by configuring the ECG sensor according to the library instructions. The system then processes and filters the ECG data to obtain the heart rate and other relevant information.

5.7 Data Display and Analysis

Once all data is processed, the system displays the parameters on an LCD and stores them on the cloud in parallel. Afterwards, implemented algorithms and calculations to analyze the data, such as detecting abnormal heart rhythms or temperature variations, are carried out.

6   Result and Analysis

Adopting the system excrement was done on 250 random older adults aged between 30 and 40. The sample contains 125 male and 125 female individuals. The experiment was repeated five times at one and a half hours each to test the system's accuracy. Following results are obtained from the experiment.

The individuals showed a variation in body temperature, as shown in Fig. 5. It is observed that the temperature at these intervals falls within the normal range for human body temperature in Celsius, which is typically considered to be between 36.1 °C (97°F) and 37.2 °C (99°F).

While observing the pulse rate, the interval variations are described in Fig. 6. It is to be noted that the recorded pulse rate ranges between 60 and 100 beats per minute (bpm) which is considered normal. However, it is essential to note that individual variations exist, and factors such as age, fitness level, and overall health can influence the resting pulse rate. Athletes or individuals with a high level of cardiovascular fitness may have resting pulse rates below 60 bpm, which can still be considered normal. No such case was found in our experiment.

Oxygen saturation measures the percentage of hemoglobin in the blood saturated with oxygen. It is often measured using a pulse oximeter. Using the system, Fig. 7 displays the results of the recorded data from the experiment. The oxygen levels recorded in the experiments are within the normal range for oxygen saturation generally, i.e., between 95% and 100% indicating healthy individuals in the sample.

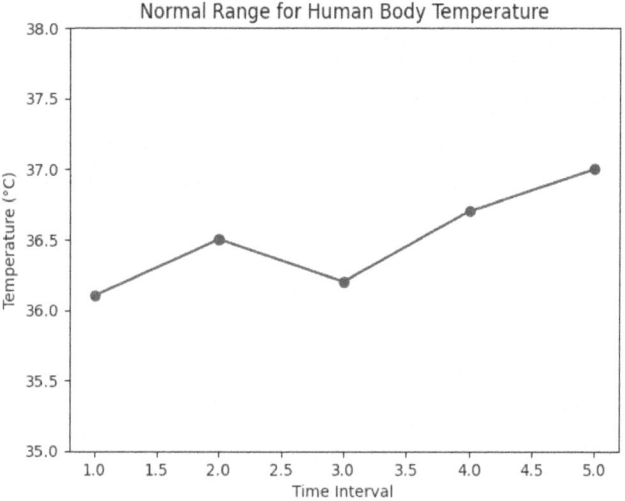

Fig. 5. Human Body Temperature

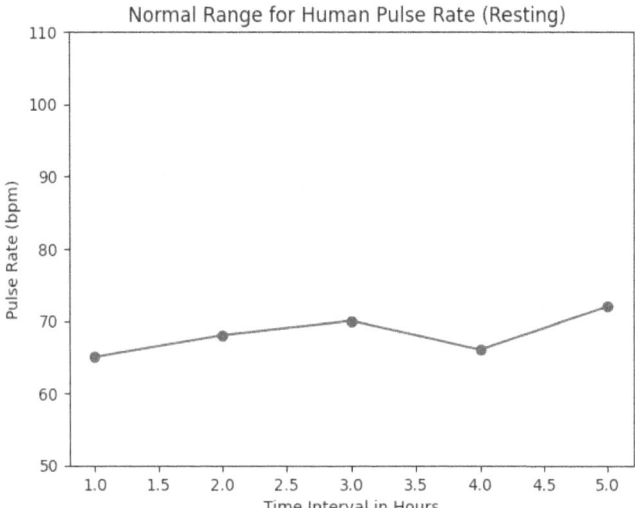

Fig. 6. Pulse Rate

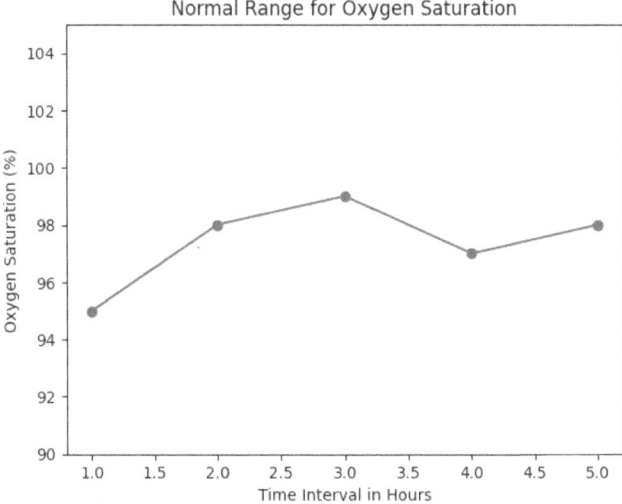

Fig. 7. Oxygen Saturation (SpO2)

The excrement also included ECG monitoring, and the following are the results. The recorded data by the system shows the P-wave duration representing atrial depolarization of the individuals was between 80 and 120 ms (Fig. 8).

While for the QRS complex, representing ventricular depolarization, the system recorded a duration between 70 and 100 ms which is considered normal among adults (Fig. 9).

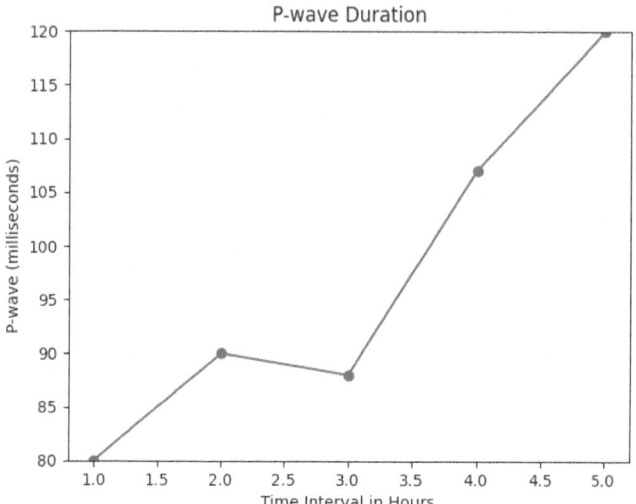

Fig. 8. P-wave Duration

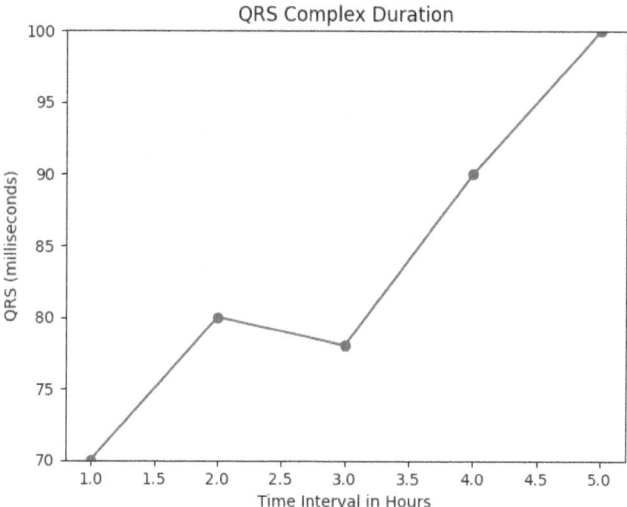

Fig. 9. QRS Complex Duration

For the QT interval, measuring the total duration of ventricular depolarization and repolarization, the recorded data shows that within the individuals in the sample, it varies between 250 to 440 ms which is considered normal for adults, including those aged 30 to 40 years (Fig. 10).

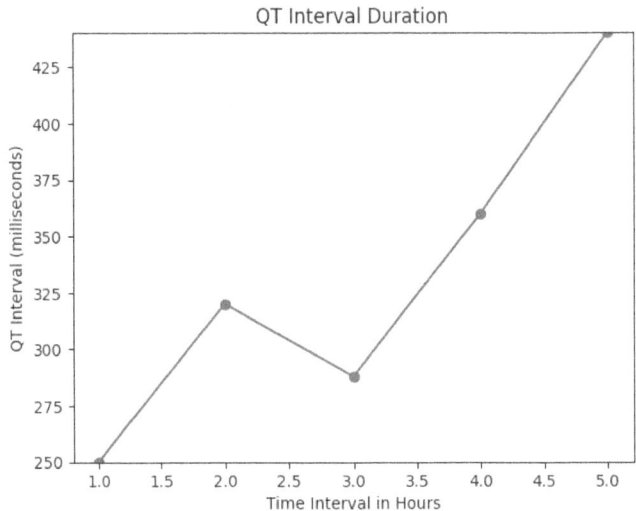

Fig. 10. QT Interval Duration

7 Conclusion

In conclusion, the implementation of an Arduino-based wearable health monitoring system for elderly care in the IoT, using sensors such as DS18B20 (temperature sensor), MAX30102 (pulse oximeter sensor), and AD8232 (ECG sensor), holds great potential for enhancing healthcare management and improving the quality of life for elderly individuals by integrating of these sensors with Arduino enables continuous monitoring of vital signs, including body temperature, heart rate, blood oxygen saturation (SpO2), and electrocardiogram (ECG). By leveraging the capabilities of Arduino and IoT technologies, several vital benefits are achieved, as listed below:

The wearable health monitoring system allows for real-time, continuous monitoring of vital signs, enabling early detection of abnormalities and prompt interventions. Elderly individuals can be monitored remotely, allowing them to live independently while ensuring their well-being. The gathered data from the sensors can be analyzed and processed to derive meaningful insights about the individual's health status. It facilitates personalized healthcare approaches, enabling healthcare providers to deliver tailored interventions and improve disease management strategies. Integrating sensors like the MAX30102 and AD8232 enables the detection of falls and abnormalities in heart rhythm, allowing immediate alerts and assistance. It enhances the safety and well-being of elderly individuals, reducing the risks associated with falls and cardiac events.

The Arduino platform and the selected sensors provide a cost-effective and versatile solution for wearable health monitoring. Arduino's affordability, ease of use, and extensive community support make it accessible for widespread adoption in healthcare applications. However, it is essential to note that successful implementation requires proper calibration, validation, and adherence to established standards. Calibration procedures specific to each sensor should be followed to ensure accurate measurements.

Appropriate data processing algorithms and analysis techniques must also be applied to the collected data for meaningful interpretation and decision-making.

In summary, implementing an Arduino-based wearable health monitoring system for elderly care in the IoT, utilizing sensors like DS18B20, MAX30102, and AD8232, showcases the potential to revolutionize healthcare management for elderly individuals. By providing continuous monitoring, personalized interventions, and enhanced safety features, such a system can significantly improve the quality of life and promote independent living for the elderly.

References

1. Gubbi, J., Buyya, R., Marusic, S., Palaniswami, M.: Future Generation Computer Systems Internet of Things (IoT): a vision, architectural elements, and future directions. Future Gener. Comput. Syst. **29**, 1645–1660 (2013)
2. Yao, W., et al.: Study and application of an elevator failure monitoring system based on the internet of things technology. Sci. Program. **2022** (2022). https://doi.org/10.1155/2022/251 7077
3. Aranda, F.J., Parralejo, F., Álvarez, F.J., Paredes, J.A.: Performance analysis of fingerprinting indoor positioning methods with BLE. Expert Syst. Appl. **202**, 117095 (2022). https://doi.org/10.1016/j.eswa.2022.117095
4. Kher, R.K., Patel, D.M.: A comprehensive review on wearable health monitoring systems. Open Biomed. Eng. J. **15**(1), 213–225 (2022). https://doi.org/10.2174/187412070211501 0213
5. Jain, A., Sarkar, A., Ather, D., Raj, D.: Temperature based automatic fan speed control system using Arduino. SSRN Electron. J. (2022). https://doi.org/10.2139/ssrn.4159188
6. Khan, R.L., Priyanshu, D., Ather, D., Allataifeh, H.: An implementation of internet of things-based live temperature and humidity monitoring system. In: Proceedings of the 2022 11th International Conference on System Modeling and Advancement in Research Trends, SMART 2022 (2022). https://doi.org/10.1109/SMART55829.2022.10046723
7. Chuah, Y.D., et al.: Fall detection of elderly people in bathroom: a complement method of wearable device. Int. J. Appl. Eng. Res. **11**(6) (2016)
8. Wan, J., et al.: Wearable IoT enabled real-time health monitoring system. EURASIP J. Wirel. Commun. Netw. **2018**(1) (2018). https://doi.org/10.1186/s13638-018-1308-x
9. Yang, Z., Zhou, Q., Lei, L., Zheng, K., Xiang, W.: An IoT-cloud based wearable ECG monitoring system for smart healthcare. J. Med. Syst. **40**(12) (2016). https://doi.org/10.1007/s10 916-016-0644-9
10. Almarashdeh, I., et al.: Real-time elderly healthcare monitoring expert system using wireless sensor network. SSRN Electron. J. (2019). https://doi.org/10.2139/ssrn.3415732
11. Mahapatra, S., Singh, A.: Application of IoT-based smart devices in health care using fog computing. Fog Data Analytics for IoT Applications: Next Generation Process Model with State of the Art Technologies, pp. 263–278 (2020)
12. Noury, N., et al.: Fall detection - principles and methods. In: Annual International Conference of the IEEE Engineering in Medicine and Biology - Proceedings (2007). https://doi.org/10.1109/IEMBS.2007.4352627
13. Shi, Q., et al.: Progress in wearable electronics/photonics—Moving toward the era of artificial intelligence and internet of things. InfoMat **2**(6), 1131–1162 (2020). https://doi.org/10.1002/inf2.12122
14. Darwish, A., Hassanien, A.E.: Wearable and implantable wireless sensor network solutions for healthcare monitoring. Sensors **11**(6), 5561–5595 (2011). https://doi.org/10.3390/s11060 5561

Sentiment Analysis of Twitter Data Using Supervised Machine Learning

Akanksha Singh Solanki$^{(\boxtimes)}$ and Puneet Nema$^{(\boxtimes)}$

Computer Science Department, Laxmi Narain College of Technology
Bhopal, Bhopal, Madhya Pradesh, India
S99.akanksha@gmail.com, puncetn@lnct.ac.in

Abstract. Finding and labelling expressions of opinion or emotion in written content is the primary focus of sentiment analysis. Now more than ever, people turn to online communities to share their perspectives and vent their emotions. As a result, there is a mountain of data being produced each day that may be successfully mined for insights. Conducting sentiment analysis on this kind of data can help generate a holistic perspective on certain items. Sentiment research on Twitter can be difficult because of the widespread use of slang and misspellings. The rapid influx of fresh words also makes it harder to analyse and compute the sentiment than it would be with more conventional sentiment analysis methods. The maximum number of characters allowed in a tweet is 140. Therefore, another challenge is overcoming the limitations of short communications to learn crucial details. Sentiment analysis of tweets can greatly benefit from knowledge-based techniques and machine learning. As individuals adapt to new ways of interacting on social media platforms like Snapchat, Instagram, Twitter, etc., the amount of data they generate increases at an exponential rate. There are literally billions of fresh pieces of content uploaded every day, including text, music, and video. This is due to the fact that numerous people frequent the website in question. These individuals wish to express their views on whatever subject they feel fits. These entries are meant to express the thoughts of a single individual on a particular subject. The goal of this research is to analyze these posts and identify the feelings that motivated their creation. We've settled on Twitter as the venue for this effort. The changes made to this social networking site are referred to as "tweets." In this research, we look at how Twitter users feel about specific businesses. Critical feedback on the company's products from people all around the world would be offered by generating a basic sentiment score and then classifying it as positive or negative.

Keywords: Twitter · Sentiment analysis · Machine Learning · LSTM model

1 Introduction

When it comes to reaching a wide audience with short messages, Twitter is one of the most popular options. Twitter is a popular platform for individuals to share and discuss their thoughts and ideas on a wide range of topics. Marketing initiatives benefit from customers' willingness to voice their opinions on a wide range of topics related to brands and goods through blogs. Researchers also use online data to do Sentiment Analysis on how the general public feels about a product or topic .Sentiment analysis is used to detect and extract private details from a text (P. K. K. K. Sentamilselvan, 2020). Whether the underlying sentiment is favourable, negative, or neutral, sentiment analysis is the most commonly used text categorization approach [1].

Challenges in Twitter sentiment analyses:

- Some tweets are often written in unpleasant language, whereas other brief messages lack emotional tone indicators.
- URLs, Hashtags, emojis, abbreviations, and acronyms are utilised often on Twitter.

Sentiment analysis is the process of identifying and labelling emotional overtones in written content. S. Santhosh Baboo [2022] argues that social media platforms like YouTube, Twitter, and Facebook can be crucial in pandemic situations. Twitter is a popular, effective, and all-encompassing social media site where millions of individuals discuss various issues [2]. Massive amounts of information may be found in these systems. Ninety percent of these sources are either texts or media. Using a technique called "Sentiment Analysis," one may determine the intensity and direction of emotions like happiness, sadness, grief, anger, and adoration in a piece of writing, review, or online remark. Opinion mining, also known as sentiment analysis (B. V. Gupta, 2022), is a technique for gauging the overall tone of a text with respect to a certain information source. Sentiment analysis is a challenging and rapidly developing field of research because of the prevalence of slang, misspelt words, truncated forms, different characters, regional accents, repeated characters, and incoming emojis. Sentiment analysis is useful in many contexts, including social media [3]. [2016] (P. Nakov) The ubiquitous 5-point "HIGHLY POSITIVE," which replaces older 2- or 3-point scales, has become the de facto standard.

A scale is used in the business world anywhere expert opinion is needed. Examples include the five-point scale used by Trip Advisor, Amazon, and the "help" section of several websites. In the field of machine learning, moving from a two- or three- point category scale to a five-point ordered scale is called an "ordinal" categorization shift "(also known as ordinal regression).

1.1 Twitter Sentiment Analysis

One application of text mining is Sentiment Analysis, which attempts to determine a text's underlying emotional tone, such as a tweet. You may share any honest emotion or viewpoint on Twitter. The problem with sentiment analysis is that it would be unnatural for any algorithm to promise 100% accuracy or prediction. Problems with the sentence and document levels may affect the sentiment analysis results on Twitter data. Searching for positive and negative phrases in a sentence, for example, is inefficient since the flavor of a text block is largely reliant on its surrounding context. There are 1.3 billion people on Twitter, with 330 million actively using the platform every month and 145 million using it daily. Since Twitter is the most understandable source of real-time public discussions, its data may be useful for gauging customer sentiment as individuals and markets respond to corporate actions.

1.2 Sentiment Analysis

Both sentiment analysis and opinion mining fall under the text mining umbrella. It's a method for anticipating how people will feel about a certain issue. It appears in a wide variety of contexts, including online reviews, social media, and news articles. Market research and governmental policymaking are two other applications for sentiment analysis findings. Many businesses rely heavily on customer feedback since it improves service quality and boosts deliveries, not to mention it helps shape policy. The primary goal of sentiment analysis is to ascertain whether or not a given text exhibits a positive or negative attitude. Conversations, forums, and weblogs are all great sources for the data collected via sentiment analysis. Sentiment analysis is becoming more common due to social media's growing significance. To gauge public reaction to its products and policies, a company may commission polls and surveys of its customers. Consumers compared prices and researched products using consumer sentiment analysis. Once upon a time, marketers would employ sentiment analysis to learn more about their target market and their experiences with their products and services. One reason is that so many websites are hosted on the web today, and that number is only expected to grow, making it impossible to collect opinion data from them all. Another is that there aren't enough resources to do so. The need for a standardized method to get the same results is evident. In addition, the internet's text corpora provide useless and useful data that might be used in the study's analysis. There is often a fine distinction between these two types of data, which creates extra work for analysts. Human readers have a hard time sifting through online content to get what they need and then summarizing what they've found [5].

1.3 Machine Learning

Algorithms, which are collections of mathematical procedures used to determine links between variables, constitute backbone of machine learning techniques that are currently available. In this work, we will present the training process and evaluate an algorithm to determine whether or not a breast tissue sample contains cancerous cells based on the characteristics of sample itself. While the specifics of how various algorithms work may vary, there is consistent pattern in their development. For all their apparent obscurity, machine learning (ML) algorithms often bear more than a passing resemblance to more familiar forms of statistical analysis. Given their similarities, the line between statistical and machine learning methods can be difficult to draw. One method for distinguishing different techniques is examining their basic objectives. The objective of statistical techniques is to make inferences, more especially, to draw inferences about populations or to gain scientific insights from data acquired from a sample representative of that group. The requirement to make inferences about the correlations between variables motivates the use of several statistical processes, including logistics and linear regression, which can generate predictions for new data. To develop a model that captures the association between clinical characteristics and mortality following organ transplant surgery, for instance, it is necessary to understand elements that distinguish low mortality risk from high mortality risk. By doing so, we can create interventions to boost success rates and lower mortality. Therefore, understanding the connections between the different variables should be the goal of any statistical conclusion.

In contrast, the fundamental concern in machine learning is an accurate prediction, the "what" rather than the "how." For instance, if the prediction is accurate in image recognition, the relationship between individual characteristics (pixels) and conclusion is of minimal importance. Because the link between many inputs, including pixels in a video or image and geo-location, is generally complex and non-linear, this is an essential component of machine learning methodologies. When there are a large number of predictors, each of which only adds a small amount to the model, and when the interactions between the predictors and outcomes are non-linear, it is extremely difficult to characterize the links between the predictors and outcomes in a rational fashion.

Supervised Learning: In a nutshell, labeled training manuals are essential to the success of supervised learning. Supervised learning is a powerful classification technology that has shown highly encouraging results when used for opinion classification. Sentiment analysis typically employs the supervised classification methods of SVM (Support Vector Machine), NB (Naïve Bayes) and ME (Maximum Entropy).

- **Naïve Bayes**
- This probabilistic machine learning model is used to solve classification issues, and Bayes theorem is the model's underlying theoretical framework. (Gandhi, 2018). Below is the general Bayes equation.

$$P(B) = \frac{P(A)P(B)}{P(B)}$$

If B has already happened, then it is possible to calculate the likelihood of A happening. Evidence (B) and a working hypothesis (x) It is assumed that the features do not interact with one another in any way. Naive Bayes classifiers come in a variety of flavours, including the multinomial, the Bernoulli, and the Gaussian. The more comprehensive Statistical n-Gram Modeling can be derived from this classifier. Compared to previous modeling methods, this one is superior for classifying texts. This method takes into account the relationships between neighboring words.

- **Support vector machine**

Because it provides a good method for text classification, SVM (Support Vector Machine) is a discriminative classifier that is taken into consideration. It is a technique of classification that is used in statistics. Several nonlinear mappings are utilised within the SVM to transform the input (actually-valued) feature vectors into a higher-dimensional feature space. The concept of structural hazard minimization is the basis for the development of SVMs. Finding a hypothesis (h) with the lowest error probability is the goal of structural risk minimization. On the other hand, traditional pattern recognition learning algorithms are predicated on empirical risk minimization, an approach that aims to maximize the efficiency of the training data. Structural risk minimization aims to identify a working hypothesis (h) with the lowest error rate. When computing the hyper aircraft to split the statistics points, such as Education and SVM, problems with quadratic optimization arise. Because its class complexity is not dependent on the function space's dimensionality, SVM can study a greater number of pattern combinations and can scale more effectively. SVM has ability to replace the teaching styles dynamically each time there is a new sample throughout the category.

- **Maximum entropy**

The weights constitute the defining characteristic of maximum entropy. It is also known as the exponential classifier due to the fact that it functions by first extracting some features from the input, combining those features linearly, and then applying the result as the sum of the exponents.

Unsupervised Learning: In contrast to the supervised learning technique, this technique does not involve the creation of any predefined labels for the classes. Compared to supervised learning techniques, these methods are more difficult to implement and require more time, yet they yield fairly similar results to those methods. Unsupervised learning is a method that can be accomplished through the use of neural networks.

Semi-supervised: Because it employs both labelled and unlabeled data, semi-supervised learning can be regarded as a hybridization of unsupervised and supervised approaches that have been examined up to this point. This is because semi-supervised learning uses labelled and unlabeled data, which is why this result is. This is because semi-supervised learning makes use of both labelled & unlabeled data. Consequently, it is a mode of education that falls somewhere in the middle of "learning without supervision" and "learning with supervision." In the real world, labelled data may be few in a number of scenarios, whereas unlabeled data are plentiful; as a result, semi-supervised learning is effective in these kinds of circumstances. The result of any semi-supervised learning model should be a prediction that is more accurate than the one that could be made by utilising the model's labelled data alone. This is the ultimate goal of any semi-supervised learning model. Semi-supervised learning has applications in many different areas, such as identifying fraudulent activity, labelling data, categorization of text, and machine translation.

2 Literature Review

Tokenization, stop-word elimination, and stemming are all used in Lexicon-based Sentiment Analysis, which has been developed by a number of researchers. However, accuracy is lower than in prior machine learning-based studies. Following these steps, we can identify if a person is feeling neutral, positive, or negative. In this work (Aldinata) [2023], researchers looked at LGBT Sentiment Analysis, a highly discussed but divisive issue in modern society. Basic processing is done on Tweets from all 50 US states before their emotions are determined. Logistic Regression, Naive Bayes, XG Boost, Linear Support Vector Machine, Pattern Analyzer, and Text Blob are among the five sentiment categorization techniques tested, along with raw and cleaned data. The most important discovery was that using Logistic Regression without text preprocessing yields the greatest F1-score (70.87%). Using their sentiment classifier on tweets from the United States concerning the LGBT population, they discovered that the great majority of posts landed in the "neutral" category [5]. Public opinion was surveyed (T. H. Jaya Hidayat, 2022) and sorted into three groups: pro, con, and neutral, in order to better understand how people feel about this new information. Against, indifferent, and in favour The distributed model and the distributed bag of words are used as Doc2Vec models. Logistic regression and support vector machines are used as classifiers in this system. The results demonstrate that almost all of the models and classifiers are against the development of Rinca Island, and their accuracy rates are higher than 75% [6]. The proposed architecture for encoding news material is presented in this paper (R. Das and T. D. Singh, 2012). The text branch encodes semantic content information with consideration for news semantic data. The observable parts of a news picture are extracted and encoded simultaneously. The next step is to use a fusion layer to combine visual and textual data for joint sentiment classification in a multimodal framework [7]. To organise the user's emotions, this framework was created. To successfully fuse cross-modal data for prediction of ultimate emotion, a decision-level fusion approach is used for all three models in the final stage. Using an internal Assamese dataset, they conducted trials and found that contextually integrated multimodal features outperformed the best unimodal features by 9.3 percentage points. In order to correctly interpret the sentiment of trending tweets in the data stream given by the Twitter API, the authors of this paper (A. Motz) [2022] used many algorithms to reach a consensus. Multiple methods, including support vector machines, Naive Bayes, textblobs, and the Lexicon Approach, were used. They believe that by combining the two methods, they will have more success [8]. The examination of their model using a labelled dataset shows that the integration of these four methods yielded the greatest overall accuracy (68.29%) of all methods examined. In order to identify sad Twitter users, the author of this work (F. Azam) builds a machine learning model to examine users' linguistic patterns. Using both Random Forest and Support Vector machine training, they created diagnostic models but found that Random Forest was more effective [9]. They think the study's results may be used to create a new technique for discovering depressed people on social media. Based on research (F. E. Ayo, 2020), To further address problems with Twitter's data streams, we also built a comprehensive metadata framework for the categorization of hate speech on the platform. In terms of identifying hate speech, the built-in generic metadata architecture outperformed all previous methods, with an accuracy of 0.95, an F1- score of 0.93, a recall of 0.92, and a precision of 0.93. Also, a generic metadata architecture for classifying the tone of hate speech achieved a 91.5% F1 score, which was significantly higher than that of competing methods [10]. Here we investigate the use of SL (single linkage), CL (complete linkage), and AL (average linkage) hierarchical clustering methods for Twitter sentiment analysis (M. Bibi, W. Aziz, 2020). By putting together a cooperative framework made up of AL, SL, and CL, we are able to operationalize the idea of picking the best cluster for tweets by majority vote. They evaluate hierarchical clustering strategies against k- means and two additional contemporary classifiers (support vector machines and Nave Bayes). Accuracy and efficiency in terms of time are used as evaluation criteria for clustering and classification [11]. The results of the experiments reveal that majority-voting-based cooperative clustering yields high- quality clusters, albeit at the cost of time efficiency. The Naive Bayes Algorithm in RapidMiner tools helped the researchers of V. A. Fitri [2019] reach an accuracy of 86.43% in their tests, which was much higher than the accuracy they obtained using Random Forest and other methods.

Both the C4.5 and the Decision Tree models [12] achieved an accuracy of 82.91%. Using a dataset of tweets obtained from a GitHub repository, the authors of this study (Y. Gupta and P. Kumar) [2019] discuss the training of multiple deep learning and machine learning models. By analysing Twitter sentiment towards various candidates, the proposed approach was put to the test in a case study that correctly predicted the outcome of elections in Punjab in February 2017 [13]. The suggested SA system instantaneously does sentiment analysis and displays the results in rhythm with Twitter posts. A user-friendly dashboard displaying the results has been built. Minutely, the screen refreshes with the latest tweets and data visualisations. Tweets are also saved in a comma-separated values (CSV) file for use in making projections. (M. R. Islam, 2018) looked into Facebook users' feelings of melancholy. They used the KNN algorithm to spot people feeling down on Facebook. Their study achieved accuracy levels between 60% and 70% across a range of matrices [14]. This research by S. Rahman and J. N. Hemel [2018] uses machine learning methods to propose a relationship between changes in Bitcoin's price and the emotional state of its users. They have detailed their plan for carrying it out, including the results of their analysis and the criteria they used to determine costs. They used a sentiment-based strategy to tackle the issue of foreseeing fluctuations in the price of bitcoin, therefore establishing the relevance of public opinion in the cryptocurrency industry. In addition, the results of this study suggest a novel approach to making use of data collected from social media platforms [15].

This study (E. S. Tellez, 2017) aims to determine the best word n-grammetokenizer and token-weighting technique to use with a Support Vector Machine classifier trained on two Spanish datasets, using text transformations including entity removal, stemming, and lemmatization. In order to determine what features are shared by the most effective classifiers, the process involves analysing every possible combination of text changes and the variables associated with them. We also introduce a unique approach to combining n-grammes with words and q-grammes with characters. This innovative combination of words and characters yields a classifier that improves upon previous word-based combinations by 11.17 and 5.62 percentage points, respectively, when applied to the INEGI and TASS'15 datasets [16].

After outlining Twitter data sentiment analysis, current sentiment analysis tools, relevant work approaches, and a case study showcasing effort, the authors of this work (P. Mishra and R. Rajnish) [2017] move on to provide their findings. The data shows that although some respondents were enthusiastic, others were not, and still others were ambivalent.

Connected business insights in the telecommunications sector are the focus of another recent study (N. A. Vidya, 2017). The NBR scale ranges from 0 to 40, with PT XL AxiataTbk receiving 32.3, PT TelkomselTbk receiving 19.0, and PT Indosat receiving 10.9.Tbk, correspondingly, after a comprehensive assessment of these five items [18].

The purpose of [22] was to develop and train a model that can accurately and auto-matically classify customer twitter reviews of leading cell phone brands. For obtaining the polarity values of each tweet review in dataset, they employed the Python TextBlob module. In addition, they utilized Logical Regression, Nave Bayes, RF SVM, and DT, as well as TF-IDF vectorizers and Bag of Words, to train and construct a model which will classify customer tweet reviews into 5 opinion groups: Weakly Negative, Strongly Negative Strongly Positive and Neutral, Weakly Positive. They found that employing the Bag of Words vectorizer, LR, and SVM methods outperformed other algorithms with an accuracy of 88% while utilizing TF-IDF vectorizer, SVM outperformed other algorithms with an accuracy of 87%.

This study [23] suggested a machine-learning model for classifying Twitter postings as either negative, positive or neutral. They applied their approach to a database consisting tweets from six different US-based airlines. They began their model with preprocessing procedures in which they cleaned tweets and extracted characteristics to represent them as feature vectors, and then they constructed their Bag of Words (BoW) model. Tweets were classified using six ML approaches throughout the classification stage: DT, LR, SVM, RF, and XgBoost. For the goal of testing and validating the data, they utilized the k-fold cross-validation method after dividing the data into 70% training and 30% testing during the validation stage. After evaluating each classifier, they calculated its Precision, Accuracy, Recall, and F1 Score. After looking over results from every classifier, they settled on SVM as having the highest accuracy (83.33%).

In this study, [24] Using Twitter's textual data and a sentiment analysis method, the suggested model analyzes the views of Indian populace on plastic ban, and results show that a machine classifier based on a learning method can achieve an accuracy of 77.94% in its classifications across a variety of datasets. The outcome will assist determine how efficient and successful India's ban on polybags would be when fully implemented.

[24] One study used a variety of features, such as TF-IDF, Bag of Words, emoticon lexicons, and N-gram, in conjunction with DL and ML algorithms to deduce sentiment from airline tweets. This study shows that the emotional tone of a message is most strongly influenced by the emoticons used in it. The superiority of DL algorithms over ML algorithms has also been established.

The purpose of [20] is to use machine-learning approaches to do an in-depth ordinal regression-based sentiment analysis of tweets. In the suggested method, tweets are first subjected to pre-processing, after which a feature extraction approach is used to generate an effective feature set. The scoring and balancing of these aspects can then be found under various different classes. With the findings of sentiment analysis as input, the proposed system uses classification techniques RF, DTs, Support Vector Regression, and Multinomial Logistic Regression to assign ratings to statements. The developers of system rely on a Twitter dataset that is made available by NLTK corpus resources in order to put together the version of the system that is actually functional. Experiments validate the effectiveness of the proposed strategy for detecting ordinal regression with ML methods. In addition, the data shows that among all the algorithms tested, DTs performs the best.

[22] The findings of this opinion poll are expected to be useful for all parties involved, but especially for Go-Jek. The SVM algorithm is dependable for classification and regression since it can handle high-dimensional data sets, classification problems, kernel linear and non-linear regression. However, there is a problem with SVM when trying to optimise it by picking the right parameters. The issue of picking appropriate parameters in support machine vector approach can only be solved by using a genetic algorithm.

3 Methodology

The study's issue statement, methodology, and steps taken to complete the study are all discussed here. An algorithm with excruciatingly specific instructions and a comprehensive flowchart of the whole study process are also included here.

3.1 Problem Statement

Unfortunately, not all data mining problems can be solved with the same machine learning approach. Data mining cannot be done without experimentation. As a consequence, they ranked every component of a wide variety of Tweet classification methods into several ontologies to determine which was most suitable. They hope to create a technique that takes a phrase or sub-sentence as input and outputs an emotion score for that particular chunk of text. Building a second mechanism that can break down a text (like a tweet) into as many subsentences as it includes ontologies is necessary for this to happen. This is now being researched alongside this dissertation at our institution.

Proposed Methodology
The Twitter API is used to collect data, and then the tweets are classified as either positive or negative. The Natural Language Toolkit's (NLTK) corpora repository makes the dataset available to the public for use in their own studies. NLTK is extensively utilised in a variety of fields of study. There are a total of 10,000 tweets in the database, 5,000 of which are bad and 5,000 of which are favourable. Next, we employed a number of supplementary libraries, including numpy, pandas, tensorflow, sklearn, genism, and seaborn. All null values, emoticons, URLs from tweets, Twitter handles, punctuation, extra spaces, digits, and special characters have been deleted from the data during preprocessing. The LSTM model was then used to obtain a performance matrix including measures of precision, accuracy, recall, and F1-Score. This section details the tactics and procedures that will be used to accomplish the set goals. The anticipated outcomes of these methods are also discussed. What follows are explanations of a few of these tactics:

a) *Data Collection*
   In this work, we have utilized a dataset from Kaggle data repository that identifies Twitter sentiment analysis.
b) *Data Preprocessing*
   Data preprocessing is a crucial aspect of the data mining and analysis process because it transforms raw data into a form that can be read and analysed by computers and machine learning. Text, images, videos, and other forms of real- world information are chaotic. It is not always well-written or coherent, and it often lacks essential details or has an illogical, inconsistent structure. Machines are best able to analyse data when it is well organised and presented as a string of zeros and ones. This makes it easy to do calculations on structured data like integers and percentages. However, before any analysis can be performed, any textual or visual data that has not been preprocessed in a consistent manner must be cleaned and structured.

Feature Extraction

"Feature extraction" means converting raw data to a set of manipulable numerical attributes without altering the meaning or structure of the data itself. Using machine learning on unprocessed data does not yield the same outcomes.

Tokenizer: This technique breaks the given text into tokens (small parts) and removes any punctuation from textual data. This study employed nltk. Tokenization is performed through Tokenize methods (a built-in function of nltk toolkit).

Data Splitting

"Split" data means to separate it into two or more distinct groups. When data is divided in half, typically one half is used to test the model and the other is used to train it. The data set was split in half for this investigation. We used 75% of the training set and 25% of the testing set.

Classification

One subset of Supervised Learning algorithms is the classification algorithm, which takes in data and figures out how to categorise it. We needed some time to figure things out. Before an algorithm can accurately categorise new data, it must be trained on an existing collection of observations. Use an LSTM model for machine learning-based categorization and ordinal regression on tweets.

LSTM Model

Long short-term memory (LSTM) Models may help reveal enduring connections. The dynamic LSTM/LBU gadget eliminates the gradient issues. While the LSTM model's performance is comparable to that of more complicated networks, it has fewer nodes. A Constant Error Carousel (CEC) occurs when the flaws of one mechanism are combined with the weight of another. Target timing moves forward or backward as interval width is increased or decreased. The phasing in of CEC serves two purposes: the first is the introduction of multiplicative units; the second is the introduction of past experience

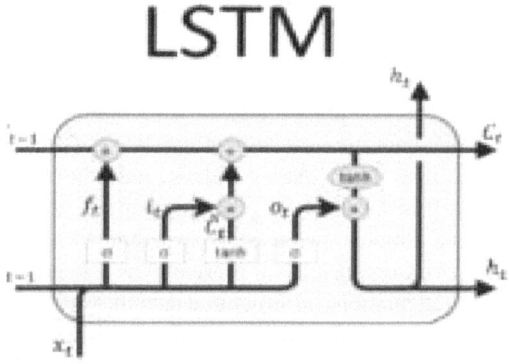

Fig. 1. LSTM Model

recall and the identification of the mental state that produces it. When it comes to modeling long- term dependencies, standard LSTMs have been shown to perform better than Recurrent Neural Networks (LSTMs) (Fig. 1).

Data Pre-processing
The process of cleaning, converting, and arranging raw data into a format that can be efficiently evaluated and used for a given purpose is referred to as data preprocessing. This process can also be referred to as data preparation. The purpose of preprocessing of data is to get it in a more accurate, consistent, and useable state so that it may be used later on in the analysis process. This step is critical for obtaining meaningful results from data analysis and machine learning algorithms. Following processes were done:

Removing Null Values
In the process of data pre-processing, it is usual practise to get rid of any values that are null. It entails finding and dealing with any data points in a dataset that are either missing or incomplete in some way. It is essential to get rid of null values in order to guarantee the correctness and dependability of the data.

Removing URLs
In NLP (natural language processing), one of the most common preprocessing steps involves removing URLs, also known as Uniform Resource Locators, from text input. URLs are not helpful for natural language processing activities since they do not provide any information that is meaningful for the analysis of the text. Eliminating them helps to reduce noise in the data, which in turn improves the NLP algorithms' overall performance.

Removing Emojis
The elimination of emojis from text data is a common step in the pre-processing phase of natural language processing (NLP) projects. Emojis are graphical characters or icons that are commonly used in online forums, text messages, and social media platforms to communicate a range of emotions. Emojis are not considered significant text for natural language processing (NLP) tasks; hence, deleting them can help minimize noise in the data. Despite the fact that they can sometimes be helpful.

Removing Twitter Handles
Preprocessing text data for use in natural language processing (NLP) tasks sometimes involves removing Twitter handles. Each user on Twitter is given a "handle," which is a special reference to their username. The "@" character, which is commonly used to preface handles, is not parsed as meaningful text by natural language processing tools. Data noise is reduced and NLP algorithm efficiency is increased when they are eliminated.

Removing Punctuations
Natural language processing (NLP) tasks sometimes necessitate the removal of punctuation from text data as a preprocessing step. NLP algorithms can be hampered by

punctuation marks like commas, periods, and exclamation points because they are not part of the text's meaning. Data noise can be reduced by removing them.

Removing Extra Space

Commonly performed in NLP preprocessing is removal of unnecessary spaces from text data. Because of formatting discrepancies, text data may contain unnecessary gaps, which might hinder the efficiency of NLP algorithms. Cleaning up the data and making it easier to process by NLP algorithms is facilitated by getting rid of unnecessary gaps.

Removing Numbers and Special Characters

In NLP tasks, removing numbers and special characters from text data is a common preprocessing step. Numerals and special characters, such as symbols, punctuation marks, and numbers, are not regarded as meaningful text for NLP tasks and can cause NLP algorithms to malfunction. Eliminating them lessens data noise and enhances the effectiveness of NLP algorithms.

Removing Stop Words

The performance of NLP algorithms can be improved by removing stop words from the text input in order to make the data set smaller overall. Stop words are words which are utilized frequently in a language but don't have much of an impact on meaning of language as a whole. Examples of stop words are "a," "the," "and," "is," and others. These terms appear somewhat frequently in the text data, but they are not likely to convey any information that is relevant.

Removing Single Quotes New Line Characters

Text data may contain newline characters or single quotes as a result of inconsistent formatting, which might hinder the efficiency of natural language processing (NLP) algorithms. By removing them, we can clean up the data and boost the efficiency of algorithms.

3.2   Proposed Algorithm

Input: NLTK dataset **Output:**Get classified outcomes **Strategy:**
Step 1: Start implementation process Step
2: Import set of data (NLTK)
Step 3: Processing of raw data before it is used

- Check null value

- Remove Punctuation

- Remove emoji

- Remove Twitter handles

- Remove newline character Step 4: Apply feature extraction techniques
- Tokenizer

Step 5: Split dataset into training& testing set that divided into 75:25

- Training set (75%)
- Testing set (25%)

Step 6: A proposed machine learning model

- LSTM memory

Step 7: Evaluation Parameters (Recall, Accuracy, Precision and F1-score)
Step 8: Get results. Step 9: End

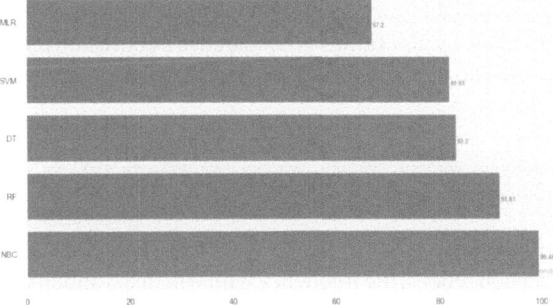

4   Results Illustrations

In the previous study use four machine learning techniques to perform sentiment analysis of twitter data based on ordinal regression. The twitter dataset is made publicly available by the NLTK, preprocessing and feature extraction technique and scoring and balancing the data. The Previous accuracy of multinomial logistic regression, SVM, Random Forest & Decision Tree are 67.2%, 81.95%, 83.2% & 91.81%. In this work I proposed the model Naïve Bayes Classifier from NLTK to achieve the accuracy 99.5%

4.1   Dataset Description

In the future, we plan to improve our approach by attempting to use bigrams and trigrams. Furthermore, we intend to investigate different machine learning techniques and deep learning techniques, such as Deep Neural Networks, Convolutional Neural Networks, and Recurrent Neural Networks (Fig. 2).

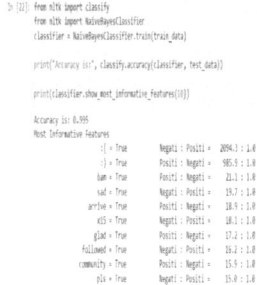

Fig. 2. Twitter sentiment analysis graph

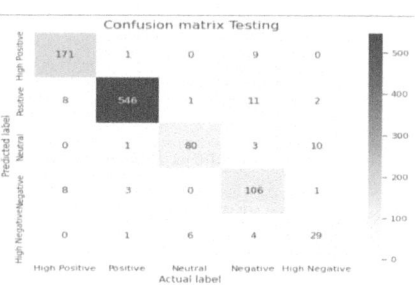

Fig. 3. Confusion Matrix on LSTM

4.2 Performance Metrics

1) Accuracy
In context of sentiment analysis on Twitter data, accuracy is a commonly used metric to assess performance of a The above Fig. 3 shows confusion Matrix of proposed LSTM model utilizing NLTK dataset. In graph, axis shows the actual label and y-axis shows predicated label each label shows. The number of successful and unsuccessful classifications made by algorithms is tabulated in a confusion matrix. "True Positive" classification model. The formula for accuracy is the same as general classification problem:

$$\text{Accuracy} = \frac{|TN| + |TP|}{|TN| + |FN| + |TP| + |FP|} \tag{1}$$

2) Precision
Precision reflects the frequency with which the anticipated result of a classifier is accurate when it represents true. The preciseness formula is:

$$\text{Precision} = \frac{|TP|}{|FP| + |TP|} \tag{2}$$

3) Recall
A recall is a metric used to assess performance of a classification model, particularly in context of imbalanced datasets. It is the ratio of a total number of accurate occurrences in training data to fraction of correct predictions. The formula for recall in the context of sentiment analysis on Twitter data is:

$$\text{Recall} = \frac{|TP|}{|FN| + |TP|} \tag{3}$$

4) F1 Score

F1 score is the answer to the problem of inaccurate results on unbalanced data (Hawar Sameen Ali Barzenji) [2021]. When data is unbalanced, we use the F1 score to help us out. The F1 score is a combination of the recall and accuracy scores [19]. The F1 rule of score is means both projected and actual values are positive. The confusion matrix that we have here has a true positive value of171, and a true negative value indicates that both the actual and anticipated values are negative. Here, in this confusion matrix, the True negative value is 29.

$$F - measure = 2 \times \frac{Recall \times Precision}{Recall + Precision} \tag{4}$$

Table 1. Comparison of proposed and base models using NLTK dataset

Parameters	Proposed Classifier	Base Classifier		
	LSTM	RFC	DT	Multinomial linear regression
Accuracy (%)	0.93	86.00	84.17	80.71
Precision (%)	0.93	82.92	80.63	75.87
F1-score (%)	0.93	68.99	80.67	71.45
Recall (%)	0.93			

As shown in Table 1 compares the dataset performance of numerous classifiers. LSTMs, RFCs, DTs, and MLRs are being compared. Each classifier's recall, precision, accuracy, and F1score are listed. A classifier's accuracy is its ability to accurately classify data items. LSTM has the best accuracy at 93%, followed by RFC at 86%, DT at 84.17%, and Multinomial Linear Regression at 80.71%.Precision refers to percentage of "yes" data that was accurately identified. LSTM classifier has the highest precision at 93%, followed by the RFC, DT, and Multinomial Linear Regression. F1-scoreindicates a classifier's precision-recall balance. In this table, the LSTM classifier has the greatest F1-score (93%), followed by the DT (80.67%), Multinomial Linear Regression (71.45%), and RFC (68.99%).Recall, also known as sensitivity or (True Positive Rate) TPR, is proportion of positive data items correctly detected by the classifier.

5   Conclusion

The purpose of NLP and ML-based sentiment analysis on Twitter is to determine a tweet's overall tone by locating and extracting subjective information (positive, negative, or neutral). Many disciplines will find this work useful since it streamlines previously time-consuming processes like equation solution and analysis. With this work, the proposed study on categorising and analysing Twitter spam streams is complete. To improve classification accuracy and use machine learning classifiers to identify spam tweets, we describe a feature extraction strategy based on a combination of random forest, linear regression, and PCA (principal component analysis). Simulation results reveal that the suggested work has a greater detection ratio when compared to other current efforts. Results from this proposed study demonstrate a significant improvement over existing classifiers in terms of accuracy, recall, precision, and F1-score when using a large amount of data. With some tweaks to the proposed algorithm for identifying positive and negative tweets, this hybrid approach is also used for sentiment classification, with favourable results. Opinion and text mining have an area dedicated to analysing tweets for sentiment. It examines sentiments expressed in tweets and calibrates a machine learning model for future use based on the results. The approach consists of five stages: sentiment detection, text pre- processing, data collection, sentiment categorization, model training, and testing. Improvements in this field over the past decade have allowed for model efficiencies of 85–90%. However, there is still a lack of nuanced data. The widespread usage of slang and acronyms further complicates their practical use. Many analyzers perform poorly when the number of classes grows. The generalizability of the concept outside the context of the current debate has not been properly explored. As a result, sentiment analysis has enormous development potential in the future.

References

1. Sentamilselvan, P.K.K.K., Aneri, D., Athithiya, A.C.: Twitter sentimental analysis using machine learning techniques. Int. J. Innov. Technol. Explor. Eng. **9**(3), 2249–8958 (2020). https://doi.org/10.35940/ijeat.c6281.029320
2. Santhosh Baboo, S., Amirthapriya, M.: Sentiment analysis and automatic emotion detection analysis of twitter using machine learning classifiers. Int. J. Mech. Eng. **7**(2), 974–5823 (2022)
3. Gupta, B.V.: Comparison of sentiment analysis algorithms using twitter and review dataset. Int. J. Res. Appl. Sci. Eng. Technol. **10**(4), 2299–2304 (2022). https://doi.org/10.22214/ijraset.2022.41785
4. Nakov, P., Ritter, A., Rosenthal, S., Sebastiani, F., Stoyanov, V.: SemEval-2016 task 4: Sentiment analysis in twitter (2016). https://doi.org/10.18653/v1/s16-1001
5. Aldinata, Soesanto, A.M., Chandra, V.C., Suhartono, D.: Sentiments comparison on Twitter about LGBT. Procedia Comput. Sci. **216**, 765–773 (2023). https://doi.org/10.1016/j.procs.2022.12.194
6. Hidayat, T.H.J., Ruldeviyani, Y., Aditama, A.R., Madya, G.R., Nugraha, A.W., Adisaputra, M.W.: Sentiment analysis of twitter data related to Rinca Island development using Doc2Vec and SVM and logistic regression as classifier. Procedia Comput. Sci. **197**, 660–667 (2022). https://doi.org/10.1016/j.procs.2021.12.187
7. Das, R., Singh, T.D.: A multi-stage multimodal framework for sentiment analysis of Assamese in low resource setting. Expert Syst. Appl. **204**, 117575 (2022). https://doi.org/10.1016/j.eswa.2022.117575
8. Motz, A., Ranta, E., Calderon, A.S., Adam, Q., Alzhouri, F., Ebrahimi, D.: Live sentiment analysis using multiple machine learning and text processing algorithms. Procedia Comput. Sci. **203**, 165–172 (2022). https://doi.org/10.1016/j.procs.2022.07.023
9. Azam, F., Agro, M., Sami, M., Abro, M.H., Dewani, A.: Identifying Depression among Twitter Users using Sentiment Analysis (2021). https://doi.org/10.1109/ICAI52203.2021.9445271
10. Ayo, F.E., Folorunso, O., Ibharalu, F.T., Osinuga, I.A.: Machine learning techniques for hate speech classification of twitter data: state-of-the-art future challenges and research directions. Comput. Sci. Rev. **38**, 100311 (2020). https://doi.org/10.1016/j.cosrev.2020.100311

11. Bibi, M., Aziz, W., Almaraashi, M., Khan, I.H., Nadeem, M.S.A., Habib, N.: A cooperative binary-clustering framework based on majority voting for twitter sentiment analysis. IEEE Access **8**, 68580–68592 (2020). https://doi.org/10.1109/ACCESS.2020.2983859

12. Fitri, V.A., Andreswari, R., Hasibuan, M.A.: Sentiment analysis of social media twitter with case of anti-LGBT campaign in Indonesia using Naïve Bayes, decision tree, and random forest algorithm. Procedia Comput. Sci. **161**, 765–772 (2019). https://doi.org/10.1016/j.procs.2019.11.181

13. Gupta, Y., Kumar, P.: Real-Time Sentiment Analysis of Tweets: A Case Study of Punjab Elections (2019). https://doi.org/10.1109/ICECCT.2019.8869203

14. Islam, M.R., Kamal, A.R.M., Sultana, N., Islam, R., Moni, M.A.: Detecting Depression Using K-Nearest Neighbors (KNN) Classification Technique (2018). https://doi.org/10.1109/IC4ME2.2018.8465641

15. Rahman, S., Hemel, J.N., Anta, S.J.A., Al Muhee, H., Uddin, J.: Sentiment analysis using R: an approach to correlate cryptocurrency price fluctuations with change in user sentiment using machine learning. In: 2018 Joint 7th International Conference on Informatics, Electronics & Vision (ICIEV) and 2018 2nd International Conference on Imaging, Vision & Pattern Recognition (icIVPR), pp. 492–497 (2018). https://doi.org/10.1109/ICIEV.2018.8641075

16. Tellez, E.S., Miranda-Jiménez, S., Graff, M., Moctezuma, D., Siordia, O.S., Villaseñor, E.A.: A case study of Spanish text transformations for twitter sentiment analysis. Expert Syst. Appl. **81**, 457–471 (2017). https://doi.org/10.1016/j.eswa.2017.03.071

17. Mishra, P., Rajnish, R., Kumar, P.N.: Sentiment analysis of Twitter data: case study on digital India (2017). https://doi.org/10.1109/INCITE.2016.7857607

18. Vidya, N.A., Fanany, M.I., Budi, I.: Twitter sentiment to analyze net brand reputation of mobile phone providers. Procedia Comput. Sci. **72**, 519–526 (2015). https://doi.org/10.1016/j.procs.2015.12.159

19. Barzenji, H.S.A.: Sentiment Analysis of Twitter Posts using Machine Learning Algorithms, pp. 980–983 (2021)

20. Zervoudakis, S., et al.: OpinionMine: a Bayesian-based framework for opinion mining using twitter data. Mach. Learn. Appl. **3**, 100018 (2021). https://doi.org/10.1016/j.mlwa.2020.100018

21. Zhao, H., et al.: A machine learning-based sentiment analysis of online product reviews with a novel term weighting and feature selection approach. Inf. Process. Manage. **58**(5), 102656 (2021). https://doi.org/10.1016/j.ipm.2021.102656

22. Windha Mega, P.D., Haryoko: Optimization of parameter support vector machine (SVM) using genetic algorithm to review go-jek's services. In: 2019 4th International Conference on Information Technology, Information Systems and Electrical Engineering, ICITISEE 2019 (2019). https://doi.org/10.1109/ICITISEE48480.2019.9003894

23. Wongkar, M., Angdresey, A.: Sentiment analysis using naive bayes algorithm of the data crawler: twitter. In: Proceedings of 2019 4th International Conference on Informatics and Computing, ICIC 2019 (2019). https://doi.org/10.1109/ICIC47613.2019.8985884

24. Ullah, M.A., et al.: An algorithm and method for sentiment analysis using the text and emoticon. ICT Exp. **6**(4), 357–360 (2020). https://doi.org/10.1016/j.icte.2020.07.003

Automated Drone Detection for Surveillance and Security Enhancement

M. Kalidas[1], T. Priya[2], V. Ansal[3], S. Mayakannan[4], P. K. Dhal[5], S. Sathish Kumar[6], and Kibebe Sahile[7(✉)]

[1] Department of MCA, Chaitanya Bharathi Institute of Technology, Gandipet, Hyderabad 500075, India
mkalidas_mca@cbit.ac.in

[2] Department of Mathematics, NPR College of Engineering and Technology, Natham, Dindigul, Tamilnadu, India

[3] Department of Electrical and Electronics Engineering, NIT Goa, Farmagudi, Ponda, Goa 403401, India

[4] Department of Mechanical Engineering, Vidya Vikas College of Engineering and Technology, Tiruchengode, Namakkal, Tamilnadu, India

[5] Department of Electrical and Electronics Engineering, Vel Tech Rangarajan Dr Sagunthala R&D Institute of Science and Technology, Avadi, Chennai, Tamilnadu 600062, India

[6] Department of Electrical and Electronics Engineering, M. Kumarasamy College of Engineering, Karur, Tamilnadu 639113, India

[7] Department of Chemical Engineering, College of Biological and Chemical Engineering, Addis Ababa Science and Technology University, Addis Ababa, Ethiopia
kibebe.sahele@aastu.edu.et

Abstract. In recent years, there has been a surge in interest in a few features of high-tech commercial drones. Small drones have a higher potential for being used for illegal operations due to their ability to evade ground security while transporting payloads. Security breaches like this may only be avoided with drone tracking and monitoring. Recognizing drones in surveillance footage can be challenging due to the similarity between small drones and birds, especially against complex backdrops. Keeping an eye out for drones and other flying objects manually is a time-consuming and difficult task. Therefore, it is necessary to employ a mechanical means of telling drones apart from birds. In this research, we create a system for drone identification using focus measure operators (FMOs). On every video frame, the five FMO parameters are calculated. Drone identification begins with a feature ranking to determine which features are most important and then continues with a classification using a random forest (RF) classifier. The Workshop on Small-Drone Surveillance, Detection, and Counteraction Techniques (WOS-DETC), with funding from the Safe Shore Consortium, provides the data used to assess the suggested method's efficacy in the Drone-vs-Bird Detection Challenge at IEEE AVSS2021. The suggested method presents to identify drones with drone present (DP) vs neither drone nor bird present (NDNBP) (two class), DP vs both bird and drone present (BBDP) vs NDNBP (three class), DP vs BP vs BBDP vs NDNBP (four class) with average acc 94.15% with sensitivity 96.69%, acc 93.60% with sensitivity 96.20%, acc 92% with sensitivity 95%, respectively for Moving Camera (MC) recordings. For drone identification, the average acc for

A. Gupta et al. (Eds.): ICAIA 2023, CCIS 2308, pp. 120–135, 2025.
https://doi.org/10.1007/978-3-031-84394-5_10

a two-, three-, and four-class classification method using a moving and a stationary camera is 96.24%, 94.12%, and 95%, respectively.

Keywords: Drone · bird · Surveillance · Unmanned aerial vehicle · Camera · detection · security

1 Introduction

Manufacturers of unmanned aerial vehicles (UAVs) are cranking out new models as technology allows. UAVs are employed for various tasks due to their portability and versatility [1, 2]. Some examples include package delivery, prescription distribution, surveying, public space monitoring, geographical mapping, assistance delivery, news reporting, disaster management, wildlife monitoring, and farming. However, criminal activities like smuggling contraband over borders, restricted zones, and even into prisons rely on unauthorized drone flights [3]. Identifying UAVs can be challenging because of their tiny size and resemblance to birds in flight.

There is a risk due to the proliferation and quick development of UAVs, which can be used for spying and interfering with airborne aircraft [4]. To solve this problem, sophisticated drone-detecting technologies are needed. Identifying UAVs with any certainty is a major obstacle for surveillance systems. Extensive study is being undertaken in this field, creating several autonomous drone surveillance systems [5]. The key issue is accurately identifying UAVs when they are hovering over valuables. Drone scanning duties are complicated because, at great distances, in particular, drones seem like birds. The complexity, detection range, and capabilities of the many created surveillance and detection systems differ. Drone monitoring becomes more challenging as a result [6]. Drone identification becomes more difficult when drones of varying sizes and low top speeds become more widely available. Various methods for detecting and pinpointing UAVs on the ground include radar, audio, video, and radio frequency monitoring. Several anti-drone systems employing these or similar methods have been created [7]. The technology and use cases that go into these systems ultimately define how well they work. In addition, there is currently no foolproof way to identify and locate all drones simultaneously, let alone defend against them. Authors [8] present a visual technique for detecting and localizing drones, complete with a 3D location tracking and prediction system.

A neural network is utilized to extract features from trajectories to categorize flying objects as unmanned aerial vehicles (UAVs) or birds. However, if the UAV is flying at an angle to the lens plane, the trajectory detection approach will not work [9]. Designed for identifying minuscule objects outside the Camera's field of view, the region-founded convolutional neural network (CNN) with ResNet (residual networks) technique described is difficult and tough to deploy in real-time applications. The median background removal and CNN utilized for UAV recognition can identify the image region containing an object, regardless of whether the Camera is stationary or moving. We need to decrease the frequency of false positives to implement an early warning detection system that yields reliable findings. By doubling the resolution, the super-resolution approach may be utilized to spot the UAV [10]. Using a neural network trained on the characteristics extracted from generic Fourier descriptors, we can distinguish between drones and birds as airborne targets [11]. Drone detection is accomplished by combining visual and auditory characteristics, but audio noise hinders performance.

Methods for detecting drones include video, audio, radar, and RF surveillance [12]. In contrast, the radar surveillance equipment moves slowly, rises rather high, and has a small footprint. There are two primary categories of video surveillance: MC-based and SC-based. Drone surveillance with stationary cameras has a visibility problem. There is a lot of background noise and obstructions in the way of RF surveillance [13]. As a result of these challenges, a focus-measure per-frame-based approach for drone surveillance has been created. Recent work on the subject has focused on deep neural networks and CNN area recommendations (faster R-CNN), both of which have demonstrated greater performance compared to previous methods but need a large amount of data for model training[14–16]. The tiny drone may be detected and identified by the suggested method in both moving and stationary cameras and in environments with varying backgrounds (such as the sky or a hilly landscape). We employed an RF classifier to accurately categorize drone-related frames in each video.

2 Dataset

The dataset used in this research was sourced from the "Drone-Vs-Bird Detection Challenge," which had been provided by the organizers of the 17th IEEE International Conference on Advanced Video and Signal-based Surveillance's (AVSS) "4th International WOSDETC2021." This dataset comprised 77 videos that were carefully selected for the specific purpose of the Drone-Vs-Bird Detection Challenge [17]. These videos had various durations, spanning from 16.67 s to 1.7 min. Each video was characterized by its distinct camera angle and background, ensuring a well-balanced representation of both terrestrial and aquatic settings. The primary objective of including these videos was to create scenarios where flying drones and birds were observable against diverse backgrounds, thereby introducing intricacy and potential distractions to the detection task.

3 Experimental Framework

In the beginning, the 45 videos are split into two categories: those with a moving camera and those with a stationary one. There are further subdivisions of these two types. This study completes the prescribed experimental design, broken down into three distinct types of video content: SC, MC, and MSC.

3.1 Moving Camera Video (MC-V)

To get the best shots of the birds and drone, the Camera isn't pointing in one certain direction during the recording session. These recordings are made in the evening when the sky is clear. Drones are constantly in the air during filming, and birds may be seen flying into and out of frame at any time. Forty films in all, 14 of which are drone-only. An example categorization task involving these video collections is as follows:

MC video-based on

a) **Two classes (MC-V2):** There are two types of frames in this set: those with drone's present (DP) and those with neither drones nor birds (NDNBP). This course has 20 videos.

b) **Three classes (MC-V3):** This set has three distinct types of images: those with a DP, those with BBDP, and those with NDNBP. The MC-V3 curriculum consists of five videos.

c) **Four classes (MC-V4):** This set has three distinct types of images: those with a DP, those with BBDP, and those with NDNBP. The MC-V3 curriculum consists of five videos.

3.2 SC Video (SC-V)

The video camera in a steady position is pointed straight up in the sky. There's a hazy, dim background. After 5 s of flying to the left, the drone finally comes into view, accompanied by a small flock of birds. After a short time of flying together, the birds take off, and the drone begins. Out of a total of 37 videos in this category, 18 include just drone footage. Therefore, the following three classification tasks are built using a total of 19 videos:

a) **Two classes (SC-V2):** Each frame is classified as either "drone present" (DP) or "neither drone nor bird present" (NDNBP). Thirteen videos may be found here.

b) **Three classes (SC-V3):** Here, we classify images as either having a drone in them (DP), having both birds and drones present in them (BBDP), or having neither (NDNBP). The SC-V3 playlist currently has five videos.

c) **Four classes (SC-V4):** There are four distinct types of frames in this group: those in which only a DP, only BP, BBDP, and NDNBP. The SC-V4 category currently only has one video.

3.3 MSC Video (MSC-V)

Drone detection falls into the same category as Moving Camera -V and SC-V, intended to work independently of camera motion. Therefore, three distinct classification problems are formed from combining SC-V and MC-V frames regardless of camera movement. The following three categories of categorization issues are formed from the frames of 77 videos. Combining the SC-V and MC-V frame types results in the following three categories:

I. **MSC video-based two classes (MSC-V2):** Drone presence (DP) and NDNBP are used here. Currently, this section contains 33 videos.

II. **MSC video-based three class (MSC-V3):** Here, we classify images into three distinct groups: those in which a drone is present (DP), those in which both birds and drones are present (BBDP), and those in which neither is present (NDNBP). This MSC-3 course has a total of 10 videos.

III. **MSC video-based four class (MSC-V4):** Here, we classify images as either having only a DP, only BP, BBDP, or NBNDP. In this MSCV4 course, you will see two videos.

All of the clarified input has been classified in preparation for creating a reliable mechanism for identifying drones. Table 1 displays the sum of frames and movies for each class combination (SC (V2 to V4), (MC (V2 to V4), and MSC (V2 to V4)). It is shown here that 2,777 frames are averaged from multiple MSC conditions.

Table 1. Frames and videos in the WOSDETC data, organized by class

Class	Stationary Camera – Video	Moving Camera – Video	MSC-V
Two (Frames)	451	1080	1531
(Videos)	13	20	33
Three (Frames)	572	842	1414
(Videos)	5	5	10
Four (Frames)	687	855	1542
(Videos)	1	1	2

4 Methodology

Figure 1 depicts the proposed approach for drone detection. The system is fed the incoming video stream. These movies are then rendered into still images. The FMOs are added to the frames once they are complete. The FMO is a data representation of a frame's focus information. Values of FMO change depending on the types of content included in the frames, such as DP, BP, NDNBP, and BBDP. Drones present at a given time are categorized based on this value. As an ensemble learning technique for regression, classification, and other issues, the RF classifier is employed to complete the classification [18]. The computed attributes are graded to further evaluate the efficacy of features for drone recognition. The bounding boxes depicted in Fig. 1 offer visual context and clarity to the proposed drone detection approach. By outlining the drone and bird using bounding boxes, the illustration provides a clear visual representation of how the system identifies and categorizes objects in the frames. These bounding boxes serve as indicators of the presence of drones and birds within the frames, aligning with the categories defined by the FMO values. The bounding boxes make it easier for readers to comprehend how the proposed approach operates, as they directly depict the objects of interest and their locations within the frames.

4.1 Focus Measure-Based Features

Images' degrees of focus or sharpness can be calculated using features based on focus measure operators (FMOs). Based on their operational philosophy, FMOs may be divided into six classes. We calculate 19 unique features for each video based on the Feature Motion Object. There is extensive information about FMOs and their applications [19]. With the help of feature rating, the nineteen features are narrowed down to the top five.

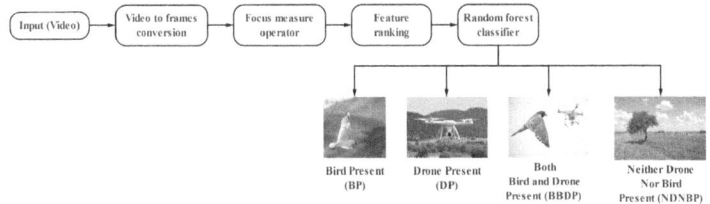

Fig. 1. Identification of Drones Using Proposed Method

Histogram entropy (HE), Gray-level variance (GLV), Helmli's mean (HM), Absolute central moment (ACM), Normalized GLV (NGLV), and technique are the five characteristics that are more significant throughout all 77 films. The following is a comprehensive breakdown of these FMOs:

Absolute Central Moment (ACM): The author proposed the operator for auto-focus. The ACM relative to the local mean quantifies the image's grayscale range. The histogram of the picture and other statistical metrics are key. The absolute value operator is put into an ACM picture without changing the average grey level [20, 21]. For a drone-captured picture of the form Im(u,v), the histogram may be expressed mathematically as follows [22]:

$$ACM = \sum_{i=1}^{N} |i - \mu_m| F_i \tag{1}$$

whereas

μ_m = mean intensity value of HistofI$_m$(u,v).

N = Number of gray levels in the images.

F_i = relative frequency of the ith gray level.

Gray-Level Variance (GLV): Auto-focus can be achieved by using the grayscale differences between imagesTo put another way, in the context of a picture's local window Im(u, v), it's the blur measurement for the pixel located at (x, y). Sharpness is greater in the areas with higher gray-level values. As an example of how to express the GLV for Im(u,v), see [23]:

$$GLV = \sum_{(i,j)\epsilon(x,y)} (I_m(u, v) - \mu)^2 \tag{2}$$

where μ = gray level mean within acceptable limits (x,y).

NGLV: The GLV is normalized by μ_g to arrive at this value. NGLV is a metric that may be used to distinguish between clear and blurry images.

$$NGLV = \frac{\sum_{(u,v)\epsilon(x,y)} (I_m(u, V) - \mu_g)^2}{\mu(x, y)} \tag{3}$$

where $\mu(x,y)$ is the average over the window of the nearby coordinates.

Histogram Entropy (HE): A sharpness measure showing information variety. Because the information richness of a focused image is so great, it is also employed to compute focus measures. The image's HE equals $-\sum R_k \log(R_k)$). Where R_k is the k_{th} grey level's relative frequency.

HM Method: By calculating the ratio of the tiny window's mean grey value to its centre grey value, the mean technique identifies the image's sharp parts. Helmli and Scherer developed an equation for determining the local contrast of images by determining the ratio between the level of intensity of each pixel in the image Im(u,v) and the average grey level of neighbouring value of (u,v), as shown in the following equation.

$$HM(\mu, v) = \begin{cases} \frac{\mu_m(u,v)}{I_m(u,v)}, & if \ \mu_m(u.v) \geq I_m(u,v) \\ \frac{I_m(u,v)}{\mu_m(u,v)}, & oherwise \end{cases} \tag{4}$$

To determine $\mu_m(u,v)$, we look inside a region centred at (u,v) (u,v).

4.2 Classification

The prediction process of the Random Forest (RF) classifier involves using previously labelled training data as a reference to determine the appropriate classification for new images. The RF classifier constructs an ensemble of decision trees based on this training data, with each tree individually making predictions. When presented with a new image, each decision tree in the ensemble predicts a classification based on the characteristics of the image that are similar to instances in the training data. The final classification for the new image is determined by combining the predictions of all decision trees through a voting mechanism, where the majority class is selected as the predicted classification. A full model is constructed using the random forest classifier in the suggested procedure. To increase prediction performance and limit overfitting, random forest classifiers employ saver ageing on multiple tree classifiers trained on independent dataset samples [24]. The Random Forest (RF) classifier is a key component of the drone detection process described in the approach. The RF classifier serves as an ensemble learning technique capable of handling a range of tasks, including regression and classification. In this context, the RF classifier is employed to perform the classification of drones based on the categorization provided by the FMO values. By utilizing the RF classifier, the system can make accurate determinations about the presence or absence of drones in specific frames. The classifier leverages the collective knowledge of multiple decision trees to enhance the accuracy and robustness of the drone detection process.

When opposed to using a single decision tree, the RF's usage of several decision trees results in a more robust and reliable conclusion. It uses trees from the training dataset to construct rules automatically at each node. The randomness in RF is split into two parts: Each decision tree is grown using a random subset of the training dataset, which is generated in the first layer. At each node, the optimal strategy for data partitioning and class label assignment is determined by factoring in the second random component [25]. The ultimate classification of each image pixel is decided by a vote in an RF, even if various trees may produce various class labels for the same data point. The F1-score, Acc, Sen, and Pre are measures of categorization performance. The next section elaborates on the suggested method's findings.

4.3 Feature Ranking

FMO-based ranking characteristics are investigated for their potential. Applying a two-sample t-test, classify the features into binary categories. The suggested technique divides groups into four categories. As a result, we use binary combination techniques for feature rating. Initially, the six potential feature combinations are ranked. We calculate the list index (Li) and actual value for every possible combination of criteria. The resulting Z values were then averaged to determine the final ranking, which determines the order of characteristics. When choosing features, the relief algorithm is applied [26–30]. For supervised models that rely on the distance between pairs of observations to predict an outcome, distance-based methods provide the most accurate method for evaluating the significance of individual features.

5   Result and Discussion

Forty-five films with a bird and a drone in various settings are used to verify the accuracy of the suggested technique. Each frame of the video is analyzed by computing the five FMO parameters. The method's efficacy is evaluated by filming a series of tests with varying configurations of drones and birds in the frame [30–35]. The MC-V2 category's performance in classifying 20 videos is first calculated and displayed in Table 2. Each video's Acc, Sen, Pre, and F1-score are shown here. In each table of results, the chosen films are denoted by the notation Vid_i, where i is the corresponding video number. There is a wide variation in Acc throughout the considered films, from 90% to 99% (see Table-Table:2), with an avg of 958.5% and a standard deviation (std) of 2.345%. Table 2 shows the same data for the MC-V3 group, with an Acc of 90%-96% and an average of 92.802.58. Table 2 displays the results of the moving camera-V4 category Accuracy, which shows a score of 96.1% accuracy and an F1-score of 89.1%.

Table 3 displays the findings from permanently installed camera systems and the results of computing SC-performance V2s in categorical categorization for 13 videos.

The average Acc for all the movies under consideration is 94.15 ± 3.601%, while the average Pre Acc-is 66.23%, as shown in the table above. Table 3 also lists the Acc for the SC-V3 category, which ranges from a min of 81% to a maximum of 99.999%, with an average of 93.60 ± 7.197%.

Classification efficiency is tracked when the Camera is in motion and at rest. The MSC-V2 classification and performance metrics for a subset of 33 films are displayed in Table 4. The table displays that the minimum Acc for the films under consideration is 87%, the maxi accuracy is 99%, and the average accuracy is 96.24% with an std of 2.17%. Table 4 also lists the accuracy for the MSC-V3 category, which ranges from a low of 81.0% to a high of 99.0%, with an average of 94.12 ± 5.11%. As a further note, the Accuracy standard deviation for the MSC-V4 category is 2.82 percentage points higher.

The comparison between capturing images with a stationary camera (SC) and a moving camera (MC) is a central aspect of our research. This comparison is essential to understanding the impact of camera motion on the quality and effectiveness of image detection, particularly in scenarios involving drones and birds. By exploring the differences between these two camera setups, we gain insights into how motion-induced

128   M. Kalidas et al.

Table 2. Avg Accuracy, Prediction, and other outcomes for Moving Camera-V2

S. No	Video Number	Accuracy (%)	Sen (%)	Prediction (%)	F1-Score (%)
1	Vid_1	95	98	65	78
2	Vid_2	97	98	93	95
3	Vid_3	99	99	91	95
4	Vid_4	97	94	98	96
5	Vid_5	97	98	95	96
6	Vid_6	90	97	33	50
7	Vid_{10}	97	98	33	50
8	Vid_{11}	96	98	19	31
9	Vid_{13}	98	100	14	25
10	Vid_{14}	97	98	33	49
11	Vid_{15}	97	98	47	63
12	Vid_{16}	97	98	91	94
13	Vid_{18}	96	98	90	92
14	Vid_{19}	96	98	86	92
15	Vid_{20}	96	100	86	72
16	Vid_{22}	94	96	57	65
17	Vid_{23}	90	95	50	79
18	Vid_{24}	94	97	68	57
19	Vid_{25}	97	98	41	84
20	Vid_{26}	98	99	73	80
	Avg	95.85	97.75	63.15	72.15
	Minimum	90	94	14	25
	Maximum	99	100	98	96
(MC-V3)					
1	Vid_8	95	97	88	93
2	Vid_9	90	91	87	89
3	Vid_{12}	99	100	97	98
4	Vid_{17}	92	92	90	91
5	Vid_{21}	91	94	80	86
	Average	92.8	94.20	88.4	91.4
	Minimum	90	91	80	86
	Maximum	96	97	90	93

(continued)

Table 2. (*continued*)

S. No	Video Number	Accuracy (%)	Sen (%)	Prediction (%)	F1-Score (%)
(MC-V4)					
Vid$_7$	96%	100%	80%		89%

Table 3. Identification of drones using a classification system of the SC-V2 type:

Sr. No	Video Number	Accuracy (%)	Sen (%)	Prediction (%)	F1-score (%)
1	Vid$_{27}$	87	91	70	79
2	Vid$_{28}$	91	96	18	30
3	Vid$_{30}$	90	93	80	86
4	Vid$_{31}$	97	98	89	93
5	Vid$_{32}$	93	97	16	28
6	Vid$_{35}$	96	98	75	85
7	Vid$_{36}$	96	97	88	93
8	Vid$_{37}$	93	97	61	75
9	Vid$_{39}$	96	99	90	90
10	Vid$_{41}$	99	99	87	93
11	Vid$_{42}$	98	98	75	85
12	Vid$_{43}$	97	99	96	97
13	Vid$_{45}$	91	95	16	28
	Average	94.15	96.69	66.23	67.07
	Minimum	87	91	16	28
	Maximum	99	99	90	97
SC-V3					
1	Vid$_{33}$	96	98	78	87
2	Vid$_{34}$	95	100	71	83
3	Vid$_{38}$	97	99	14	24
4	Vid$_{40}$	81	85	76	81
5	Vid$_{44}$	99	99	83	90
	Average	93.60	96.2	64.4	73
	Minimum	81	85	14	24
	Maximum	99	100	83	90
SC-V4					
Vid$_{29}$		92%	95%	50%	65%

Table 4. The proposed system's classification results for Moving Camera, Stationary Camera, and Moving and Stationary Camera videos

Class	Acc	SC-V (%)	MC-V (%)	Moving and Stationary Camera videos (%)
Two	Mean Accuracy	94.1 ± 3.6	95.8 ± 2.3	95.1 ± 2.1
	Man Sen	96.6 ± 2.4	97.7 ± 1.4	95.2 ± 4.7
	Minimum Accuracy	87	90	87
	Maximum Accuracy	99	99	99
	F1-score	67.07	72.15	70.15
Three	Mean Accuracy	93.6 ± 7.1	92.8 ± 2.5	93.2 ± 5.1
	Man Sen	96.2 ± 6.0	94.2 ± 2.7	93.2 ± 5.1
	Minimum Accuracy	81	90	81
	Maximum Accuracy	99	96	99
	F1score	73	91.4	82.2
Four	Mean Accuracy	92 ± 2.3	96 ± 1.9	94 ± 2.8
	Mean Sen	995 ± 1.4	100	97.5 ± 3.5
	F1-score	65	89	77

variations influence the performance of detection algorithms. This analysis is critical for enhancing the robustness and accuracy of detection systems in scenarios where camera motion is a factor, contributing to the broader objectives of improving the detection of drones and birds in complex visual environments.

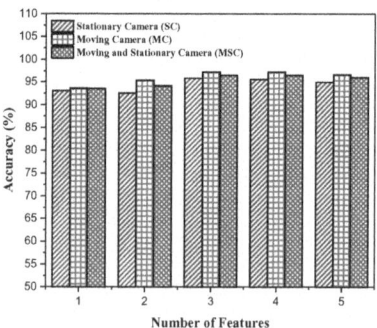

Fig. 2. Feature vs. Accuracy in Two-Class Classification with Moving Camera, Stationary Camera, and MSC

In addition, the F1-score is calculated for the chosen clips in each scenario to demonstrate the equilibrium between drone and bird detection. For all of the tested films, the created system averaged an MC success rate of 96%, an SC success rate of 94.39%, and an MSC success rate of 95.32%. Class-wise average accuracy versus feature count for the SC-V, Moving Camera -Video, and MSC-V are displayed in Figs. 2, 3, and 4, respectively. High Acc is attained while employing three characteristics for two- and three-class categorization. In addition, a high Acc can only be achieved in dynamic and static camera footage by using all five criteria for four-class categorization.

Table 5 compares the suggested strategy to several commonly used alternatives. A convolutional neural network (CNN) based approach to drone identification is provided. Three SC films are used to test the approach and get an Acc. of 90.6%.

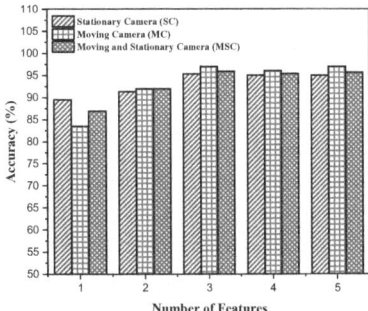

Fig. 3. Comparison of features and accuracy for Moving Camera, Stationary Camera, and MSC-based three-class classification

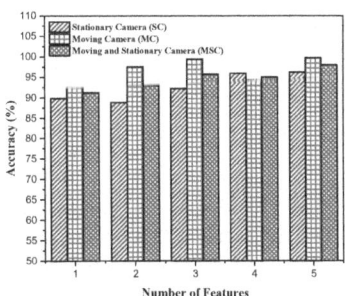

Fig. 4. Classification with four classes using a Moving Camera, Stationary Camera, and MSC system: features versus accuracy

A CNN-based technique is offered with 97.5% Accuracy, but it has only been tested on six movies. The suggested approach was tested on 45 videos shot by mobile and fixed cameras, with an average Acc. of 94.77%.

In this study, we provided an FMO-based strategy for using drones for surveillance. The probability of drone identification with a single camera is studied, and the results demonstrate a high detection rate. Since the proposed technique can be applied to drone monitoring using several cameras, it could one day be used to track moving drones for decision-making in security [36–42].

Table 5. The Result of the Proposed Method

Author	Method	Video type	No. of videos	Accuracy (%)
Proposed method	RF and FMOs classifier	Moving Camera	20	95.42
		Stationary Camera	19	93.89
		Moving Camera and Stationary Camera	45	94.77

6 Conclusion

To create a drone surveillance system, this study delves into the parameters of the focal measure operator (FMO), specifically the absolute central moment, the GLV, the NGLV, the Gramme entropy, and Helmli's mean technique. Frame-based motion operators (FMOs) are calculated for every video frame, and their unique characteristics are uncovered. Therefore, a system based on random forest classifiers is created to automatically recognize drones. Dataset 2021 from the "Drone vs Bird Detection Challenge", is utilized to put the proposed method through its paces, consisting of 45 videos. The suggested approach is preferable because of its high accuracy and sensitivity. A system with three to five characteristics may outperform deep learning and data-dependent alternatives. It is possible to build a multicamera-based intelligent system for drone recognition in a wide field of view using the proposed technology.

References

1. Singha, S., Aydin, B.: Automated drone detection using YOLOv4. Drones **5**(3) (2021). https://doi.org/10.3390/drones5030095
2. Pansare, A., et al.: Drone detection using YOLO and SSD a comparative study. In: 2022 International Conference on Signal and Information Processing, IConSIP 2022 (2022). https://doi.org/10.1109/IConSIP49665.2022.10007489
3. Sommer, L., Schumann, A.: Deep learning-based drone detection in infrared imagery with limited training data. In: Proceedings of SPIE - The International Society for Optical Engineering, vol. 11542 (2020). https://doi.org/10.1117/12.2574171
4. Tiwari, R., Dubey, A.K.: Detection of camouflaged drones using computer vision and deep learning techniques. In: Proceedings of the Confluence 2022 - 12th International Conference on Cloud Computing, Data Science and Engineering, pp. 380–383 (2022). https://doi.org/10.1109/Confluence52989.2022.9734191
5. Jain,R., Nagrath, P., Thakur, N., Saini, D., Sharma, N., Hemanth, D.J.: Towards a smarter surveillance solution: the convergence of smart city and energy efficient unmanned aerial vehicle technologies. Stud. Syst., Dec. Control **332**, 109–140 (2021). https://doi.org/10.1007/978-3-030-63339-4_4
6. Kumar, N.M., Sudhakar, K., Samykano, M., Jayaseelan, V.: On the technologies empowering drones for intelligent monitoring of solar photovoltaic power plants. Procedia Comput. Sci. **133**, 585–593 (2018). https://doi.org/10.1016/j.procs.2018.07.087

7. Soderlund, G.: Introduction to 'charting, tracking, and mapping: new technologies, labour, and surveillance. Soc. Semiot. **23**(2), 163–172 (2013). https://doi.org/10.1080/10350330.2013. 777589

8. Saini, D.K., Bala, S., Sharma, A.K., Zia, K.: Emerging technologies for pandemic and its impact. Intell. Syst. Ref. Libr. **209**, 291–310 (2021). https://doi.org/10.1007/978-981-16-2972-3_14

9. Chang, K.-J., Chuang, C.-W., Chiu, J.-T., Chen, J.-Y.: Flying watchdog: a drone with edge AIoT for residential safety and fall detection by face and posture recognition. In: APWCS 2022 - 2022 IEEE VTS Asia Pacific Wireless Communications Symposium, pp. 181–185 (2022). https://doi.org/10.1109/APWCS55727.2022.9906504

10. Sriram, R., Vamsi, A., Vigneshwari, S.: Telemetry-based autonomous drone surveillance system. Lect. Not. Electr. Eng. **708**, 419–427 (2021). https://doi.org/10.1007/978-981-15-8685-9_43

11. Zhai, X., Liu, K., Nash, W., Castineira, D.: Smart autopilot drone system for surface surveillance and anomaly detection via a customizable deep neural network. In: International Petroleum Technology Conference 2020, IPTC 2020 (2020). https://doi.org/10.2523/iptc-20111-ms

12. Dilshad, N., Hwang, J., Song, J., Sung, N.: Applications and challenges in video surveillance via drone: a brief survey. In: International Conference on ICT Convergence, vol. 2020-October, pp. 728–732 (2020). https://doi.org/10.1109/ICTC49870.2020.9289536

13. Minhas, M.S., Zelek, J.: Defect detection using deep learning from minimal annotations. In: VISIGRAPP 2020 - Proceedings of the 15th International Joint Conference on Computer Vision, Imaging and Computer Graphics Theory and Applications, vol. 4, pp. 506–513 (2020). https://www.scopus.com/inward/record.uri?eid=2-s2.0-85083563004&partnerID=40&md5=e918203fa3ab51a5c42eae0dc3a9a2ed

14. Siewert, S., et al.: Image and information fusion experiments with a software-defined multispectral imaging system for aviation and marine sensor networks. In: AIAA Information Systems-AIAA Infotech at Aerospace, 2017 (2017). https://doi.org/10.2514/6.2017-0877

15. Roh, S.-B., Oh, S.-K., Pedrycz, W., Seo, K.: Development of autofocusing algorithm based on fuzzy transforms. Fuzzy Sets Syst. **288**, 129–144 (2016). https://doi.org/10.1016/j.fss.2015.08.029

16. Shim, S.-O., Aziz, W., Banjar, A., Alamri, A., Alqarni, M.: Improving depth computation from robust focus approximation. IEEE Access **7**, 20144–20149 (2019). https://doi.org/10.1109/ACCESS.2019.2897744

17. Mannan, S.M., Malik, A.S., Choi, T.-S.: Reducing intricacy of 3D space for 3D camera. In: Proceedings of the International Symposium on Consumer Electronics, ISCE (2008). https://doi.org/10.1109/ISCE.2008.4559551

18. Shim, S.-O., Malik, A.S., Mahmood, M.T., Choi, T.-S.: Estimation of depth map based on focus adjustment. In: Proceedings of SPIE - The International Society for Optical Engineering, vol. 7073 (2008). https://doi.org/10.1117/12.798191

19. Caruso, S., Bonaque-González, S., Oliva-García, R., Rodríguez-Ramos, J.M.: Relative multiscale deep depth from focus. Signal Process. Image Commun. **99** (2021). https://doi.org/10.1016/j.image.2021.116417

20. Shete, S., Bhavsar, A., Sao, A.K.: Enhancing shape from focus-measure-fusion and sparse representation. In: ACM International Conference Proceeding Series, vol. 14 (2014). https://doi.org/10.1145/2683483.2683542

21. An, Y., Kang, G., Kim, I.-J., Chung, H.-S., Park, J.: Shape from focus through Laplacian using a 3D window. In: Proceedings of the 2008 2nd International Conference on Future Generation Communication and Networking, FGCN 2008, vol. 2, pp. 46–50 (2008). https://doi.org/10.1109/FGCN.2008.139

22. Choi, W.-J., Choi, T.-S.: Fast three-dimensional shape recovery in TFT-LCD manufacturing. In: Proceedings of SPIE - The International Society for Optical Engineering, vol. 7073 (2008). https://doi.org/10.1117/12.798232
23. Zhao, M., Zhou, J.: Detection method of stone surface roughness based on CCD camera. Guangxue Jishu/Optical Tech. 44(3), 310–314 (2018). https://www.scopus.com/inward/record.uri?eid=2-s2.0-85055925070&partnerID=40&md5=edf28d89b026634e35d800a5e4f80ad4
24. Rusinol, M., Chazalon, J., Ogier, J.-M.: Combining focus measure operators to predict OCR accuracy in mobile-captured document images. In: Proceedings - 11th IAPR International Workshop on Document Analysis Systems, DAS 2014, pp. 181–185 (2014). https://doi.org/10.1109/DAS.2014.11
25. Sayar, M.S., Akgül, Y.S.: Depth from moving apertures. In: Lecture Notes in Electrical Engineering, LNEE, vol. 264, pp. 189–197 (2014). https://doi.org/10.1007/978-3-319-01604-7_19
26. Pavliček, P.: Measurement of the shape of objects by shape from focus. In: Proceedings of SPIE - The International Society for Optical Engineering, vol. 9524 (2015). https://doi.org/10.1117/12.2189187
27. Abbas, Q., Ibrahim, M.E., Khan, S., Baig, A.R.: Hypo-driver: a multiview driver fatigue and distraction level detection system. CMC-Comput. Mat. Continua 71(1), 1999–2017 (2022). https://doi.org/10.32604/cmc.2022.022553
28. Akhtar, M.M., Zamani, A.S., Khan, S., Shatat, A.S.A., Dilshad, S., Samdani, F.: Stock market prediction based on statistical data using machine learning algorithms. J. King Saud Univ.-Sci. 34(4), 101940 (2022). https://doi.org/10.1016/j.jksus.2022.101940
29. AlAjmi, M.F., Khan, S., Sharma, A.: Collaborative learning outline for mobile environment. In: Paper presented at the 2014 International Conference on Issues and Challenges in Intelligent Computing Techniques (ICICT) (2014)
30. Alfaifi, A.A., Khan, S.G.: Utilizing data from twitter to explore the UX of "Madrasati" as a Saudi e-learning platform compelled by the pandemic. Arab Gulf J. Sci. Res. 39(3), 200–208 (2022). https://doi.org/10.51758/AGJSR-03-2021-0025
31. Khan, J., et al.: Secure smart healthcare monitoring in industrial internet of things (IIoT) ecosystem with cosine function hybrid chaotic map encryption. Sci. Programm., (Article ID 8853448), 22 (2022). https://doi.org/10.1155/2022/8853448
32. Khan, S.: Data visualization to explore the countries dataset for pattern creation. Int. J. Online Biomed. Eng. 17(13), 4–19 (2021). https://doi.org/10.3991/ijoe.v17i13.20167
33. Khan, S.: Visual data analysis and simulation prediction for COVID-19 in Saudi Arabia using SEIR prediction model. Int. J. Online Biomed. Eng. 17(8) (2021). https://doi.org/10.3991/ijoe.v17i08.20099
34. Khan, S.: Business intelligence aspect for emotions and sentiments analysis. In: Paper presented at the 2022 First International Conference on Electrical, Electronics, Information and Communication Technologies (ICEEICT) (2022)
35. Khan, S., Alfaifi, A.: Modeling of coronavirus behavior to predict it's spread. Int. J. Adv. Comput. Sci. Appl. 11(5), 394–399 (2020). https://doi.org/10.14569/IJACSA.2020.0110552
36. Khan, S., Alghulaiakh, H.: ARIMA model for accurate time series stocks forecasting. Int. J. Adv. Comput. Sci. Appl. 11(7), 524–528 (2020). https://doi.org/10.14569/IJACSA.2020.0110765
37. Fazil, M., Khan, S., Albahlal, B.M., Alotaibi, R.M., Siddiqui, T., Shah, M.A.: Attentional multi-channel convolution with bidirectional LSTM cell toward hate speech prediction. IEEE Access 11, 16801–16811 (2023)
38. Keshta, I., et al.: Energy efficient indoor localisation for narrowband internet of things. CAAI Trans. Intell. Technol. (2023)

39. Khan, S., Ch, V., Sekaran, K., Joshi, K., Roy, C.K., Tiwari, M.: Incorporating deep learning methodologies into the creation of healthcare systems (2023)
40. Khan, S., et al.: Transformer architecture-based transfer learning for politeness prediction in conversation. Sustainability **15**(14), 10828 (2023)
41. Khan, S., Moorthy, G.K., Vijayaraj, T., Alzubaidi, L.H., Barno, A., Vijayan, V.: Computational intelligence for solving complex optimization problems (2023)
42. Khan, S., et al.: Manufacturing industry based on dynamic soft sensors in integrated with feature representation and classification using fuzzy logic and deep learning architecture. Int. J. Adv. Manuf. Technol., 1–13 (2023)

IoT-Driven Dynamic Behavior Intervention Model for Sustainable Hygiene Practices: Insights from Household Water Consumption

R. Srinivasan[1], K. Vimala Devi[2], David Winster Praveenraj[3], S. Venkatasuvrananian[4], K. Subramani[5], V. S. Prasanth[6], and Babu Rao Gaddala[7(✉)]

[1] Department of Mechanical Engineering, Sri Krishna College of Technology, Kovaipudur, Coimbatore 641042, India
[2] School of Computer Science and Engineering, Vellore Institute of Technology, Vellore 631014, India
vimaladevi.k@vit.ac.in
[3] CHRIST (Deemed to be) University, Bangalore, India
david.winster@christuniversity.in
[4] CSBS, Saranathan College of Engineering, Trichy, India
veeyes@saranathan.ac.in
[5] School of Business and Management, CHRIST (Deemed to be University), Bangalore Yeshwanthpur Campus, Bangalore 560 073, India
subramani.k@christuniversity.in
[6] Department of Electronics and Communication Engineering, Madanapalle Institute of Technology and Science, Madanapalle, Andhra Pradesh 517 325, India
[7] Department of Chemical Engineering, University of Technology and Applied Sciences (Higher College of Technology Muscat), Muscat, Oman
baburao.gaddala@utas.edu.om

Abstract. IoT-enabled technologies have advanced so that smart sensor systems can observe and recognize human behavior in various contexts, including energy consumption and healthcare, with remarkable efficiency and effectiveness. One example is using the Internet of Things (IoT) technology to better comprehend human water consumption behavior and establish and maintain clean environments. Static models have typically been used to model the behavior intervention process throughout time. While these static approaches perform adequately when predicting general human behavior, they fall short when tracking and reacting to shifts in behavior in IoT settings. The authors of this study proposed a dynamic behavior intervention model to forecast the hygiene-related water-use habits of individual households. This model takes its cues from the structure equation model method and the notion of control engineering, which originated in the expanded theory of planned behavior (ETPB). The current ETPB dynamic behavior model with system parameter estimation using an artificial neural network (ANN) is assessed for its intervention trend using a residential water use case study. It has been shown that the ETPB dynamic model helps the process of intervening in people's behavior.

Keywords: Internet of Things (IoT) · ETPB · dynamic behavior · artificial neural network (ANN)

© The Author(s), under exclusive license to Springer Nature Switzerland AG 2025
A. Gupta et al. (Eds.): ICAIA 2023, CCIS 2308, pp. 136–154, 2025.
https://doi.org/10.1007/978-3-031-84394-5_11

1   Introduction

Public health issues, including obesity, sexual risk-taking, cancer prevention, and poor cleanliness, have all been demonstrated to benefit from a change in individual behavior [1]. Predicting and monitoring specific human actions is crucial for effective illness prevention, diagnosis, and treatment. There has been a lot of focus from the business and academic worlds on the detrimental effects of poor sanitation and water on infectious diseases and plagues in impoverished countries [2]. Diarrhoea, parasite infections, skin illness, and eye disease are preventable with proper hygiene.

Monitoring behavior and translating medical advice into intervention guidance are two of the most difficult challenges facing clinicians and decision-makers in water hygiene, which is part of the larger field of general healthcare practices research [3]. Many treatments aimed at changing people's behavior regarding water use have not been empirically studied because they do not lend themselves to investigation within the currently dominant research paradigm [4].

A family's water consumption is strongly affected by the family's water habits and behavior (physical, behavioral pattern) and the response of the household to water-related initiatives [5]. Drinking water preferences about the social-psychological and geographical factors of water scarcity were examined. He discovered that factors such as money, education, and personality shape the drinking water preferences of Shanghai residents [6]. Most healthcare-related IoT systems involve physical actions/behaviors using RFID (Radio-frequency identification), sensor devices, or other approaches.

The research developed an updated version of the Theory of Planned Behaviour (TPB) structure by incorporating considerations of individual beliefs and actions toward water conservation. The updated concept of planned behavior is shown in Fig. 1.

H1. Positive changes in public opinion on water conservation have led to a rise in households planning to reduce water use and boost conservation efforts.

H2. The purpose of preserving water and implementing conservation activities has increased as the subjective norm towards water conservation has grown.

H3. The higher the public backing for water conservation, the more likely people will take steps to reduce their water usage at home.

H4. When people feel more capable of conserving water, they are more likely to prioritise water conservation in their homes.

H5. The correlation between attitude, PBC, subjective norms, and self-efficacy shows the influence of previous behaviour on present and future water-saving intentions.

The underlying connections between TPB components will also be explored beyond the direct implications of previous behaviour on water conservation intent. Thus, we conjectured.

H6. Attitude, PBC, subjective norms, and self-efficacy all moderated the impact of intention on water conservation behaviours among earlier generations.

H7. Current water conservation intentions can be influenced by past Behaviour via TPB factors (attitude, PBC and Subjective norms).

However, to put professional advice and knowledge into practice, managers, doctors, and policymakers face a challenging problem: predicting and understanding human water usage behaviour to improve sanitation standards. Hygiene education programmes focused on behavioural interventions have often been and still are mostly conducted

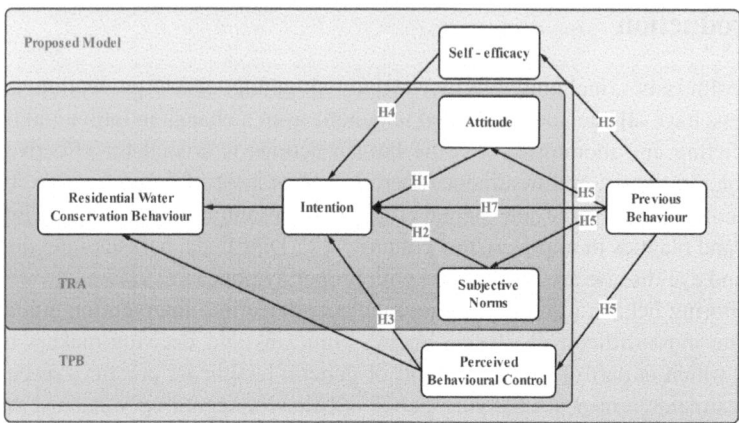

Fig. 1. An enhanced TRA and TPB model for residential water-saving behaviour expansion.

face-to-face. In-person counselling as a means of intervening in dominant behaviour is expensive and time-consuming, and it's unrealistic to expect every user to have constant access to counsel [7, 8]. Recently, IoT-enabled healthcare technologies have allowed for delivering health promotion information to a broad population while catering to each person's specific patterns of behaviour [9]. The Internet of Things enables smart water systems to monitor and analyse people's water consumption patterns to promote sanitary lifestyles. Environmental hygiene includes personal cleanliness practices, proper food storage and handling, environmental hygiene around the home, and good trash management [10].

To model and predict hygienic behaviours, experts often turn to conventional fixed behaviour interventions, in which the same treatment plan is employed for all participants regardless of their characteristics [11]. This study aims to learn more about an innovative online behaviour intervention that makes suggestions that evolve based on the user's unique IoT setting. An online drug misuse prevention programme, for instance, might be adjusted somewhat for each participant based on demographic information (age, gender, education and income level, etc.), and participants' reactions to the program's intended outcomes change over time [12, 13]. Interventions in past studies have consistently been based on a theory or model for influencing behaviour change. Human behaviour theories and models like the TPB, the Stage of Change (Transtheoretical model - TTM), the Health Belief Model (HBM), and so on have been employed in quantitative studies of online intervention tactics [14]. Static data analysis, such as that employed by the SEM, forms the backbone of most models that describe behaviour change. These static methods have shown some success in predicting the problem above, but they still have a ways to go before they can adequately model behaviour treatments or fully grasp the temporal dynamics of the process by which people's behaviours change [15].

Researchers explored the potential of dynamic behaviour intervention models for predicting individual families' water use patterns about personal cleanliness [16]. This model's intervention technique is grounded in control engineering principles and features a time lag and system coefficients. This dynamic modelling, which employs a fluid-analogy-based technique to simulate large psychological systems with the control engineering paradigm, can provide light on the time-dependent nature of the behaviour intervention process [17]. ETPB provides the theoretical foundation for our dynamic system, which is first expressed mathematically using a structural equation method and then posed as a fluid analogy to include elements of a dynamic system [18]. This research not only identifies three types of parameters in a dynamic system but also presents a model and an ETPB questionnaire to aid in collecting behavioural data from online users. Data gathered in China was used in a case study that focused on how individual households handled their water supply [19]. An ANN model of water consumption was given, and the effectiveness of the dynamic behaviour model was evaluated by analysing three ways of parameter selection to highlight the possibility for different tendencies in intervention.

This paper's main contributions are (i) the development of an IoT-based procedure for collecting ETPB data and (ii) treatment result analysis and prediction through the creation of a dynamic model of behaviour intervention. These enhancements to our knowledge of family water consumption behaviour are crucial to making strides toward more sanitary conditions in urban homes.

The remaining parts of this research are structured as follows. In Sect. 2, there is a look back at the various models used to study the dynamics of behaviour modification. Section 3 on online behaviour change technique in an IoT scenario covers the theoretical underpinnings, behaviour modelling, data collection, and experimental simulation. The paper concludes and recommends further research after assessing the dynamic ETPB model in Sect. 4.

2 Literature Review

The topic of utilising the internet to influence people's behaviours is well-documented. Researchers to find out which theories and models of behaviour change are most effective at encouraging people to alter their negative online behaviours. Theories are the most popular tool for influencing behaviour change because they help researchers identify the most important factors in shaping people's actions. Popular models and theories include TPB, TTM (the theory of thought experiments), HBM (the theory of mind), the social cognitive model, and many others. Researchers agree that statistical approaches are useful for analysing behaviour change models [20]. Common statistical methods in behaviour research include multiple linear regression and structural equation modelling.

Researchers in [21] used conventional multiple regression to assess an attitude-behaviour hypothesis, which they then used to display TPB timelines and path diagrams. The authors thoroughly assessed exploration, data visualisation, and assumption-checking strategies for the behavioural sciences. Multiple binary logistic regression analyses were used to determine the correlation between various traits and personalities (self-efficacy, self-identity etc.). Despite its widespread use in research, multiple linear regression has shown limited success in simulating dynamic systems and conducting in-depth analyses of complex systems. For an SEM analysis of the mediating conceptual framework, a method (sedentary behaviour and physical activity). However, the SEM-based statistical approach has increasingly shown less ability to model intervention systems but continues to be beneficial for modelling behaviour modification.

Dynamical systems in the behavioural sciences, like time-varying adaptive intervention, have been shown to benefit from engineering control concepts [22]. The concepts of "dynamic system" (meaning a multivariate time-varying method), "tailoring variable", and "process analysis" are all applicable to models for behaviour modification, and they can be found in the fields of control engineering and adaptive intervention. However, no studies established a complete and generic structure for gathering data on behaviour change and modelling it as a dynamic process, despite showing the utility of simulating behaviour change with control engineering principles like resistance, time delay, etc. The effectiveness of dynamic modelling and the consequences for accurate prediction of outcomes have not been the subject of a systematic review.

Therefore, in this study, we develop a dynamic process model for behaviour change that can facilitate the development of a thorough methodology for behaviour change, evaluation of performance, and forecast of specific problems.

3   Proposed Methodology

This process has many moving parts because altering people's behaviour is difficult. The online behaviour change technique in an IoT setting is depicted in Fig. 2 and might be used to tackle such a complex problem by gathering data on behaviours and offering feedback. The ETPB Questionnaire, drawn from the TPB, supplements our primary data collection techniques. The information gathered will then be put through a simulation of an intervention procedure for a specific problem behaviour using the ETPB dynamic model [23]. The methodology's game mechanism can guide users with persuasion techniques, and the motivation system can serve as a safety net to guarantee the simulated feedback leads to the intended results. The accessibility of your online behaviour modification project will improve with each of these feedback methods. Furthermore, to validate individual intervention, it is also vital to collect data through a well-planned follow-up system.

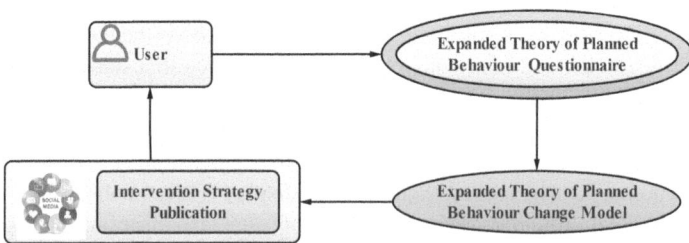

Fig. 2. Methodology for an Internet-of-Things-based Online behaviour intervention system

3.1   ETPB

The ETPB details the theory behind the behaviour change strategy. Despite its relevance to psychology studies, TPB has been the focus of the current controversy about whether it is sufficient and whether other external variables might help enhance intention prediction [24]. The results of the studies above have suggested new variables that can be used to increase the predictive capacity of TPB for both intentions and actions. As external dimensions of TPB, prior behaviour and self-efficacy have been highlighted by ETPB as having substantial implications on TPB. Figure 3 shows the theoretical foundations upon which ETPB is built. The ETPB hypothesis offers that an individual's intention reveals their readiness to engage in a certain behaviour and that one's behaviour acts as a quantifiable reaction to a stated goal. The extent to which one expects positive or negative consequences from behaviour is what is meant by "attitude toward the behaviour." When we talk about someone's prior behaviour, self-efficacy, or perceived capacity to carry out a specific behaviour, we're referring to past activities that may have influenced their decision to engage in that behaviour.

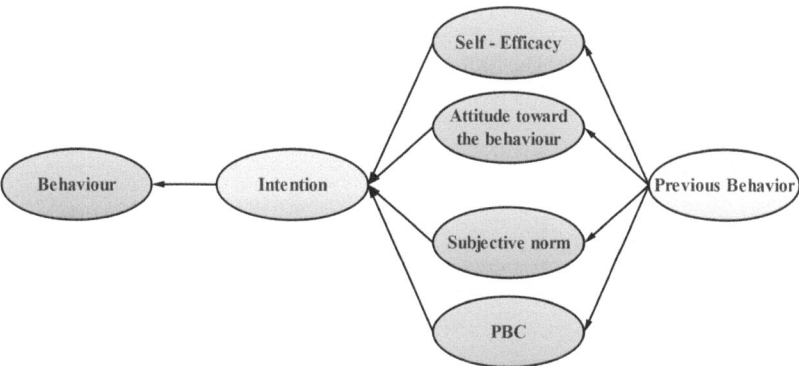

Fig. 3. Conceptual structure of ETPB

3.2 Behaviour Change Modelling

This strategy is best described as a modelled version of the behaviour intervention process. By including an SEM time delay and related coefficient, we can explicitly characterise the modelling process, establishing the concept of control engineering to simulate intervention in Behaviour across Time.

Using an SEM containing latent variables, we looked into the connections between alterations to human behaviour and the mechanisms by which that behaviour is altered. Despite its popularity in the social sciences, SEM's features get murky when the model structure becomes complex [25–35]. Here, we solely consider path analysis using SEM, assuming that the independent variables were measured accurately and that the problem variables were easily observable. Since no conclusive evidence exists between Behaviour and SN, this variable has been ignored during simulations. Figure 4 shows the SEM variable definitions and the ETPB flowchart.

We hypothesised and compared inventory management in distribution chains to a fluid analogue to bring dynamic implications to the ETPB SEM model, as the classic SEM model can only explain a static system. Figure 5 depicts the ETPB SEM inventory

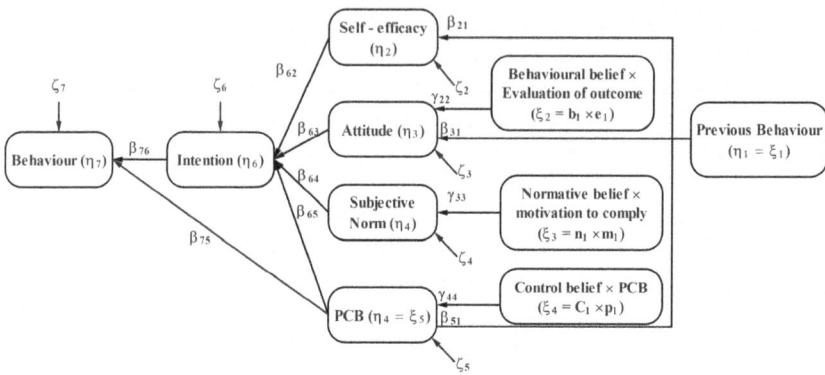

Fig. 4. Path analyses and variable definitions for the ETPB model

for each endogenous variable. The following equation can be used as a general expression for ETPB behaviour intervention based on the principle of mass conservation.

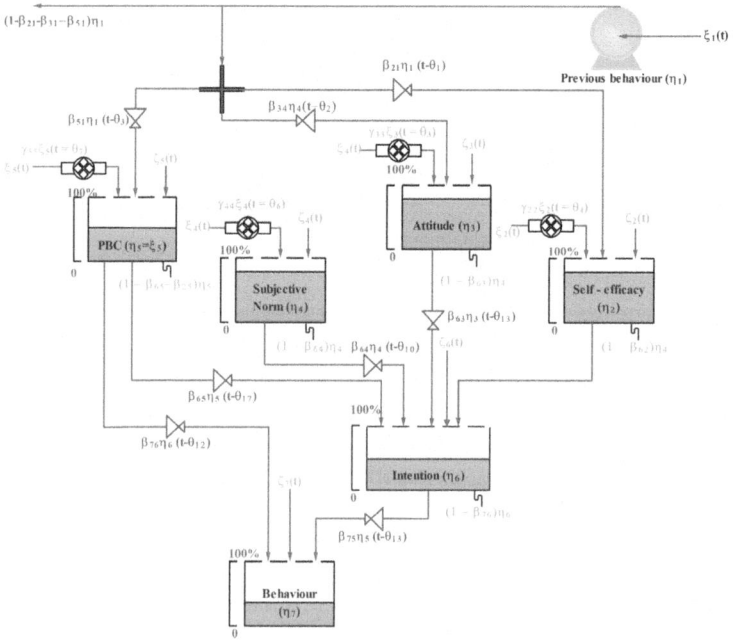

Fig. 5. Fluid analogy for ETPB model

$$'Inventory'\ Accumulation = Fluid\ Inflow - Fluid\ Outflow \qquad (1)$$

Time and mass transaction description in a system was built using three parameters: the inventory transaction time delay θ_i, the inventory mass relation (β_{ij} and g_{ij}), and the inventory capacity constant (t_i). Specifically, t_i is consistent, representing the capacity of each inventory, j_i denotes uncontrollable elements, and θ_i is the time delay in transporting inventory. Regression weight factors β_{ij} and g_{ij} are matrices. The jth vector of disturbance variables represents a stochastic signal in the system with zero mean. Equations (3) and (4) on the bottom page show the accumulation of inventory (η_i) in the fluid analogy system based on (1).

$$\eta_1 = \xi_1(t)$$

$$= \beta_{21}\eta_1(t - \theta_1) + \beta_{31}\eta_1(t - \theta_2) + \beta_{51}\eta_1(t - \theta_3) + (1 - \beta_{21} - \beta_{31} - \beta_{51})\eta_1(t) \quad (2)$$

$$T_2\frac{d\eta_2}{dt} = \beta_{21}\eta_1(t - \theta_1) + \gamma_{22}\xi_2(t - \theta_4) - \beta_{62}\eta_2(t - \theta_8) - (1 - \beta_{62})\eta_2(t) + \xi_2(t)$$

$$= \beta_{21}\eta_1(t - \theta_1) + \gamma_{22}\xi_2(t - \theta_4) - \beta_{62}\eta_2(t) + \xi_2(t) \quad (3)$$

$$T_3\frac{d\eta_3}{dt} = \beta_{31}\eta_1(t - \theta_2) + \gamma_{33}\xi_2(t - \theta_5) - \beta_{63}\eta_3(t - \theta_9) - (1 - \beta_{63})\eta_3(t) + \xi_3(t)$$

$$= \beta_{31}\eta_1(t - \theta_2) + \gamma_{33}\xi_3(t - \theta_3) - \beta_{63}\eta_3(t) + \xi_3(t) \quad (4)$$

$$T_4\frac{d\eta_4}{dt} = \beta_{44}\xi_4(t - \theta_6) - \beta_{64}\eta_4(t - \theta_{10}) - (1 - \beta_{64})\eta_4(t) + \xi_4(t)$$

$$= \beta_{44}\eta_4(t - \theta_3) - \beta_{64}\eta_4(t) + \xi_4(t) \quad (5)$$

$$T_5\frac{d\eta_5}{dt} = \beta_{51}\eta_1(t - \theta_3) - \gamma_{55}\xi_5(t - \theta_7) - \beta_{65}\eta_5(t - \theta_{11}) - \beta_{75}\eta_4(t - \theta_{13}) -$$
$$(1 - \beta_{65} - \beta_{75})\eta_5(t) + \xi_4(t)$$

$$= \beta_{51}\eta_1(t - \theta_3) - \gamma_{55}\xi_5(t - \theta_7) - \eta_5 + \xi_5(t) \quad (6)$$

$$T_6\frac{d\eta_6}{dt} = \beta_{62}\eta_2(t - \theta_8) + \gamma_{63}\eta_3(t - \theta_9) + \beta_{64}\eta_4(t - \theta_{10}) + \beta_{65}\eta_5(t - \theta_{11}) - \beta_{76}\eta_6(1 - \eta_{12}) - (1 - \beta_{76})\eta_6(t) + \xi_6(t)$$

$$= \beta_{62}\eta_5(t - \theta_8) + \beta_{63}\eta_3(t - \theta_9) + \beta_{64}\eta_4(t - \theta_{10}) + \beta_{65}\eta_5(t - \theta_{11}) - \eta_6(t) + \xi_6(t) \quad (7)$$

$$T_7\frac{d\eta_7}{dt} = \beta_{76}\eta_6(t - \theta_{12}) + \beta_{75}\eta_5(t - \theta_{13}) - \eta_7(t) + \xi_4(t) \quad (8)$$

where, $\eta_1(t) = j_1(t)$, $j_2(t) = b_1(t) \times e_1(t)$, $j_3(t) = n_1(t) \times m_1(t)$, $j_4(t) = c_1(t) \times p_1(t)$ and $\eta_5(t) = j_5(t)$ according to the SEM model (t). At the steady state condition of $d_i/dt = 0$, the dynamic system reduces to an SEM model (Eqs. (6)–(8)). There is only a first-order relationship between time t and the pace at which stocks fluctuate in a system.

3.3 Data Collection

It is crucial that this mechanism for intervening in undesirable behaviour collect and exchange data from more than just websites due to the pervasive nature of the internet and IoT-enabled devices (such as smartphones, tablets, personal computers, e-readers, etc.). As part of health psychology, information regarding human behaviour change may be gleaned from these interconnected products via IoTs. In the realm of health psychology, IoT-enabled devices offer valuable insights into human behavior change. These interconnected products, like smartphones and tablets, can provide rich data about individuals' behaviors, patterns, and interactions. This information can be harnessed to better comprehend the dynamics of behavior change, facilitating the development of effective interventions and strategies. The wake-up record of a specific sensor can be analysed to reconstruct past behaviour. The "wake-up record" refers to the recorded data generated by a specific sensor when it becomes active or wakes up from a dormant state. By analyzing the wake-up records of sensors, we can reconstruct past behavior patterns. For instance, in the context of a smart device, this data could include timestamps and usage patterns that shed light on an individual's past activities and interactions. In addition, we propose using data from IoT resources like GPS, online availability, sensor networks, and sensor identifiers to substantiate behavioural change investigations. The proposal to utilize data from various IoT resources, such as GPS, online availability, sensor networks, and sensor identifiers, serves to enhance the credibility of behavioral change investigations. These diverse data sources provide a more comprehensive and accurate picture of an individual's behaviors and interactions. By combining data from multiple IoT resources, we can validate and reinforce our findings related to behavioral changes.

Based on the traditional TPB questionnaire, the ETPB questionnaire collects data/objective parameters about behaviour and includes questions about one's attitudes, goals, past behaviour, moral norms, descriptive norms, and confidence in one's ability to change. We recommend using a seven-point Likert scale, with one representing strong disapproval and seven representing high approval. Two questions assess the reporter's attitude toward the targeted behaviour (good, dreadful, harmful, advantageous), and two questions assess the degree to which the reporter feels social pressure to partake in or refrain from the behaviour (subjective norm [SN]).

4 Case Study and Evaluation

Three parameters were identified in this study and applied to a case study focusing on the water-use habits of individual households. In this part, we compared two models to illustrate the variations in intervention behaviour trends and assess the performance of the dynamic ETPB method. We begin with a model that predicts residential water consumption from survey-calculated behaviour (where survey BV is a degree to signify the attainment of a planned objective) and other factors. The dynamic ETPB model is the focus of the second model. We compare these two models under specific conditions to anticipate the normal water use behaviour of urban apartments.

4.1   ANN Water Consumption Model

In this study, we employ an ANN model based on a multilayer network architecture. This MLP uses data from the ETPB questionnaire to identify HS, NORs, month, EL, IL, and BV, feeds that information into two hidden perceptron layers, and then uses that information to generate Water Consumption Data (WCD) in units of tonnes per month. Two nodes are used for the hidden layer. The regression equation used to determine BV (questionnaire calculated BV) is based on research into how to encourage water-saving behaviours [36–40]. The following formula is used to derive Behaviour from ETPB survey data.

$$BV = (observed\ intention * 0.69) + (observedPBC * 0.27) \tag{9}$$

Apartments, one of the most common housing types in China's main cities, and their shared gardens were the primary focus of the model's development, limiting its scope to internal demand. As household leakage is highly variable and unclear, and the forecast of leakage is tied to many other variables not included in the modelling questions, it was disregarded.

The following form expresses the ANN for residential water use:

$$WCD = F(W_2 F(W_1 R + \psi_h I) + \psi_0 I) \tag{10}$$

This method uses the sigmoid function F(x), the threshold layer weight vectors W1, W2, h, and o, and the unit matrix I to predict water usage (where WCD is the standard water usage O/P vector and R is the standard survey I/P vector comprising NOR, HS, IL, EL, M, and Behaviour Value). This ANN model was tested by comparing its predictions to the original data fifty times. Figures 6 a and b display the reliability of this model's testing results. The ANN water consumption method has an average error of <0.8% when comparing observed and predicted values.

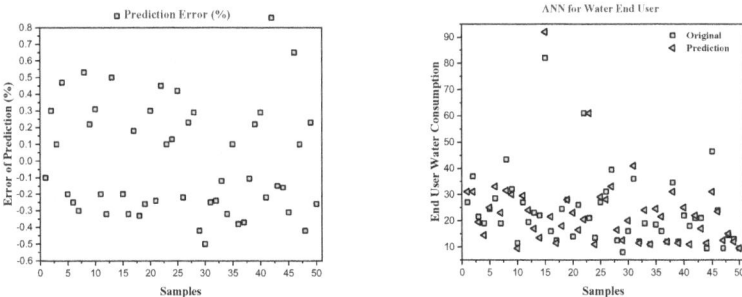

Fig. 6. (a) Prediction error test and (b) Artificial Neural Network water consumption method

4.2   ETPB Model Evaluation

The two most important parts of evaluating a model: estimating system parameters and analysing intervention behaviour patterns. We compared the performance of an Artificial

Neural Network water consumption model and an ETPB behaviour change method, each of which was trained with data from a different set of households (EL, NOR, etc.) that could result in a distinct pattern of water consumption or behaviour related to the household's use of water, to determine the efficacy of the dynamic ETPB model.

Since the ETPB questionnaire was used to directly measure all inventory values, it was possible to assume that the inflow resistances of all inventories were equally small, or g22-g44 $= 1$. In this scenario, inventory coefficients would serve as an interpretation for inventory output transfer resistances. TPB variable coefficients and regression equation analysis in the study yield transfer resistances of $21 = 0.3$, $31 = 0.2$, and $51 = 0.4$, which comprehensively investigate the regression and correlation analysis of parameters related to behavioural change. The starting mass input of every inventory is set with the mean parameter values of b0 $= 3$, e0 $= 3$, n0 $= 1$, m0 $= 5$, c0 $= 5$, and p0 $= 1$ to represent a typical reporter. The ETPB questionnaire score is the first direct inventory measurement. Hence, the time constant ti is defined as the average rate between the first indirect and the first direct inventory measurements. When we model the system with time delay parameters of (Time $(1 - 7) = 0$, time $(8 - 13) = 2$), (Time $(1 - 7) = 1$, time $(8 - 13) = 3$), and (Time $(1 - 7) = 3$, time $(8 - 13) = 4$), (unit of delay is in 'month'), we discover that the three-intervention strategy may rapidly alter the writer's attitude, SN, self-efficacy and PBC. It will take four months to finish all of the intervention procedures.

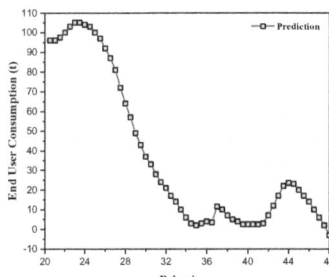 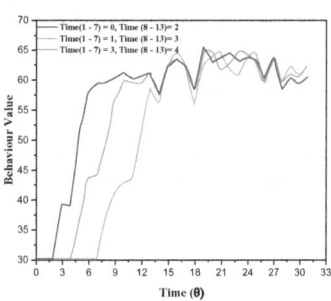

Fig. 7. ANN Water Consumption Model and ETPB Behavioural Intervention of a) Artificial Neural Network b) External influence on behaviour modification in ETPB

The ANN-predicted water consumption trend is then compared to the ETPB-predicted behaviour intervention trend. These values are used as starting points for the ANN: HS $= 2$, EL $= 3$, NORs $= 3$, and IL $= 2$. A household with a yearly income of between 100 and 200 thousand RMBs is indicative of an 87 square metre flat with three residents older than 12 years old, a representative with a bachelor's degree (who will typically denote the landlord's EL), and a representative with a bachelor's degree.

The outcome of ETPB's Behaviour is the amount of water that was conserved. If the rising water savings trend in Fig. 7a and the falling water consumption trend in Fig. 7b can be reconciled, then it can be concluded that an ETPB dynamic model can provide the proper behaviour intervention trend, given the first-order relationship between BV and ETPB Behaviour (water saved). Figure 7a depicts the ETPB model's curve, which begins steadily before spiking sharply in the centre and then levelling off toward the end. Water consumption drops precipitously in the middle of the curve in Fig. 7b before levelling out. However, as shown in Fig. 7a, the ETPB model has demonstrated less capability to replicate the growth of the amount of water conserved at the beginning of the intervention. A first-order variation in the inventory equation makes it impossible to model an inverse reaction in a dynamic system simulation [41–43].

Results from an evaluation of the proposed ETPB model's ability to anticipate patterns of intervention in family behaviour based on data from an ANN household water consumption model are shown in Figs. 7a and b. The pattern of behaviour intervention may follow the same general trajectory as the ANN's input zone water consumption trend. However, these two approaches cannot be a perfect fit for the initial stages of implementing behavioural interventions to reduce indoor water use. Further study into a dynamic system with a second-order deviation is needed to boost the proposed model's prediction accuracy.

4.3 Data Collection

The population at large in this scenario lives in the heart of Shenzhen, China. In 2002, the Estate Company began development on it. All 1480 residents here live in apartment buildings seven stories tall, making this a classic example of a mature Chinese urban neighbourhood. The sampling neighbourhood is conveniently located near public transportation hubs, commercial districts, and educational institutions. Only 128 of the 200 apartment homes selected randomly in January 2013 returned the survey with a return rate of 64%. A multivariate outlier test using the Marhalanobis distance as a criterion excluded nine outliers with a P value of less than 0.001. Water Utilities supplied monthly records on water consumption based on readings taken from metres set in each residence in the studied area. The Cronbach's alpha for this study of water usage reliability was 0.782. In addition to ETPB variables and water usage data, other factors such as house size (HS), NORs (number of persons), number of inhabitants' education levels, etc., are collected for use in ANN simulation results. The data analysis findings are shown in Table 1.

Table 1. Results from reliability tests and data collecting

parameters	reliability	format	min	max	mean	sd
Attitude	Scale α = 0.996 calculated	7-point Likert scale	1.1	7.1	5.996	1.383
Perceived Behavioural Control	Scale α = 0.913 calculated	7-point Likert scale	2.767	7.1	5.206	1.275
SN	Scale α = 0.784 calculated	7-point Likert scale	1.767	7.1	5.133	1.398

(*continued*)

Table 1. (*continued*)

parameters	reliability	format	min	max	mean	sd
self-efficacy	Scale α = 0.881 calculated	7-point Likert scale	1.1	7.1	5.198	1.511
Previous Behaviour	Scale α = 0.839 calculated	7-point Likert scale	1.372	5.736	3.925	0.917
Intention	Calculated average correlation (r) = 0.927	7-point Likert scale	2.1	7.1	5.712	1.340
No. of Residents	Scale α = 0.954 calculated	Residents under 12 years old were ignored	1.1	11.1	4.2179	2.206
HS	Scale α = 0.995 calculated	(58, 87.114,167 m × m)	1.1	4.1	2.360	0.963
household income	Scale α = 0.847 calculated	(30–100k RMB/Y) (100–200k RMB/Y) (200–500K RMB/Y) (greater than 5,00 000 K RMB/Y)	1.1	4.1	1.889	0.919
Education	Scale α = 0.956 calculated	(High school or lower) (Bachelor's degree) (Master's degree or higher)	1.1	3.1	1.906	0.807

5   Conclusion

In conclusion, this study introduces a new approach to understanding how behavior interventions work, specifically in the context of the Internet of Things (IoT). By using a theoretical framework called ETPB and data collection methods, we can better study and model psychological and behavioral interventions. Our findings suggest that the ETPB model, which mimics behavior change, aligns well with our ANN model. The research also presents a dynamic model that simulates how household water usage changes in response to interventions. This approach isn't limited to just hygiene; it can be applied to areas like reducing carbon emissions and water consumption.

However, there are some limitations to this study. The ETPB model doesn't capture all the intricacies of behavior change mechanisms, and incorporating more psychological research and techniques could enhance the accuracy of simulations. Additionally, the dynamic ETPB model relies on long-term observation data to determine intervention strategies. Lastly, while various stochastic technologies are proposed for prediction, this study doesn't test those methods. More research in these areas could help overcome these limitations. In essence, this research offers a fresh perspective on behavior interventions using IoT, providing insights into household water usage changes and potential applications in various fields. While there are constraints, further studies can explore solutions and refine these approaches.

1. Y. Fu and W. Wu, "Behavioural informatics for improving water hygiene practice based on IoT environment," *J. Biomed. Inform.*, vol. 78, pp. 156–166, 2018, doi: https://doi.org/10.1016/j.jbi.2017.11.006.
2. S. Kato, M. Ando, T. Kondo, Y. Yoshida, H. Honda, and S. Maruyama, "Lifestyle intervention using Internet of Things (IoT) for the elderly: A study protocol for a randomised control trial (the BEST-LIFE study)," *Nagoya J. Med. Sci.*, vol. 80, no. 2, pp. 175–182, 2018, doi: https://doi.org/10.18999/nagjms.80.2.175.
3. J. Gong, "Juvenile Crime Monitoring and Characteristic Analysis Based on the Internet of Things and Grid Management," *Mob. Inf. Syst.*, vol. 2022, 2022, doi: https://doi.org/10.1155/2022/5141745.
4. Z.-C. Ding, J. Yan, and J. Fu, "Internet and Mobile Phone Addiction Self-Control Mediate Physical Exercise and Subjective Well-Being in Young Adults Using IoT," *Mob. Inf. Syst.*, vol. 2021, 2021, doi: https://doi.org/10.1155/2021/9923833.
5. C. S. Frempong, V. O. Charles-Unadike, J. A. Anaman-Torgbor, and E. E. Tarkang, "Correlates of intention to practice good food hygiene among street food-vendors in Kadjebi District, Ghana," *Int. J. Heal. Promot. Educ.*, 2022, doi: https://doi.org/10.1080/14635240.2022.2075428.
6. Y. A. B. Buunk-Werkhoven and A. Dijkstra, "Gender variations in determinants of oral hygiene behaviour: A secondary analysis based on the theory of planned behaviour," in *Planned Behaviour: Theory, Applications and Perspectives*, 2014, pp. 37–54. [Online]. Available: https://www.scopus.com/inward/record.uri?eid=2-s2.0-84949512320&partnerID=40&md5=19de430d71a2070e11214f1f92c0a5d0
7. B. S. Jorgensen, J. F. Martin, M. W. Pearce, and E. M. Willis, "Predicting Household Water Consumption With Individual-Level Variables," *Environ. Behav.*, vol. 46, no. 7, pp. 872–897, 2014, doi: https://doi.org/10.1177/0013916513482462.
8. "Remote Sensing for Agriculture, Ecosystems, and Hydrology XXIV," in *Proceedings of SPIE - The International Society for Optical Engineering*, 2022, vol. 12262. [Online]. Available: https://www.scopus.com/inward/record.uri?eid=2-s2.0-85142878931&partnerID=40&md5=0b3cb46ce00042c937abeef40e2f4952
9. K. Rathnayaka, H. Malano, M. Arora, B. George, S. Maheepala, and B. Nawarathna, "Prediction of urban residential end-use water demands by integrating known and unknown water demand drivers at multiple scales II: Model application and validation," *Resour. Conserv. Recycl.*, vol. 118, pp. 1–12, 2017, doi: https://doi.org/10.1016/j.resconrec.2016.11.015.
10. K. Rathnayaka et al., "Factors affecting the variability of household water use in Melbourne, Australia," *Resour. Conserv. Recycl.*, vol. 92, pp. 85–94, 2014, doi: https://doi.org/10.1016/j.resconrec.2014.08.012.
11. I. B. Addo, M. C. Thoms, and M. Parsons, "Household Water Use and Conservation Behaviour: A Meta-Analysis," *Water Resour. Res.*, vol. 54, no. 10, pp. 8381–8400, 2018, doi: https://doi.org/10.1029/2018WR023306.

12. S. Wöhrle, G. Heisenberg, and R. I. Rocha Lima Filho, "Predicting household water supply using satellite imagery and deep learning," in *Proceedings of SPIE - The International Society for Optical Engineering*, 2022, vol. 12262. doi: https://doi.org/10.1117/12.2636152.

13. M. S. Rahim, K. Anh Nguyen, R. A. Stewart, D. Giurco, and M. Blumenstein, "Predicting Household Water Consumption Events: Towards a Personalised Recommender System to Encourage Water-conscious Behaviour," in *Proceedings of the International Joint Conference on Neural Networks*, 2019, vol. 2019-July. doi: https://doi.org/10.1109/IJCNN.2019.8851868.

14. L. Tawalbeh, F. Muheidat, M. Tawalbeh, M. Quwaider, and G. Saldamli, "Predicting and Preventing Cyber Attacks during COVID-19 Time Using Data Analysis and Proposed Secure IoT layered Model," in *2020 4th International Conference on Multimedia Computing, Networking and Applications, MCNA 2020*, 2020, pp. 113–118. doi: https://doi.org/10.1109/MCNA50957.2020.9264301.

15. M. Mashayekhi and E. Santini-Bell, "Detection of damage-induced fatigue response based on structural health monitoring data of in-service steel bridges using artificial neural network," in *Structural Health Monitoring 2019: Enabling Intelligent Life-Cycle Health Management for Industry Internet of Things (IIOT) - Proceedings of the 12th International Workshop on Structural Health Monitoring*, 2019, vol. 1, pp. 294–301. doi: https://doi.org/10.12783/shm2019/32127.

16. N. F. Espinoza-Sepulveda and J. K. Sinha, "Design for Vibration-Based Fault Diagnosis Model by Integrating AI and IIoT," in *Lecture Notes in Mechanical Engineering*, 2022, pp. 278–285. doi: https://doi.org/10.1007/978-3-030-93639-6_23.

17. R. G. Nair and N. Kumar, "An Efficient Elderly Disease Prediction and Privacy Preservation Using Internet of Things," in *Industrial Internet of Things (IIoT): Intelligent Analytics for Predictive Maintenance*, 2021, pp. 369–392. doi: https://doi.org/10.1002/9781119769026.ch15.

18. S. Wang, B. Li, M. Yang, and Z. Yan, "Missing data imputation for machine learning," in *Lecture Notes of the Institute for Computer Sciences, Social-Informatics and Telecommunications Engineering, LNICST*, 2019, vol. 271, pp. 67–72. doi: https://doi.org/10.1007/978-3-030-14657-3_7.

19. H. Shah, D. Pandya, K. Panchal, and N. P. More, "Classification of Machine and Deep Learning Techniques for Financial Fraud Detection of Healthcare Industry," in *2022 International Conference on Futuristic Technologies, INCOFT 2022*, 2022. doi: https://doi.org/10.1109/INCOFT55651.2022.10094538.

20. S. Karthiga and A. M. Abirami, "Deep Learning Convolutional Neural Network for ECG Signal Classification Aggregated Using IoT," *Comput. Syst. Sci. Eng.*, vol. 42, no. 3, pp. 851–866, 2022, doi: https://doi.org/10.32604/csse.2022.021935.

21. M. A. de Oliveira, G. Sedrezt, and G. G. H. Cavalheiro, "ML-based Plant Stress Detection from IoT-sensed Reduced Electrodes," in *Proceedings of the International Florida Artificial Intelligence Research Society Conference, FLAIRS*, 2023, vol. 36. doi: https://doi.org/10.32473/flairs.36.133180.

22. A. Prabha *et al.*, "IoT-based Battery Management System by Deploying a Machine Learning Model," in *3rd International Conference on Smart Electronics and Communication, ICOSEC 2022 - Proceedings*, 2022, pp. 439–445. doi: https://doi.org/10.1109/ICOSEC54921.2022.9952044.

23. S. Bande and V. V Shete, "Smart flood disaster prediction system using IoT & neural networks," in *Proceedings of the 2017 International Conference On Smart Technology for Smart Nation, SmartTechCon 2017*, 2018, pp. 189–194. doi: https://doi.org/10.1109/SmartTechCon.2017.8358367.

24. J. C. P. Cheng, W. Chen, K. Chen, and Q. Wang, "Data-driven predictive maintenance planning framework for MEP components based on BIM and IoT using machine learning algorithms," *Autom. Constr.*, vol. 112, 2020, doi: https://doi.org/10.1016/j.autcon.2020.103087.

25. S. Anitha *et al.*, "Reducing Child Undernutrition through Dietary Diversification, Reduced Aflatoxin Exposure, and Improved Hygiene Practices: The Immediate Impacts in Central Tanzania," *Ecol. Food Nutr.*, vol. 59, no. 3, pp. 243–262, 2020, doi: https://doi.org/10.1080/03670244.2019.1691000.

26. N. Gold *et al.*, "Effectiveness of digital interventions to improve household and community infection prevention and control behaviours and to reduce the incidence of respiratory and/or gastrointestinal infections: a rapid systematic review," *BMC Public Health*, vol. 21, no. 1, 2021, doi: https://doi.org/10.1186/s12889-021-11150-8.

References

1. Singha, S., Aydin, B.: Automated drone detection using YOLOv4. Drones **5**(3) (2021). https://doi.org/10.3390/drones5030095

2. Pansare, A., et al.: Drone detection using YOLO and SSD a comparative study. In: 2022 International Conference on Signal and Information Processing, IConSIP 2022 (2022). https://doi.org/10.1109/IConSIP49665.2022.10007489

3. Sommer, L., Schumann, A.: Deep learning-based drone detection in infrared imagery with limited training data. In: Proceedings of SPIE - The International Society for Optical Engineering, vol. 11542 (2020). https://doi.org/10.1117/12.2574171

4. Tiwari, R., Dubey, A.K.: Detection of camouflaged drones using computer vision and deep learning techniques. In: Proceedings of the Confluence 2022 - 12th International Conference on Cloud Computing, Data Science and Engineering, pp. 380–383 (2022). https://doi.org/10.1109/Confluence52989.2022.9734191

5. Jain, R., Nagrath, P., Thakur, N., Saini, D., Sharma, N., Hemanth, D.J.: Towards a smarter surveillance solution: the convergence of smart city and energy efficient unmanned aerial vehicle technologies. In: Krishnamurthi, R., Nayyar, A., Hassanien, A.E. (eds.) Development and Future of Internet of Drones (IoD): Insights, Trends and Road Ahead. Studies in Systems, Decision and Control, vol. 332, pp. 109–140. Springer, Cham (2021). https://doi.org/10.1007/978-3-030-63339-4_4

6. Kumar, N.M., Sudhakar, K., Samykano, M., Jayaseelan, V.: On the technologies empowering drones for intelligent monitoring of solar photovoltaic power plants. Procedia Comput. Sci. **133**, 585–593 (2018). https://doi.org/10.1016/j.procs.2018.07.087

7. Soderlund, G.: Introduction to 'charting, tracking, and mapping: New technologies, labour, and surveillance.' Soc. Semiot. **23**(2), 163–172 (2013). https://doi.org/10.1080/10350330.2013.777589

8. Saini, D.K., Bala, S., Sharma, A.K., Zia, K.: Emerging technologies for pandemic and its impact. In: Kumar Bhoi, A., Mallick, P.K., Narayana Mohanty, M., Albuquerque, V.H.C.D. (eds.) Hybrid Artificial Intelligence and IoT in Healthcare. Intelligent Systems Reference Library, vol. 209, pp. 291–310. Springer, Singapore (2021). https://doi.org/10.1007/978-981-16-2972-3_14

9. Chang, K.-J., Chuang, C.-W., Chiu, J.-T., Chen, J.-Y.: Flying watchdog: a drone with edge AIoT for residential safety and fall detection by face and posture recognition. In: APWCS 2022 - 2022 IEEE VTS Asia Pacific Wireless Communications Symposium, pp. 181–185 (2022). https://doi.org/10.1109/APWCS55727.2022.9906504

10. Sriram, R., Vamsi, A., Vigneshwari, S.: Telemetry-based autonomous drone surveillance system. In: Bhoi, A.K., Mallick, P.K., Balas, V.E., Mishra, B.S.P. (eds.) ETAEERE 2020. LNEE, vol. 708, pp. 419–427. Springer, Singapore (2021). https://doi.org/10.1007/978-981-15-8685-9_43

11. Zhai, X., Liu, K., Nash, W., Castineira, D.: Smart autopilot drone system for surface surveillance and anomaly detection via a customizable deep neural network. In: International Petroleum Technology Conference 2020, IPTC 2020 (2020). https://doi.org/10.2523/iptc-20111-ms

12. Dilshad, N., Hwang, J., Song, J., Sung, N.: Applications and challenges in video surveillance via drone: a brief survey. In: International Conference on ICT Convergence, vol. 2020-October, pp. 728–732 (2020). https://doi.org/10.1109/ICTC49870.2020.9289536

13. Minhas, M.S., Zelek, J.: Defect detection using deep learning from minimal annotations. In: VISIGRAPP 2020 - Proceedings of the 15th International Joint Conference on Computer Vision, Imaging and Computer Graphics Theory and Applications, vol. 4, pp. 506–513 (2020). https://www.scopus.com/inward/record.uri?eid=2-s2.0-85083563004&partnerID=40&md5=e918203fa3ab51a5c42eae0dc3a9a2ed

14. Siewert, S., et al.: Image and information fusion experiments with a software-defined multispectral imaging system for aviation and marine sensor networks. In: AIAA Information Systems-AIAA Infotech at Aerospace, vol. 2017 (2017). https://doi.org/10.2514/6.2017-0877

15. Roh, S.-B., Oh, S.-K., Pedrycz, W., Seo, K.: Development of autofocusing algorithm based on fuzzy transforms. Fuzzy Sets Syst. **288**, 129–144 (2016). https://doi.org/10.1016/j.fss.2015.08.029

16. Shim, S.-O., Aziz, W., Banjar, A., Alamri, A., Alqarni, M.: Improving depth computation from robust focus approximation. IEEE Access **7**, 20144–20149 (2019). https://doi.org/10.1109/ACCESS.2019.2897744

17. Mannan, S.M., Malik, A.S., Choi, T.-S.: Reducing intricacy of 3D space for 3D camera. In: Proceedings of the International Symposium on Consumer Electronics, ISCE (2008). https://doi.org/10.1109/ISCE.2008.4559551

18. Shim, S.-O., Malik, A.S., Mahmood, M.T., Choi, T.-S.: Estimation of depth map based on focus adjustment. In: Proceedings of SPIE - The International Society for Optical Engineering, vol. 7073 (2008). https://doi.org/10.1117/12.798191

19. Caruso, S., Bonaque-González, S., Oliva-García, R., Rodríguez-Ramos, J.M.: Relative multiscale deep depth from focus. Signal Process. Image Commun. **99** (2021). https://doi.org/10.1016/j.image.2021.116417

20. Shete, S., Bhavsar, A., Sao, A.K.: Enhancing shape from focus-measure-fusion and sparse representation. In: ACM International Conference Proceeding Series, vol. 14 (2014). https://doi.org/10.1145/2683483.2683542

21. An, Y., Kang, G., Kim, I.-J., Chung, H.-S., Park, J.: Shape from focus through Laplacian using a 3D window. In: Proceedings of the 2008 2nd International Conference on Future Generation Communication and Networking, FGCN 2008, vol. 2, pp. 46–50 (2008). https://doi.org/10.1109/FGCN.2008.139

22. Choi, W.-J., Choi, T.-S.: Fast three-dimensional shape recovery in TFT-LCD manufacturing. In: Proceedings of SPIE - The International Society for Optical Engineering, vol. 7073 (2008). https://doi.org/10.1117/12.798232

23. Zhao, M., Zhou, J.: .Detection method of stone surface roughness based on CCD camera. Guangxue Jishu/Opt. Tech. **44**(3), 310–314 (2018). https://www.scopus.com/inward/record. uri?eid=2-s2.0-85055925070&partnerID=40&md5=edf28d89b026634e35d800a5e4f80ad4

24. Rusinol, M., Chazalon, J., Ogier, J.-M.: Combining focus measure operators to predict OCR accuracy in mobile-captured document images. In: Proceedings - 11th IAPR International Workshop on Document Analysis Systems, DAS 2014, pp. 181–185 (2014). https://doi.org/10.1109/DAS.2014.11

25. Sayar , M.S., Akgül, Y.S.: Depth from Moving Apertures. In: Gelenbe, E., Lent, R. (eds.) Information Sciences and Systems 2013. LNEE, vol. 264, pp. 189–197. Springer, Cham (2013). https://doi.org/10.1007/978-3-319-01604-7_19

26. Pavliček, P.: Measurement of the shape of objects by shape from focus. In: Proceedings of SPIE - The International Society for Optical Engineering, vol. 9524 (2015). https://doi.org/10.1117/12.2189187

27. Abbas, Q., Ibrahim, M.E., Khan, S., Baig, A.R.: Hypo-driver: a multiview driver fatigue and distraction level detection system. CMC-Comput. Mater. Continua **71**(1), 1999–2017 (2022). https://doi.org/10.32604/cmc.2022.022553

28. Akhtar, M.M., Zamani, A.S., Khan, S., Shatat, A.S.A., Dilshad, S., Samdani, F.: Stock market prediction based on statistical data using machine learning algorithms. J. King Saud Univ.-Sci. **34**(4), 101940 (2022). https://doi.org/10.1016/j.jksus.2022.101940

29. AlAjmi, M.F., Khan, S., Sharma, A.: Collaborative learning outline for mobile environment. Paper Presented at the 2014 International Conference on Issues and Challenges in Intelligent Computing Techniques (ICICT) (2014)

30. Alfaifi, A.A., Khan, S.G.: Utilizing data from twitter to explore the UX of "Madrasati" as a Saudi e-learning platform compelled by the pandemic. Arab Gulf J. Sci. Res. **39**(3), 200–208 (2022). https://doi.org/10.51758/AGJSR-03-2021-0025

31. Khan, J., et al.: Secure smart healthcare monitoring in industrial internet of things (IIoT) ecosystem with cosine function hybrid chaotic map encryption. Sci. Program. **2022**(Article ID 8853448), 22 p. (2022). https://doi.org/10.1155/2022/8853448

32. Khan, S.: Data visualization to explore the countries dataset for pattern creation. Int. J. Online Biomed. Eng. **17**(13), 4–19 (2021). https://doi.org/10.3991/ijoe.v17i13.20167

33. Khan, S.: Visual data analysis and simulation prediction for COVID-19 in Saudi Arabia using SEIR prediction model. Int. J. Online Biomed.Eng. **17**(8) (2021). https://doi.org/10.3991/ijoe.v17i08.20099

34. Fazil, M., Khan, S., Albahlal, B.M., Alotaibi, R.M., Siddiqui, T., Shah, M.A.: Attentional multi-channel convolution with bidirectional LSTM cell toward hate speech prediction. IEEE Access **11**, 16801–16811 (2023)

35. Khan, S.: Business intelligence aspect for emotions and sentiments analysis. Paper Presented at the 2022 First International Conference on Electrical, Electronics, Information and Communication Technologies (ICEEICT) (2022)

36. Khan, S., Alfaifi, A.: Modeling of coronavirus behavior to predict it's spread. Int. J. Adv. Comput. Sci. Appl. **11**(5), 394–399 (2020). https://doi.org/10.14569/IJACSA.2020.0110552

37. Khan, S., Alghulaiakh, H.: ARIMA model for accurate time series stocks forecasting. Int. J. Adv. Comput. Sci. Appl. **11**(7), 524–528 (2020). https://doi.org/10.14569/IJACSA.2020.011 0765

38. Keshta, I., et al.: Energy efficient indoor localisation for narrowband internet of things. CAAI Trans. Intell. Technol. (2023)

39. Khan, S., Ch, V., Sekaran, K., Joshi, K., Roy, C. K., Tiwari, M.: Incorporating deep learning methodologies into the creation of healthcare systems (2023)

40. Khan, S., et al.: Transformer architecture-based transfer learning for politeness prediction in conversation. Sustainability **15**(14), 10828 (2023)

41. Khan, S., Moorthy, G.K., Vijayaraj, T., Alzubaidi, L.H., Barno, A., Vijayan, V.: Computational intelligence for solving complex optimization problems (2023

42. Khan, S., et al.: Manufacturing industry based on dynamic soft sensors in integrated with feature representation and classification using fuzzy logic and deep learning architecture. Int. J. Adv. Manuf. Technol. 1–13 (2023)

43. Khan, S., Altayar, M.: Industrial internet of things: Investigation of the applications, issues, and challenges. Int. J. Adv. Appl. Sci. **8**(1), 104–113 (2021). https://doi.org/10.21833/ijaas.2021.01.013

44. Khan, S., AlSuwaidan, L.: Agricultural monitoring system in video surveillance object detection using feature extraction and classification by deep learning techniques. Comput. Electr. Eng. **102**, 108201 (2022). https://doi.org/10.1016/j.compeleceng.2022.108201

A Comparison of the Feature Extraction and Fine-Tuning Approach to Fake News Detection Using BERT

Vansh Chaudhary and Shivani Aggarwal[✉]

Apex Institute of Technology, Chandigarh University, Mohali, Punjab, India
shivaniaggarwal.uhv@gmail.com

Abstract. The rampant increase in the spread of misinformation around the globe on the social media in the wake of the pandemic & international conflicts has urged the innovation in machine learning paradigms of the fake news detection to tackle the threat. Given that the task belongs to the text classification under the subfield of Natural Language Processing, the benchmark defying performance of the transformers-based models such as BERT inspired to use the pre-trained model for text classification using two approaches, one using the BERT for feature extraction and then classifying the text using standard classifiers, a methodology which has not been widely researched, while other involving fine-tuning the BERT for text classification. The various classifier models achieved remarkable accuracy up to 98%, especially Logistic Regression & Multi-layer Perceptron, which was further topped by the fine-tuned model which achieved an accuracy of 98.8% on the balanced test dataset & outperformed existing BERT-based benchmark FakeBERT on the imbalanced dataset with an accuracy of 99.21% for the task of fake news detection.

Keywords: Machine Learning · Fake News · Natural Language Processing · BERT

1 Introduction

The exponential increase in the amount of misinformation online being propagated in the form of fake news has instigated a plethora of machine learning models being applied to distinguish the fake news. Analogously around the same time, Natural Language Processing has been revolutionized with the advent of transformers & transformer-based models such as BERT (Bidirectional Encoder Representations from Transformers), GPT (Generative Pretrained Transformer) & their variants. Though various deep learning paradigms have been used for the binary classification task of the fake news from the real one, transformer-based models, despite their overwhelming accuracy, haven't been used as extensively as the conventional classifiers for the same task. To compare & analyze the performance of transformer-based models for the text classification is the primary objective of this research paper which will not only provide insights on the caliber of the fine-tuned models but can also guide which models should be preferred for

© The Author(s), under exclusive license to Springer Nature Switzerland AG 2025
A. Gupta et al. (Eds.): ICAIA 2023, CCIS 2308, pp. 155–177, 2025.
https://doi.org/10.1007/978-3-031-84394-5_12

such task with proper reasoning by comparing the performance of the various models using feature extraction as well as fine tuning approach of the transformer-based models in the task of text classification.

2 Literature Review

It is evident in the context of the modern communication infrastructure that fake news is on the verge of spiraling out of control if not checked properly with far-reaching consequences which makes it all but more significant to comprehend the background of the concept of the fake news.

Fake News which is simply a news which is fake has been on a rampant rise since the last decade [1] with the inception & mass-level acceptance of the internet which has penetrated the entire landscape of the world at an unprecedented pace. Unfortunately, the negative impacts of the fake news heavily outweigh the positive (if any) as discussed extensively in the [2]. Furthermore, the alleged manipulation & misinformation during the US Presidential Elections for targeted political propagation & campaigning [3] diabolically, during the heightened peaks of the COVID-19 causing & worsening psychological distress & disorders in the patients as well as healthcare professional [4]. The malicious intent of fake news to instigate communal riots & stock market rumors to destabilize the market were reported from throughout the globe in the last few years.

The phenomenon isn't a novel danger lurking out of nowhere in fact, in Nigeria in 2014, the infodemic paired with Ebola epidemic caused irreversible damage [5] with text messages regarding Ebola Salt solution text messages shared via WhatsApp leading to hospitalizations & even 2 deaths. [6] studied the discourse model for the justification of the fake news alongside legitimization strategies such as Authorization, Moralization & Rationalization applied by the fact checkers when faced with fabricated persuasive messages & imagery.

2.1 Pre-existing Approaches Using Traditional Models

Of the many different approaches used, there have been many instances of tackling the issue using machine learning models by presenting the problem as text classification problem with varied methodologies ranging from creating new models for the specific purpose or using pre-trained models after fine-tuning them to aptly adjust to the task of the fake news classification. Fake news detection is preferably a binary classification problem but it has been formulated with a multiclass approach as well. Restricting the review to machine learning for fake news detection, the presented solutions are further divided against the lines of traditional & modern state-of-art machine learning algorithms.

To provide a clearer picture of how to get initiated with fake-news detection, de Beer et al. [7] presented the same in their paper where the systematic methods for detecting a fake news from a given piece of text are studied. The paper reviewed & concluded that how various approaches can be used for a given problem. The common approaches studied in the problems are represented in the hierarchy chart given on the next page:

Considering fake news as an online problem, Zhang and Ghorbani [8] propounded the exponential increase to the 'volume, veracity & velocity' of fake news which makes it difficult to timely detect & check its propagation. The paper suggested probing not only the content of the fake news but also conducting a wider investigation of the social context, creator & targets as well which will assist in not only the detection and but also the prediction of the fake news generation (Fig. 1).

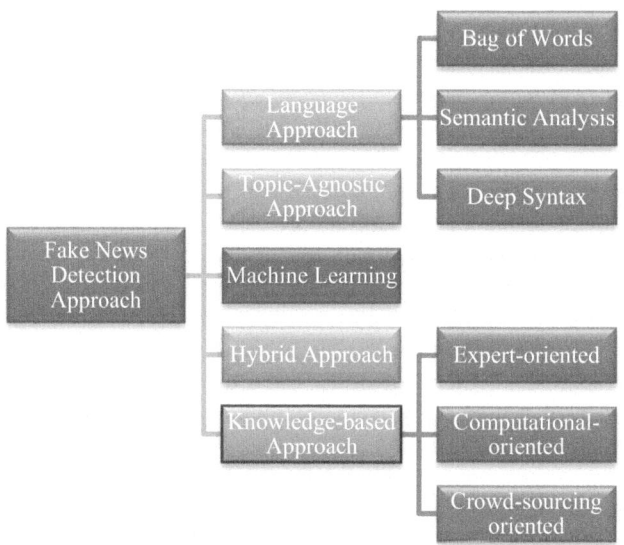

Fig. 1. Different approaches to fake news detection

Diving deeper, while the sources were differentiated as humans & non-human (social bots), the content was also classified as physical & non-physical (sentiment) & the social context constituted of the environment of dissipation of misinformation which is primarily social media platforms & online website. The comprehensive paper also evaluated already existing solutions using different measures and information for the classification & cited the findings of noted research papers reinforcing the value of auxiliary information regarding the fake articles. Turning to the machine learning aspect, the authors enlisted and contrasted nearly all the datasets available over the web using the features available which are integral to the success of a supervised learning models. Finally, the paper concludes with an ambitious proposal of a 'real-time' fake news prediction & detection system based on unsupervised learning approach of cluster & outlier analysis.

Ahmed et al. [9] performed a systematic literature review for research articles studying & executing traditional classifiers. The paper concluded that Naïve-Bayes classifiers with 96.08% accuracy alongside Support Vector Machines (SVM) paired with Neural Network achieving a whopping 99.90% accuracy [10] were the best performing classifiers for the social media posts.

To investigate further on the supervised machine learning algorithms, Khanam et al. [11] trained their model upon Liar & PolitiFact datasets using the concept of TFIDF (term frequency – inverse document frequency) vectorizer for n-gram features. Performing the preprocessing involving noise (punctuations, special symbols) removal, stemming & Part of Speech (POS) tagging to extract tokens, feature extraction is performed in which the tfidf vectorizer is extracted which is used to calculate the importance of particular term in a document in relation to a collection of other documents. After the train-test-split is performed, a number of classifier algorithms were implemented ranging from Random Forests to Naïve-Bayes, with highest accuracy from XGboost (Extreme Gradient Boosting) of 75% and an average accuracy of nearly 72%.

The imbalance & inconsistency in the accuracy can be attributed to the variety of factors such as using different datasets & models still, since SVM paired with neural network is the best performing algorithm so far, one should observe it in a more nuanced manner to come up with richer insights. Baarir and Djeffal [12] performed Fake News Detection specifically using SVM on a combined dataset created using datasets present on the Kaggle and then preprocessed with feature selection, where all the features are analyzed as to what extent are they affecting the target variable, i.e., fake or true & parameter tuning in which the effect of the parameters such as Cost, ε, the hyperparameter regarding the tube width of hyperplane & γ, the hyperparameter augmented for non-linear classification of gaussian kernel. One of the major distinctions of the paper is that the model is implemented in WEKA (Waikato Environment for Knowledge Analysis) using LIBSVM (Library for SVM) method and not conventional python notebooks. Upon implementing the models, it was concluded that N-gram approach performs superior to bag of words & that 'sentiment' feature barely contributes to the fake news detection.

2.2   Existing Systems Using Deep Learning Methodology

Another important division within the supervised learning paradigm are the individual learners & ensemble learners, viz., independently trained classifiers conglomerated into one classification models. To infer how the individual & ensemble models fare in the task of the fake news detection Ahmad et al. [13] trained & tested a 13 different machine learning upon 4 different datasets as well as using User Query to find out the best performing classifier models. The paper also included deep learning neural networks such as Long Short-term Memory (LSTM) & convolutional neural network as benchmark algorithms & judged the outcomes using standard performance metrics. The ensemble model Random Forest & Linear SVM paired with 5-fold cross validation were able to achieve accuracy as high as 99% for first dataset. The paper concluded that ensemble methods were ahead in terms of accuracy & other measures with bagging classifier based on the decision trees as the best average performer over all the datasets while benchmark algorithms apart from LSVM fared much poorer with lower accuracy.

Closer to a product is the implementation of the machine learning classifiers using search by Sharma et al. [14] where there are two variations namely static search & dynamic search. In static search, the user is prompted to enter the news article upon which the Logistic Regression (selected after metric computation) paired with GridSearchCV method is applied, returning a confidence score which then decides whether or not is the article fake with an accuracy of nearly 75%. The dynamic search involves three search parameters & a passive-aggressive classifier. The dynamic search method outperforms the static search with an accuracy of 93% though it's still a far cry from the behemoth websites of fake news detection already discussed by [7].

Focusing on the machine learning approach of the problem, which is usually guided by heuristic from language or knowledge approach, one must understand that Fake News Detection lies within the domain of Natural Language Processing, a subfield of the Machine Learning interlaced with linguistics which aims to impart to the machines the ability to understand, process & respond to natural language. Stating the NLP, it is important to address the fact that since the groundbreaking 'Attention is All You Need!' research paper [15] introducing "Transformers" into the field of NLP, majority of the tasks have shifted towards using pre-trained models followed by fine-tuning.

To implement the aforementioned problem with a proper integration of the all the features (content as well as context), Kaliyar et al. [16] proposed a robust model Deep-FakE paired with XGBoost classifier converging the neural network based deep learning with a more conventional random forest-based classifier applying tensor decomposition With experiments consisting of trying classifier & neural network over individual & combined forms of content & social context features, the DeepFakE model registered an accuracy of approximately 86.5% which was later evaluated using Cross Entropy Loss over the number of epochs.

Using other deep learning models for the fake news detection, Sachin et al. [17] discussed the challenges for the real-time detection due to dynamic nature of the network and studied the limitations of backtracking method applied previously for the problem. The employed models trained over PolitiFact & FakeNewsNet dataset, preprocessed & tokenized and then, using GloVe (Global Vectors used for retrieving word representation) to embed the dataset using 6 billion tokens which is then fed to the layers of the neural networks. Of the 7 neural network models, including individual & ensemble of the CNN & LSTM (bidirectional) with & without the attention mechanism, the highest average accuracy achieved was nearly 88.8% using CNN with bidirectional LSTM with attention mechanism. To check the statistical significance of the mean accuracies, the paper performed t-test using p-values to distinguish between the pairs of models having no significant difference with suggestion of tracking the sources of the fake news in long term for better prevention.

Given the prominence of the transformers for the NLP tasks, one of the model which is immediately relevant to the review is the FakeBERT model proposed by Kaliyar et al. [18] which involved fine tuning the powerful BERT model with pre-trained GloVe word embedding using 3 parallel one-dimensional convolutional neural network layers with 2 straightforward convolutional layers, all of which are max-pooled & then flattened before fed to two dense layers with the ReLu activation function & categorical cross-entropy as the loss function. Of the common machine learning algorithms, the highest accuracy of 89.97% was achieved using Multinomial Naïve-Bayes classifier while the originally proposed FakeBERT achieved accuracy of 98.90% besting existing classifier models while there was also considerable improvement in case of the other deep learning models with future direction of a hybrid multi-class classification interlacing the data from the echo-chambers rather than studying the models in isolation.

Perhaps the most innovative & creative algorithm implemented for the fake news detection is the Salp Swarm optimization and its two improved versions (using non-linear decreasing function and oscillating strategy for inertia weights respectively), proposed by Ozbay & Alatas [19], which were modelled upon the swarming behavior of Salps (jellyfish-like creatures). The methodology, further inculcating another biological algorithm – Gray Wolf Optimization (GWO), constructed a Document-term matrix (DTM) based on the term frequency after relevant features had been extracted via data preprocessing. Over all the datasets, the best mean accuracy is provided by Adaptive SSO (using Oscillating strategy) while GWO had overall best accuracy of 99.5% on the ISOT news dataset with mean accuracies observed as having statistically significant difference across all the datasets.

The more refined version of the approach taken by [17] & the most accurate of all the reviewed fake news detection model was proposed by Chauhan & Palivelia [20] with a similar model architecture having GloVe embedding & deep learning LSTM model with an embedded layer followed by two LSTM layers having tanh activation function followed by two Dense Output layers using ReLU & sigmoid activation functions respectively which are then used for categorizing the news as fake or true based leading to an astounding 99.88% accuracy which outperformed other benchmark algorithms including the Adaptive Salp Swarm Optimization (SSO), FakeBERT by a remarkable margin. The paper concluded with a future proposal of designing an automated system for websites for alert & report.

From state-of-art models to amateur datasets, Mridha et al. [21] conducted research from a multidimensional perspective on the task of fake news detection using 'deep' learning models. In the expansive paper, the vivid datasets available around the internet biased towards political news or completely based on the social media posts were scrutinized for affecting accuracy and also the composition as text only or text-image or propagation tree,TF-IDF, bag of words, to more complex GloVe & BERT embeddings. Moving forward to the deep learning approaches, the paper observed the most popular models from 2017 to 2021 in deep learning & then, explains the composition of each one in a discerned manner. Finally, the paper enlisted the challenges such as political bias in the text features & limiting fake news detection to NLP while it can be more efficiently spread using fake image & multimedia which demands the combination of Computer Vision as well & future research detection to incorporate GPT (Generative Pre-trained Transformer) alongside XAI (Explainable Artificial Intelligence) or other less used deep learning models such as SeqGAN & Deep Belief Network for the task.

3 Proposed Methodology

It is thoroughly clear that deep learning models are the outperforming other models in the task of fake news detection from the review of the research articles. Given the popularity & outperformance of the transformers in Natural Language Processing tasks in the present scenario, the proposal is to use a pre-trained transformer-based model for the purpose of the fake news detection. Contemporary research [22] indicates that 'encoder-only' transformers-based models are more suited for the Sequence classification than other paradigms of the 'decoder-only' or 'encoder-decoder' approach which are primarily best-suited for the text generation & text summarization. For using a pre-trained model for text classification (or any other NLP task) has two general procedures with similarities as well as the differences which must be understood before digressing into the details of the proposed methodology.

3.1 Feature Extraction

Rather than repurposing the pretrained model, the hidden states returned after processing & learning on the tokenized input texts are used as the features for training a simpler classifier model. It can be visualized by applying a classification filter on the output of the transformer model. The positive factors pertaining to this given approach is that it can be implemented even in the absence of GPU (Graphical Processing Unit) and that the model is trained very quickly. The models preferable for such an approach are the ones which do not require the computation of gradients over the dataset. Thus, while a conventional approach would yield a lower accuracy, using the conventional classifier over the feature matrix constructed based on the hidden states extracted using the transformer-based model can be significantly increase the accuracy if even the simpler classifiers such as Logistic Regression or Random Forest acting as a balance between conventional ML & deep learning (Fig. 2).

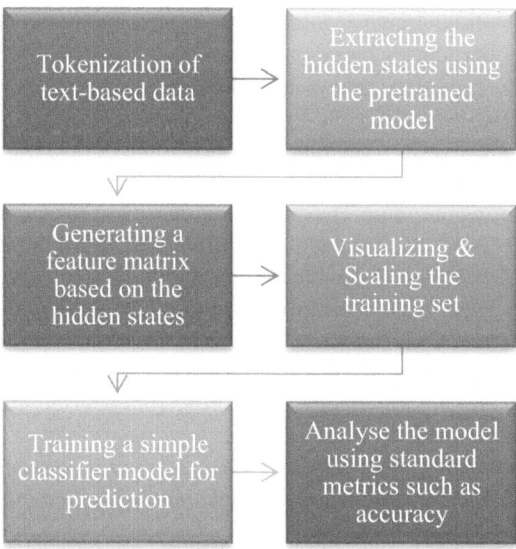

Fig. 2. Working of the feature extraction approach

The traditional machine learning models which have been selected to act as the classifier models in the feature extraction approach are standard classifiers from which have been implemented in the scikit learn library of the Python language & are all briefly discussed hereby:

- **Logistic Regression:** Reframing the classification as an estimation of probability to model the binary outcomes, logistic regression combines the linear regression, which is used to fit onto the data & then applying the logistic function with the help of the odd. Thus, logistic regression models the uncertainty between the observations of a parameter value in the training set relative to the parameter having that value in reality.
- **K-Nearest Neighbors Classifier:** The simplest of the machine learning classifiers is the one which classifies using distance (or proximity) between the points as the approach to make a prediction. K-nearest neighbours finds the nearest 'k' neighbours to the test sample from the training dataset with the majority class as selected as the label for the sample. The value of 'k' can be fine-tuned for achieving better results depending on the size of the dataset.
- **Support Vector Machine:** Popularly called as the SVM, the classifier uses a hyperplane created with a subset of samples to distinguish between different labels which acts as the decision boundary separating both the classes. The SVM are well-known for handling non-linear data points with the help of the kernel function where with the help of a suitable kernel such as polynomial or sigmoid, the data points from the training set are mapped onto a higher dimension where they can be easily separated.
- **Ensemble Models:** The ensemble method refers to particular approach where rather than using a single classifier model, multiple (similar or different) classifier models are implemented for the same training dataset & the prediction is made via voting, i.e., which label will be assigned to a sample is decided on the basis of the voting between the classifier models based on the majority. The ensemble classifiers which will be used in the feature extraction approach are:

  - *Random Forest Classifier*: The random forest is the perfect trade-off between the speed & robustness of the non-parametric models of decision trees whilst not being prone to overfitting the noise. The model is an ensemble of a number of the decision trees. Every decision tree is trained on a different subset of the features of training dataset with the purpose of having no correlation between the prediction of any decision trees. The criterion used for the attribute splitting in the implementation of the decision trees is Gini index, which is a measure of the population diversity & is aimed to be maximized.
  - *Extra Trees Classifier:* Referring to Extremely Randomized Trees Classification, it is similar to the random forest classifier with the difference of not performing bagging (bootstrap aggregating) rather going for original test sample & to undergo randomized splitting in contrast to optimal splitting in the random forest leading to a model with relatively higher accuracy but lower generalizability. The Extra Trees have a concept of the feature importance via which the feature selection is performed which is indispensable for the creation.

- **Boosting Models:** It is another ensemble modelling technique which *boosts* a series of weak classifiers to construct a single strong classifier, usually by readjusting the weights of the misclassified labels. The boosting algorithms used for the purpose of the fake news detection are:

– *AdaBoost Classifier:* Adaptive Boosting or AdaBoost is a simple & popular boosting algorithm which works by initially assigning equal weights to each of the data points & then, sequentially increasing the weights of the samples which have been classified wrongly for the next model to minimize the error in training the models; usually decision stumps are used in implementation of the AdaBoost which are finally combined together into a stronger & more robust classifier model. The re-adjustment of the sample weight occurs as per following:

– *Gradient Boosting Machine Classifier*: The GBM classifier is an improvised & robust version of the machine learning classifier which works on the intuition of hypothesis boosting of probabilistic approximation for correct parameters & deploys the weight minimization on to AdaBoost classifier with the difference that instead of fitting to the data, the model is fitted onto the residual errors of the previous model assisted by the log of the odds of target feature which is then converted to the probability, a manner similar to Logistic Regression, with hyperparameter of learning rate managing the role of each tree & how the weights will be assessed. The process is repeated over & over again until a certain threshold is met or the residual errors are negligible. The GBM is suited for the binary classification problem with one vs. all training used for multiclass problem.

– *XGBoost Classifier:* Extreme Gradient Boosting is one step ahead to the Gradient Boosting Machine by introducing the concept of parallel boosting that is instead of sequentially building the decision trees, a level-wise approach for the values of the gradient is used to find the best split for the training set out of all possible splits. The speed & performance is improved in comparison to the simple boosting algorithms with options such as regularization, parallelization & tree pruning for handling the complexity or missing values with sparsity awareness feature for handling sparse data & cross-validation to check the model from overfitting the dataset.

• **Naïve-Bayes Classifiers:** It refers to the family of classifiers which works on the concept of the Bayes theorem which uses the prior knowledge (probability) of a hypothesis to compute the posterior probability of the hypothesis, henceforth providing a probabilistic inference for the classifiers. The Naïve-Bayes model assumes that every feature present in the dataset is independent of one another & contributes equally towards the target feature. The various Naïve Bayes classifiers selected to be trained & rested for the task of fake news detection are:

– **Gaussian Naïve-Bayes Classifier:** Assuming a gaussian distribution of the values of the features. The mean & standard deviation of each feature is calculated & then, computing the parameters of the probabilistic distribution into which the new test sample is fed for the calculation of the probability. It is well-known for its scalability & the ability to work well with continuous features. Given below is the equation for a simple gaussian (normal) distribution of a random variable x:

– **Bernoulli Naïve-Bayes Classifier:** As the name suggests, the classifier assumes a Bernoulli distribution of the features & is primarily suited for binary classification problems, especially those constituting of the categorical features (discrete data). In the traditional approach of document classification, the Bernoulli Naïve-Bayes

164   V. Chaudhary and S. Aggarwal

captures the occurrence of a particular term signifying while nor focussing on the frequency of the term in the document(s) also able to model the conjecture that specific terms are not present in the model.

- **Multinomial Naïve-Bayes Classifier:** One of the most popular choices in the count vectorization based document classification problems, Multinomial Naïve-Bayes classifier is important since it involve pre-processing steps such as lemmatization & tokenization of the dataset, which involves breaking down the text into smaller & simpler components. Multinomial Naïve-Bayes is often used in conjunction with the bag-of-words approach & is preferred for its speed & low computational cost.
- **Complement Naïve-Bayes Classifier:** Another variant of Naïve-Bayes classifiers which is considered for the imbalanced data, where multinomial naïve-bayes falters, the complement Naïve-Bayes calculates the probability of a particular sample or term not being present in a class & selects the class with the lowest value of the complement probabilities represented by the formula:
- **Hyperparameter tuning of the Naïve Bayes Classifier Models using the Grid Search based on Cross-Validation Score:** The hyperparameter of the Naïve-Bayes classifiers such as var_smoothing or alpha (using sklearn library in python) can be optimized for a better performance. GridSearchCV automates the process of finding the suitable values for the hyperparameters where each parameter combination is treated as a grid point & the best values out of the specified values are selected which can be then fitted to the model with for optimal accuracy.

- **Multi-layer Perceptron Classifier:** Approaching the task of classification with deep learning, the multi-layer perceptron classifier is a feed forward neural network which is maps the input features to an output set with the help of multi-layer neutrons which are fully connected to each other. Having ReLU (Rectified Linear Unit) as activation function with Softmax for the output layer, the model learns the weights & bias from the dataset after iterating over it for a number of epochs, minimizing the loss which is governed by the learning rate hyperparameter.

3.2 Fine-Tuning the Pretrained Model

The other more robust methodology is to implement text classification by fine-tuning the pretrained model to specifically perform the downstream tasks. To understand, it is analogous to how Teflon which was originally used for the artillery and nuclear material casing due to its inertness was repurposed later as the non-stick cookware for a more commercial use [23]. In machine learning terminology, the process is referred as transfer learning a given model for a specific task. Though nearly a standard practice in the field of computer vision, it was the advent of transformers & more importantly, the exceptional ULMFit [24] regarding LSTM made it possible to adapt it into the field of Natural Language Processing where it excelled over every benchmark model in the various tasks ranging from Named Entity Recognition to Text Summarization and Question Answering. Here the entire model is trained for the process from the ground up, i.e., the hyperparameters for the pretrained model are updated as well. It isn't a simple classification filter at the output but here the input of the classification head, i.e., the extracted hidden states are also updated & adapted which leads to an overall superior performance (Fig. 3).

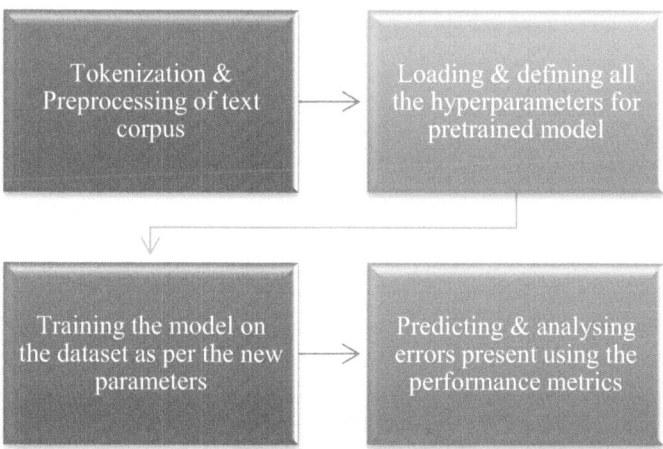

Fig. 3. Working of the fine-tuning approach

Taking into account all the significant factors, the best choice is to perform the fake news detection by fine-tuning the BERT (Bidirectional Encoding Representations from Transformers) but given the precipice that similar models have been utilized to confront the problem, the idea is to rather than using BERT the focus of the research is the performance of different classifier models in fake news detection when the BERT model is used merely for feature extraction, i.e., for designing a feature matrix using the hidden states. The approach has not been thoroughly researched since it is used only in the case of unavailability of the GPU or when the timeframe for the model training is severely limited. Thus, rather than owing to the limited resources (GPU Memory), the methodology is to capitalize upon the limitations & study the various models to understand & observe the ability of classification models for utilizing the hidden states generated by the BERT for the apt classification. Furthermore, given the affinity of BERT for smaller length text strings [25], the fake news detection will be performed on the basis of the title of the news rather than the body.

To provide a comprehensive picture of the methodology, it is important to understand the various constituents such as datasets used, models, preprocessing steps such as tokenization & the model implementation paradigm individually.

3.3 Dataset Used

The primary dataset used for training the model is the ISOT (Information Security & Object Technology) Fake News Dataset [26] provided by the University of Victoria which consists of two csv files classified as per real & fake news. Upon combination and adding the label column for the real & fake news, the various attributes of the dataset are (Table 1):

Table 1. Details of the Dataset

About Kaggle Fake News Challenge Dataset	
Shape	44898,5
Features	5 (Title, Text, Subject, Date, Label)
Null values	Absent
Labels	2 (1: True, 0: Fake)
Balance	Data Balanced

A clearer overview of the dataset can be fulfilled using the various data visualization in a more concise way from the distribution of the data in term of classes to n-gram analysis of the 'title' column since the model will be implemented onto it (Fig. 4).

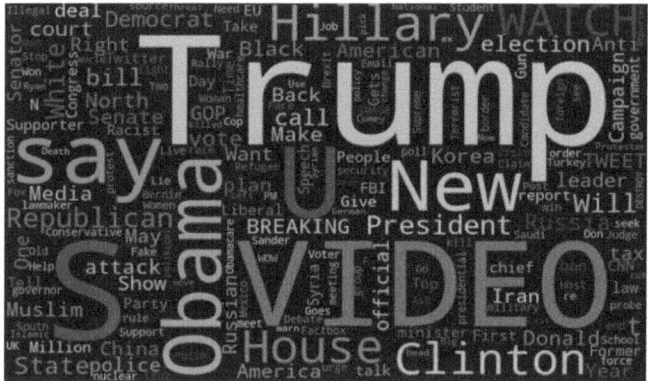

Fig. 4. Word Cloud of the Dataset

It can be observed that the US Presidential Election has heavily influenced the dataset which is very expected since it was former US President Donald Trump who catapulted the concept of fake news into the popular narrative during his electoral campaign. Having understood the datasets which are being used for the training & testing the data, it is necessary to observe & underline the standard practices which are required for formulating text data which can then be fed into the model resulting in robust performance. The noise removal, especially for the social media datasets followed by lowercasing are mandatory for every Natural Language Processing task for data cleaning.

3.4 Model Configurations

There are two paradigms for the text classification using the transformers-based models. The first one involving the feature extraction where the hidden states returned by the models are used as the features for the other simple classifiers such as Logistic Regression, Random Forest or Naïve-Bayes classifier without any changes in the pre-trained model while the other involves the fine-tuning of the model end-to-end The first approach is suited for a smaller model which is trained quickly due to the reduction in the memory required for the computation of the gradients. It differs from the more advanced approach of the updating all the hyperparameters of the pretrained model and has not been considered for much purview for the downstream task of the text classification, especially in context of fake news detection and can assist in the insight generation & model architectures in the future.After the raw text have been filtered out and preprocessed, the most crucial step for the Transformer -based models is performed, i.e., the **tokenization**, more specifically subword tokenization. Tokenization is the process of breaking a string down into smaller units which are then used by the model. Subword tokenization combines the letter & the word tokenization by keeping frequent words in the vocabulary & longer rarer words are fragmented into characters of the smaller size. Tokenization is learned from the pre-training dataset usng specific algorithms & statistics. Of the various subword tokenization algorithms available, BERT tokenizer deploy the WordPiece algorithm [27] which is demonstrated working above. As one can observe, the token [CLS] have been inserted at the end of the sentence while the token [SEP] is added at the end of the news headlin signifying the start & the end of the sequence to the model. For encoder-based models, the tokenized text is further converted into one-hot vectors known as *token encodings* whose size is constrained by the vocabulary of the tokenizer. These encodings are transformed in lower-dimensional space known as *token embeddings* which are passed onto the *encoder stack* for conversion into a sequence of embedding vectors called as *hidden states* or *context*. The entire process is illustrated in the following figure: To supplement the model with the information regarding the positions of the tokens from the text, *positional embeddings* are also provided after the token embeddings in the transformer architecture before encoding tokenized vectors.

BERT or Bidirectional Encoder Representation of Transformers, as the name suggests & aforementioned, is an encoder-only model which relies upon construing the text input into a numerical representation with computed representation dependent on the left as well as the right context of the text (Fig. 5).

BERT model was the first encoder-based model and was pretrained on BookCorups & English Wikipedia with two objectives including:

1- Masked Language Modeling: To predict the most likely candidate for randomly masked words in a text.
2- Next Sentence Prediction: To determine if one passage of the text is likely to follow another.

To fine-tune the model for the task of Fake News Detection, the encoder stack which computes the hidden states of the tokens must be briefly overviewed to guide how to handle the hyperparameters of the model to perform well for the given task.

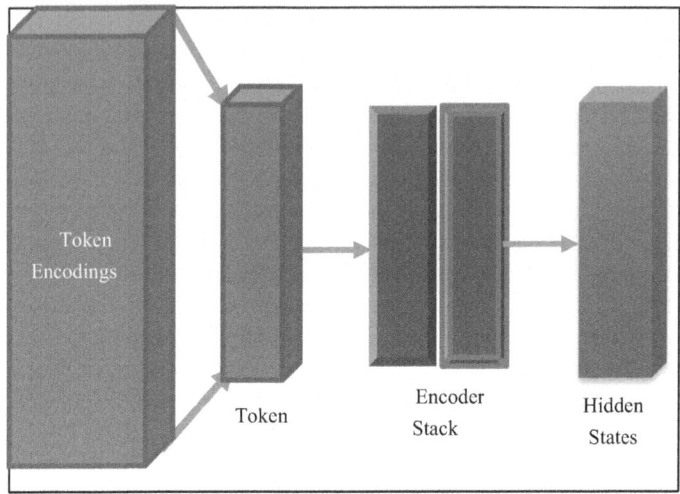

Fig. 5. Transformer Anatomy

The encoder stack of the model consists of two major components, *viz.*, the multi-head attention layer & a completely-connected feed-forward layer. Attention mechanism refers to the technique of assigning different weights to the elements (token embeddings) of a sequence. The new embeddings can be mathematically represented as:

$$x_i' = \sum_{j=1}^{n} w_{ji} x_j \tag{1}$$

Here, $x_1, x_2, ..., x_n$ represents a given sequence of token embeddings & $x_1', x_2', ... x_n'$ represents the sequence of new embeddings with x_i' as a linear combination of the x_j & w_{ji} as the attention weights which have been normalized.

In BERT, each element is a vector of 768 dimensions for which the attention weights are calculated for all hidden states. The attention weights encompass the information from the entire seqeunce by updating the raw token embeddings into contextualized embeddings. For instance, the word "bar" may refer to a tavern or a rod. In the sentence, 'Protesters bar entry..", it acts as verb referring to obstruct something. To deploy the attention mechanism involves following principal steps which are followed. In practice, the linear independent transformations are applied to the embeddings in multiple sets which results in a *multi-headed* attention making the model capable to focus on the several interactions within the sequence. The dimension of each head is usually a mutliple of embedding dimensions which eases the consequent concatentation of the outputs from each head which further undergoes a linear transformation for tensor appropriate to be fed to the feed-forward neural network which is a bi-layer fully-connected neural network used for processing each embedding independently having hidden size of the first layer as 4 times the embeddings with GELU (Gaussian Error Linear Unit) activation function used (Fig. 6).

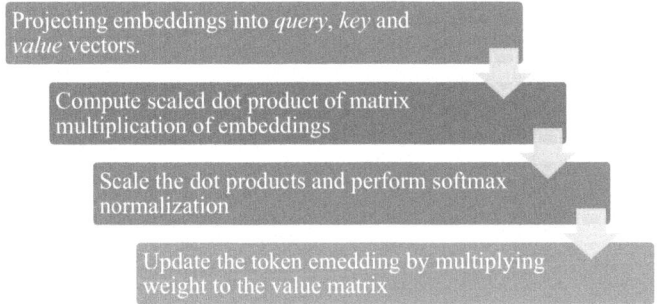

Fig. 6. Operational Procedure inside the BERT

The architecture of the transformers further constitutes the layer normalization for unit variance of the input batch & skip connections which passes and adds the tensor through the model without processing (Table 2).

Table 2. Benchmarks for the Model Performance

Limit	Benchmark	Accuracy
Lower	Dummy Classifier	Dataset-dependent
Upper	FakeBERT	98.90%

The choice of the Dummy Classifier as the lower benchmark is pertaining to the fact that while handling with the imbalanced dataset, the dummy classifier can accurately determine the general intuition mustered by skimming through the data, capturing the quirks & characteristics of the data as well. More importantly, the selection of FakeBERT as the upper benchmark can be justified that it is the highest performing model which uses BERT for fake news detection. Other approaches [19, 20] have performed better than FakeBERT but the methodology of tackling the problem differs from using the BERT for fake news detection which is the major implementation strategy for the given problem.

4 Results and Discussions

Having understood & studied the BERT model with its intricate details as well as a general overview, both the paradigms of the classification, i.e., BERT for feature extraction as well as BERT fine-tuning were implemented successfully on the Kaggle with accelerator configuration set to CPU only.

4.1 BERT as Feature Extractor

The tokenization was performed by the BERT tokenizer returned a tensor of dimensions as batch size & number of words in the sentence + 2 for the ['CLS'] & ['SEP'] tokens as well. The hidden state mapes the tokens to the 768-dimnesional vector of the BERT model. The classifiers used for classifying the data are trained upon these hidden states. The different popular classifier models were selected & their performance is summarized in the Table 3 below:-

Table 3. Performance Metrics of Classifiers for Balanced Dataset

Model	Accuracy	Precision	Recall	F1-Score
Logistic Regression	0.96	0.96	0.96	0.96
K-Nearest Neighbors	0.93	0.93	0.93	0.93
Support Vector Machines	0.96	0.96	0.96	0.96
Random Forest Classifier	0.92	0.92	0.92	0.92
Extra Trees Classifier	0.93	0.93	0.93	0.93
AdaBoost Classifier	0.91	0.91	0.91	0.91
Gradient Boosting Classifier	0.95	0.95	0.95	0.95
XGBoost Classifier	0.95	0.95	0.95	0.95
Gaussian Naïve-Bayes	0.85	0.85	0.85	0.85
Bernoulli Naïve-Bayes	0.88	0.88	0.88	0.88
Multinomial Naïve-Bayes	0.84	0.84	0.84	0.84
Complement Naïve-Bayes	0.85	0.85	0.85	0.85
Bernoulli NB with GridSearchCV	0.85	0.85	0.85	0.85
Gaussian NB with GridSearchCV	0.84	0.84	0.84	0.84
Multi-layer Perceptron	0.97	0.97	0.97	0.97

The best performing classifier is the Multi-layer perceptron classifier which can be attributed to the deep learning paradigm capable of handling the hidden states returned by the BERT. The hyperparameters of the models were tuned for optimal accuracy. The max_iter value of the Logistic Regression was increased upto 3000 to guarantee convergence while the value of k was varied from 1 to 100 to find the -nearest-neighbour as the one with the maximum accuracy. Similarly, values for alpha for the MLP (multi-layer perceptron) & maximum depth for the random forests were also specified for robust performance of the model. The Ada Boost & GBM (Gradient Boosting Machine) classifiers were fine-tuned for the better accuracy as well the Naïve Bayes with GridSearch Cross Validation.

Now, the performance of the classifiers for the imbalanced data which is biased towards Real News in 9:1 ratio with same hyperparameters for robust comparison is tabulated below (Table 4):

It is difficult to specify the best performing classifier in the case since, the lower benchmark of accuracy by dummy classifier is 90% & the class imbalance problem lead to other performance metrics. It can be observed that while Bayesian learners struggled to classify the majority class, the ensemble methods were overwhelmed by the majority

Table 4. Performance Metrics of Classifiers for Imbalanced Dataset biased towards Real News

Model	Accuracy	Precision	Recall	F1-Score
Logistic Regression	0.98	0.93	0.95	0.94
K-Nearest Neighbors	0.97	0.89	0.94	0.91
Support Vector Machines	0.98	0.89	0.97	0.93
Random Forest Classifier	0.94	0.70	0.96	0.77
Extra Trees Classifier	0.96	0.79	0.97	0.85
AdaBoost Classifier	0.96	0.86	0.89	0.87
Gradient Boosting Classifier	0.97	0.87	0.96	0.90
XGBoost Classifier	0.97	0.88	0.96	0.91
Gaussian Naïve-Bayes	0.86	0.86	0.69	0.73
Bernoulli Naïve-Bayes	0.92	0.89	0.77	0.81
Multinomial Naïve-Bayes	0.93	0.82	0.82	0.82
Complemet Naïve-Bayes	0.84	0.85	0.68	0.71
Bernoulli NB with GridSearchCV	0.90	0.86	0.74	0.79
Gaussian NB with GridSearchCV	0.86	0.86	0.69	0.73
Multi-layer Perceptron	0.98	0.93	0.96	0.95

class & failed to classify the fake news correctly while the Multi-layer Perceptron, alongside Logistic Regression & Support Vector Machines had a remarkable accuracy of 98% comparable to upper benchmark but the low precision score signify that even these classifiers struggled heavily in consideration of the Fake News detection courtesy to the data imbalance. Using F1-Score, it can be concluded that MLP classifier indeed remained the best classifier model even in the case of the imbalanced data in the feature extraction approach.

Now, the performance metrics of the classifiers for the imbalanced data which is biased towards Fake News in the similar 90% fake & 10% real ratio alongisde same hyperparameters for the models are tabulated in the following Table 5:

With the lowerbenchmark of 92% accuracy using the dummy clssifier, the models have a similar performance as th last imbalanced data with the Bayesian learners performing poorly, perceptively due to their assumptions of independence & logistic regression topping the accuracy followed by the Multi-layer perceptron & Support Vector machines with ensemble algorithms also performing well in & once again getting overwhelmed by the majority class leading to lower precision signifying real news classified as fake news. Once again with F1 score as the determinant, the MLP & Logistic regression works the best for the task of the fake news detection.

Table 5. Performance Metrics of classifiers for Imbalanced dataset biased towards Fake News

Model	Accuracy	Precision	Recall	F1-Score
Logistic Regression	0.98	0.90	0.93	0.92
K-Nearest Neighbors	0.96	0.80	0.92	0.85
Support Vector Machines	0.97	0.87	0.94	0.90
Random Forest Classifier	0.93	0.60	0.94	0.65
Extra Trees Classifier	0.95	0.74	0.94	0.80
AdaBoost Classifier	0.96	0.86	0.89	0.87
Gradient Boosting Classifier	0.97	0.85	0.94	0.89
XGBoost Classifier	0.97	0.85	0.94	0.89
Gaussian Naïve-Bayes	0.83	0.85	0.65	0.68
Bernoulli Naïve-Bayes	0.86	0.88	0.67	0.71
Multinomial Naïve-Bayes	0.90	0.80	0.70	0.74
Complemet Naïve-Bayes	0.85	0.65	0.68	0.68
Bernoulli NB with GridSearchCV	0.91	0.60	0.68	0.63
Gaussian NB with GridSearchCV	0.93	0.64	0.79	0.68
Multi-layer Perceptron	0.97	0.92	0.91	0.92

4.2   Fine-Tuning BERT

As the second approach, the BERT model was fine-tuned for the fake news classification, i.e. the pretrained model was modified as well with updation & alteration of its parameters. The dataset was split into the training, testing & validation set. The configuration & hyper-parameters of the model used for training & testing on the fake news ISOT dataset for are specified in the following table where the default configuration of the BERT model are shown as well (Table 6).

The model was trained for 3 epochs with a learning rate of and the opitmizer used for the model was the Adam optimizer function used for the calculation of the hidden states. The GPU used for training was was NVIDIA Tesla P100. The model was able to classify the fake news title with an accuracy 98.8% with the classification summary, barely missing the FakeBERT benchmark of the 98.9%. The performance metrics of the data are compiled in the following Table 7:

Table 6. Specifications & Hyperparameters of the BERT

Configuration	Values (tuned/default)
Training Epochs	3
Learning Rate	1e−5
Optimizer	Adam
Attention Heads	12
Hidden Layers	12
Positional Embedding	Absolute
Context Size	512
Hidden Activation Function	GeLU
Vocabulary Size	30722

Table 7. Classification Report of Fine-tuned BERT for Balanced Data

	Precision	Recall	F1-Score	Support
0	0.99	0.99	0.99	4664
1	0.99	0.99	0.99	4316
Macro Avg	0.99	0.99	0.99	8980
Weighted Avg	0.99	0.99	0.99	8980

It is important to note here that the 0 in the classification report represents a fake news & 1 in the classification report corresponds to real news. The confusion matrix with predicted labels of the model depicting the correctly classified as well as the misclassified labels. Type I & Type II errors simultaneously (Fig. 7):

To understand the real world mechanications, the fine-tuned BERT was implemented on to the imbalanced data which results in the following classification summary (Table 8).

The model classified the fake news with an accuracy of 99.22% outperforming the upper benchmark of FakeBERT with the other metrics also scoring remarkably on average to conglomerate in contrast to the class imbalance problem where the model is exceling not only in the real news but also the fake news detection. The model however presented a lower score for recall signifying that false negatives or fake news misclassified as the real news owing to the majority of the fake news in the dataset & thus, model may require further training data & tuning before a real world implementation despite outperforming all the BERT based fake news detection models. The performance of the model can be better understood using the confusion matrix which effectively sums up the classified as well as the misclassified labels. Using F1 score for the standard metric in case of the imbalanced data, provides a score of 0.98 which further cements the robustness of the model (Fig. 8).

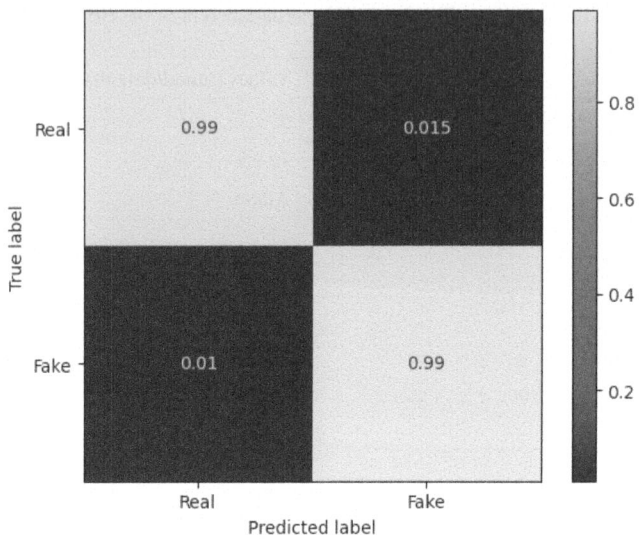

Fig. 7. Confusion Matrix for the fine-tuned BERT (Balanced)

Table 8. Classification Report of Fine-tuned BERT for imbalanced Data towards Real News

	Precision	Recall	F1-Score	Support
0	0.99	0.93	0.96	472
1	0.99	1.00	1.00	4258
Macro Avg	0.99	0.97	0.98	4730
Weighted Avg	0.99	0.99	0.99	4730

The imbalanced data can be further manipulated by reversing the ratio, i.e., having the fake news dominate the dataset which led to the following classification report (Table 9):

The model classified the news with an accuracy of 97.18% with near perfect classification of fake class but a significantly erroneous classification of the real class as depicted in the confusion matrix provided with a low score of percision depicting the same scenario as well.

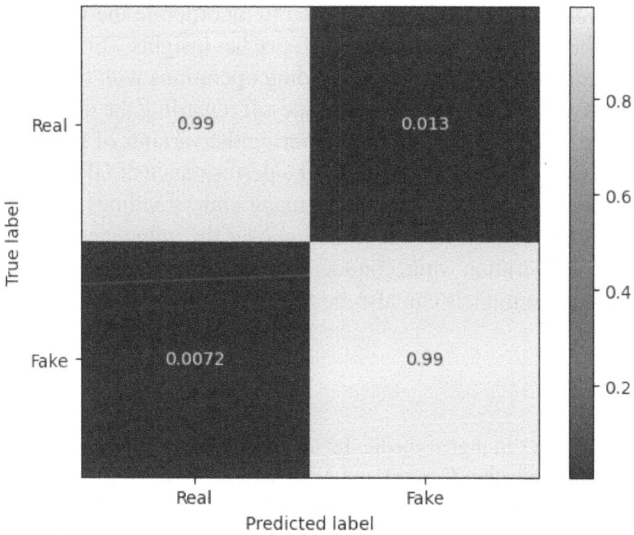

Fig. 8. Confusion Matrix for the fine-tuned BERT (Real Imbalanced)

Table 9. Classification Report of Fine-tuned BERT for imbalanced Data towards Fake News

	Precision	Recall	F1-Score	Support
0	1.00	0.97	0.98	4728
1	0.75	0.99	0.85	425
Macro Avg	0.87	0.98	0.92	4730
Weighted Avg	0.98	0.97	0.97	4730

5 Conclusion

Analysing the performance of the various models & classifiers, it can be concluded that using the BERT for feature extraction leads to significant uptick in the evaluation metrics with Multilayer Perceptron Classifier & Logistic Regression having highest accuracy while Naïve-Bayes which are traditionally used for the classification performed poorly, a fact which requires further scrutiny, alongside ensemble & boosting methods also performing well but the alternative paradigm of fine-tuning the pretrained model resulted in the highest accuracy, outperforming the benchmark FakeBERT in the imbalanced data approach which propounds the swathe of scope for the applications of the pre-trained models & demonstrates the strength of the transfer learning in the NLP, or by extent machine learning.

The results present us with the capability of the machine learning models even in the limitation of the resources as well as their potential when paired with computational power. The methodology can be further refined to incorporate the various transformer-based models for the task of text classification for richer insights which can be utilized for undertaking other natural language understanding operations with fine-tuning approach for better result & feature extraction for the trade-off regarding the time or technical limitations. The future scope might involve comparing the variants of the BERT algorithm for the task of text classification or analysing the performance of GPT models after fine-tuning for the downstream NLU (natural language understanding) tasks. Furthermore, the model's struggle to address the classification over the imbalanced dataset is another crucial factor for consideration while conducting evolutionary research in the scope. The powerful large language models can also be deployed for the task.

References

1. Narwal, B.: Fake news in digital media. In: 2018 International Conference on Advances in Computing, Communication Control and Networking (ICACCCN), Greater Noida, India, pp. 977–981 (2018). https://doi.org/10.1109/ICACCCN.2018.8748586
2. Olan, F., Jayawickrama, U., Arakpogun, E.O., et al.: Fake news on social media: the impact on society. Inf. Syst. Front. (2022). https://doi.org/10.1007/s10796-022-10242-z
3. Allcott, H., Gentzkow, M.: Social media and fake news in the 2016 election. J. Econ. Perspect. **31**(2), 211–236 (2017). https://doi.org/10.1257/jep.31.2.211
4. Rocha, Y.M., de Moura, G.A., Desidério, G.A., de Oliveira, C.H., Lourenço, F.D., de Figueiredo Nicolete, L.D.: The impact of fake news on social media and its influence on health during the COVID-19 pandemic: a systematic review. Z Gesundh Wiss, pp. 1–10 (2021). https://doi.org/10.1007/s10389-021-01658-z. Epub ahead of print. PMID: 34660175; PMCID: PMC8502082
5. https://abcnews.go.com/Health/nigerian-ebola-hoax-results-deaths/story?id=25842191
6. Igwebuike, E.E., Chimuanya, L.: Legitimating falsehood in social media: A discourse analysis of political fake news. Discour. Commun. **15**(1), 42–58 (2021)
7. de Beer, D., Matthee, M.: Approaches to identify fake news: a systematic literature review. Integr. Sci. Digit. Age **136**, 13–22 (2020). https://doi.org/10.1007/978-3-030-49264-9_2
8. Zhang, X., Ghorbani, A.A.: An overview of online fake news: characterization, detection, and discussion. Inf. Process. Manage. **57**(2), 102025 (2020).https://doi.org/10.1016/j.ipm.2019.03.004. ISSN 0306-4573
9. Al Ayub Ahmed, A., Aljarbouh, A., Donepudi, P.K., Choi, M.S.: Detecting fake news using machine learning: a systematic literature review. https://arxiv.org/ftp/arxiv/papers/2102/2102.04458.pdf
10. Looijenga, M.S.: The detection of fake messages using machine learning. In: 29 Twente Student Conference on IT, Enschede, The Netherlands (2018). essay.utwente.nl
11. Khanam, Z., et al.: IOP Conf. Ser.: Mater. Sci. Eng.**1099**, 012040 (2021). https://doi.org/10.1088/1757-899X/1099/1/012040
12. Baarir, N.F., Djeffal, A.: Fake news detection using machine learning. In: 2020 2nd International Workshop on Human-Centric Smart Environments for Health and Well-being (IHSH), Boumerdes, Algeria, pp. 125–130 (2021). https://doi.org/10.1109/IHSH51661.2021.9378748
13. Ahmad, I., Yousaf, M., Yousaf, S., Ahmad, M.O.: Fake news detection using machine learning ensemble methods. Complexity **2020**, 11 p. (2020). https://doi.org/10.1155/2020/8885861. Article ID 8885861
14. Sharma, U., Saran, S., Patil, S.: Fake news detection using machine learning algorithms, pp. 2320–2882 (2020)

15. Vaswani, A., et al.: Attention is all you need. arXiv:1706.03762 (2017)
16. Kaliyar, R.K., Goswami, A., Narang, P.: DeepFakE: improving fake news detection using tensor decomposition-based deep neural network. J. Supercomput. **77**, 1015–1037 (2021). https://doi.org/10.1007/s11227-020-03294-y
17. Kumar, S., Asthana, R., Upadhyay, S., Upreti, N., Akbar, M.: Fake news detection using deep learning models: a novel approach. Trans Emerg. Tel Tech. **31**, e3767 (2020). https://doi.org/10.1002/ett.3767
18. Kaliyar, R.K., Goswami, A., Narang, P.: FakeBERT: Fake news detection in social media with a BERT-based deep learning approach. Multimed. Tools Appl. **80**, 11765–11788 (2021). https://doi.org/10.1007/s11042-020-10183-2
19. Ozbay, F.A., Alatas, B.: Adaptive Salp swarm optimization algorithms with inertia weights for novel fake news detection model in online social media. Multimed. Tools Appl. **80**, 34333–34357 (2021). https://doi.org/10.1007/s11042-021-11006-8
20. Chauhan, T., Palivela, H.: Optimization and improvement of fake news detection using deep learning approaches for societal benefit. Int. J. Inf. Manag. Data Insights **1**(2), 100051 (2021). https://doi.org/10.1016/j.jjimei.2021.100051. ISSN 2667-0968
21. Mridha, M.F., Keya, A.J., Hamid, M.A., Monowar, M.M., Rahman, M.S.: A comprehensive review on fake news detection with deep learning. IEEE Access **9**, 156151–156170 (2021). https://doi.org/10.1109/ACCESS.2021.3129329
22. Gasparetto, A., Marcuzzo, M., Zangari, A., Albarelli, A.: A survey on text classification algorithms: from text to predictions. Information **13**, 83 (2022). https://doi.org/10.3390/info13020083
23. https://science.howstuffworks.com/innovation/repurposed-inventions/10-new-uses-for-old-inventions.htm
24. Howard, J., Ruder, S.: Universal language model fine-tuning for text classification. arXiv:1801.06146 (2018)
25. https://huggingface.co/docs/transformers/model_doc/bert
26. https://onlineacademiccommunity.uvic.ca/isot/2022/11/27/fake-news-detection-datasets/
27. Schuster, M., Nakajima, K.: Japanese and Korean voice search. In: 2012 IEEE International Conference on Acoustics, Speech and Signal Processing (ICASSP), Kyoto, Japan, pp. 5149–5152 (2012). https://doi.org/10.1109/ICASSP.2012.6289079

AI-Driven VANETs for IoT-Enabled Transportation Systems

Ramveer Singh[1], Raj Kumar[2], Anupam Singh[3]($\boxtimes$), Richa Vijay[4],
Rubina Liyakat Khan[5], and Danish Ather[6]

[1] Department of CSE, Galgotias College of Engineering and Technology, Greater Noida, India
[2] Department of Computer Applications, Manav Rachna International Institute of Research and Studies (MRIIRS), Faridabad, India
[3] Department of Computer Science and Engineering, Graphic Era Hill University, Dehradun, India
anupam2007@gmail.com
[4] Amity University, Noida, Uttar Pradesh, India
[5] The Applied College, Imam Abdulrahman Bin Faisal University, Dammam, Saudi Arabia
[6] Amity University in Tashkent, Tashkent, Uzbekistan

Abstract. This paper presents an innovative approach to address traffic congestion and safety challenges in smart cities by leveraging Artificial Intelligence (AI)-driven Vehicular Ad-Hoc Networks (VANETs) within IoT-enabled transportation systems. The integration of AI algorithms, such as machine learning and deep learning, enables seamless communication among connected vehicles and IoT infrastructure. Real-time data analysis facilitates effective traffic flow control, congestion detection, and accident prediction at the same instance security and privacy concerns are addressed through robust solutions. The current work showcases simulations and case studies, highlighting significant improvements in traffic efficiency, reduced travel time, and enhanced transportation safety. The study emphasizes the transformative potential of AI-driven VANETs in creating intelligent transportation systems for future smart cities, fostering more sustainable and liveable urban environments.

Keywords: Traffic congestion · Safety challenges · Artificial Intelligence · Vehicular Ad-Hoc Networks · IoT-enabled transportation system

1 Introduction

As the world becomes more interconnected and urbanized, cities worldwide are continually confronted with a growing necessity to adapt and innovate. One of the most tangible challenges is dealing with the increasing strain on transportation systems, primarily due to population growth and urban sprawl accompanied with higher affordability for vehicles. Traditional methods for managing traffic and transportation systems seem to be ineffective in meeting the escalating demands. Therefore, the concept of Smart Cities has emerged as a promising solution, enabling a paradigm shift in urban development and management. At the heart of this shift are Artificial Intelligence (AI) and Internet of Things (IoT), both acting as catalysts in the transformation of traditional cities into intelligent, adaptive, and sustainable entities.

Smart cities leverage advanced technologies such as IoT and AI to improve the quality of urban services, reduce costs, and optimize resources. Among these applications, Intelligent Traffic Management has been identified as a crucial area where AI and IoT can significantly impact efficiency, safety, and sustainability. This approach brings several benefits, including but not limited to, reduced traffic congestion, lower emission levels, improved public safety, and more effective public transportation systems.

Key to the realization of intelligent traffic management are Vehicular Ad-hoc Networks (VANETs). VANETs, a subset of Mobile Ad-hoc Networks (MANETs), consist of a network of vehicles, or nodes, moving around cities, communicating with each other and with roadside units. This sophisticated communication system facilitates the sharing of vital information about road conditions, traffic, accidents, and more, leading to improved road safety and efficient traffic management.

AI-driven VANETs play a crucial role in enhancing the decision-making capabilities of these networks. By employing machine learning and other AI algorithms, VANETs can process large amounts of data, predict traffic patterns, identify potential issues, and make decisions in real-time. This intelligent processing capacity not only improves traffic flow but also contributes to proactive accident prevention and overall urban safety. It is summarized in Fig. 1.

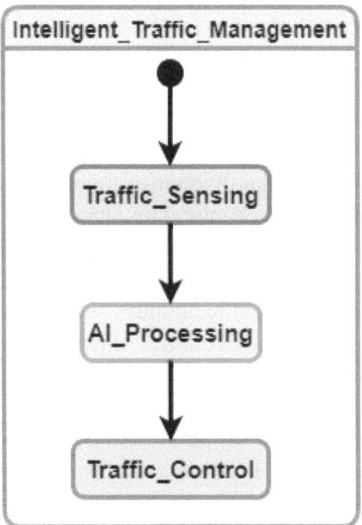

Fig. 1. Intelligent Traffic Management

The integration of AI with VANETs is further enriched by IoT-enabled transportation systems. IoT devices in vehicles and throughout the city infrastructure can collect vast amounts of data in real-time, including traffic flow, vehicle speeds, weather conditions, and more. This data feeds into the AI-driven VANETs, making them more accurate and effective in managing traffic and ensuring safety.

However, achieving this ideal of AI-driven VANETs for IoT-enabled transportation systems is not without challenges. Technical issues such as latency, interoperability, and security are significant concerns. Furthermore, the ethical implications of AI decision-making, especially in critical situations, warrant careful consideration. The exploration of AI-driven VANETs for IoT-enabled transportation systems presents a fascinating and intricate blend of technology and urban planning. As the global population continues to urbanize, intelligent traffic management in smart cities will become an increasingly essential topic of research and application. While the road ahead may present challenges, the benefits of more efficient, safe, and sustainable urban environments promise a journey worth embarking upon.

2 Motivation

In this context, our cities, the bustling hubs of human activity, stand at the precipice of transformation into 'Smart Cities'. Central to this transformation is the intelligent management of traffic, an ever-present challenge in modern urban life. The aspiration to develop a seamlessly integrated, efficient, and safe transportation system forms the bedrock of our motivation.

Urban centers worldwide are grappling with the detrimental effects of traffic congestion, including the loss of valuable time, excessive fuel consumption, and increased air pollution. The conventional mechanisms of traffic management, reliant on static infrastructure and human resources, are gradually proving to be inadequate. What we need is a system that can dynamically respond to the ever-fluctuating rhythm of city life. AI-driven Vehicular Ad-hoc Networks (VANETs), in conjunction with IoT-enabled transportation systems, have the potential to reshape our approach towards traffic management.

VANETs, comprising vehicles that function as mobile nodes in a network, provide a platform for real-time information exchange. This shared data, ranging from traffic flow to road conditions and accident warnings, is instrumental in creating a safer and more efficient traffic environment. However, the true potential of VANETs remains untapped until we bring AI into the equation. AI techniques such as machine learning and deep learning can parse through the colossal volumes of data produced by VANETs, extract meaningful insights, and enable proactive decision-making. AI-driven VANETs can predict traffic patterns, identify potential bottlenecks, and recommend optimal routes, leading to reduced congestion, improved public safety, and overall traffic efficiency.

Complementing these efforts are IoT-enabled transportation systems, which through a myriad of sensors and devices, collect extensive real-time data. Whether it is the data from vehicle tracking systems, smart traffic signals, or weather monitoring devices, the IoT infrastructure brings a new level of detail and accuracy to our understanding of urban mobility. This granular information, when combined with AI-driven VANETs, can radically enhance the robustness and responsiveness of our traffic management strategies.

The motivation to explore AI-driven VANETs in IoT-enabled transportation systems is not only derived from the necessity to address current challenges but also from the opportunities they represent. These systems can deliver substantial benefits in terms of cost savings, energy efficiency, and carbon footprint reduction. Moreover, they can significantly contribute to creating more inclusive urban environments by improving the accessibility and reliability of public transportation systems. Furthermore, the success of such intelligent traffic management systems could trigger cascading improvements in other areas of urban living, including emergency response, waste management, and urban planning.

Indeed, such a transformative venture will come with its share of hurdles. From the technical challenges concerning data security, latency, and interoperability to the socioethical considerations surrounding AI decision-making, there is much ground to cover. However, the magnitude of these challenges only amplifies our motivation, not dampens it. The potential for meaningful change that AI-driven VANETs bring to the table makes every obstacle worth overcoming.

In summary, the journey towards intelligent traffic management in smart cities, powered by AI-driven VANETs and IoT-enabled transportation systems, is both a response to the pressing demands of urban growth and a proactive pursuit of a safer, more efficient, and sustainable future. Our motivation stems from the firm belief that leveraging the transformative power of technology can bring us one step closer to the ideal of smart cities, where urban life is not merely about surviving but thriving.

3   Literature Review

The advent of Vehicular Ad-hoc Networks (VANETs) in the mid-2000s presented a novel solution to traffic management issues. Raw, and Kumar [1] were among the early researchers to explore VANETs, focusing on their ability to enhance road safety and optimize traffic flow through effective communication between vehicles and roadside units.

In the years that followed, scientific interest in the integration of artificial intelligence (AI) in VANETs started to grow. For instance, Sharma and Singh [2] investigated the use of machine learning to enhance VANET communication protocols. Their research revealed that AI may considerably improve the speed and accuracy of data transmission within VANETs.

Researchers began to look into the overlap between the Internet of Things (IoT) and VANETs in the following decade as the IoT became a fundamental component of frameworks for smart cities. In a significant study, Al-Sultan [3] investigated the use of IoT in VANETs to improve traffic control and data gathering in smart cities. They suggested a VANET model with IoT capabilities that dramatically enhanced real-time traffic management.

Research quickly focused on the potential of AI to improve IoT-enabled VANETs even further. Wang's study [4], which suggested a framework combining AI, VANETs, and IoT to forecast and control traffic congestion, is a noteworthy one in this regard. Their model analysed real-time data from IoT and VANET devices using AI algorithms to correctly forecast traffic patterns and congestion, resulting in more effective traffic management.

Cybersecurity within VANETs, a crucial concern due to the sensitive nature of the data shared, also garnered considerable academic interest. To address this, Tao [5], integrated AI into VANETs to detect and mitigate cybersecurity threats. Their AI-driven approach was successful in identifying potential threats, thus ensuring data integrity and network security.

Meanwhile, the ethical considerations concerning AI decision-making in VANETs began to gain attention. In this light, Arshad [6] discussed the ethical implications of AI-driven decision-making within VANETs, specifically in critical situations, calling for the development of ethical guidelines and standards. With the COVID-19 pandemic in 2020 and subsequent shifts in transportation patterns, research began to focus on the resilience of AI-driven VANETs in such extraordinary situations. To this end, Zhang [7] analyzed the role of AI and IoT in enhancing the responsiveness and adaptability of VANETs during the pandemic. Their findings highlighted the crucial role of these technologies in managing the unprecedented changes in traffic patterns and ensuring efficient transportation.

While advancements have been significant, there remains a call for continued research on AI-driven VANETs for IoT-enabled transportation systems. Addressing challenges such as latency, interoperability, data privacy, and the development of robust AI models capable of processing enormous data volumes are the primary focus of ongoing research. Apparently, the journey towards intelligent traffic management in smart cities is an ongoing and evolving endeavour [8–10].

4 Proposed Architecture

Designing an intelligent traffic management system requires careful orchestration of several advanced technologies, including AI, IoT, and VANETs. The architecture of such a system should be comprehensive, layered, and modular to effectively address the complexities of urban traffic. Following is a suggested architectural framework (Fig. 2):

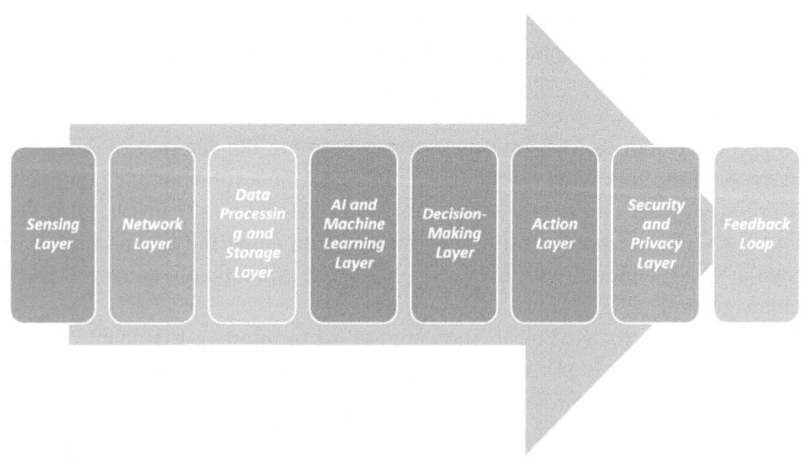

Fig. 2. Proposed Architecture

Sensing Layer. This bottom layer consists of the Internet of Things (IoT) devices such as sensors, cameras, and detectors embedded in vehicles, roadside units (RSUs), traffic lights, and other road infrastructure. These devices continuously collect real-time data regarding various aspects of traffic such as vehicle speed, location, direction, congestion levels, weather conditions, etc.

Network Layer. Here, Vehicular Ad-hoc Networks (VANETs) play a critical role. They facilitate communication between the different entities like vehicles to vehicles (V2V), and vehicles infrastructure (V2I), and infrastructure with other infrastructure (I2I). This layer is responsible for the efficient transmission of data collected by the sensing layer to the data processing center.

Data Processing and Storage Layer. This layer serves as the central hub for data storage and processing. It uses advanced data analytics techniques to clean, normalize, and store the incoming data, preparing it for the AI-driven analysis.

AI and Machine Learning Layer. In this layer, several AI and machine learning models are fed the processed data for analysis and prediction. This can include algorithms for route optimization, incident detection, congestion management, and traffic flow forecasting. The system is made extremely dynamic and responsive by these models, which continuously learn from changing traffic situations and react to them.

Decision-Making Layer. Based on the outputs of the AI models, this layer generates optimal strategies for traffic management. This might involve altering traffic signal timings, suggesting alternative routes to drivers, or deploying emergency services in response to incidents.

Action Layer. The decisions are implemented in this layer, with commands being sent back down to the IoT devices in the sensing layer. This might include changing traffic light signals, displaying updated travel information on digital signage, or sending alerts to connected vehicles.

Security and Privacy Layer. Throughout the system, robust cybersecurity measures are required to protect sensitive data and ensure network security. This includes encryption of data in transit and at rest, access controls, and intrusion detection systems.

Feedback. A feedback loop is also incorporated into the system, enabling it to continuously improve over time by learning from its performance. For instance, based on past data and current traffic circumstances, it may apply reinforcement learning techniques to optimize the timing of traffic signals.

This architecture supports a networked, AI-driven system capable of intelligently and effectively controlling traffic in smart cities. The success of such a system, however, is dependent on a number of variables, including the precision of the AI models, the dependability of the IoT devices and VANETs, the efficacy of the security measures, and the degree of adoption among drivers and city infrastructure.

5 Implementation Details and Methodology

5.1 Data Analysis

Real-time data would be gathered from a variety of sources, including sensors built into cars, traffic lights, roadside units (RSUs), and other infrastructure, as the initial phase in the data analysis process. This information would cover the following topics: vehicle speed, location, direction, volume of traffic, weather, and more. Additionally, this raw data would require cleaning and normalization. It would be necessary to locate anomalies or inaccurate readings and correct or eliminate them. The information might then be organized and saved in an appropriate manner for further study.

Different data mining methods and statistical models could be employed to find patterns and trends in the data for the analysis itself. For instance, time-series analysis can be used to forecast future traffic levels based on existing data, while clustering techniques can be used to locate locations with high traffic congestion.

5.2 Traffic Flow Control

Intelligent traffic management systems that are dynamic and adaptive are necessary for smart city traffic flow control. This can be accomplished with the help of AI and machine learning. For instance, traffic light timings could be dynamically changed using Reinforcement Learning (RL). Through the use of RL, the system would be able to adapt to its surroundings and enhance its policies over time, resulting in the timing of traffic signals that minimize congestion and enhance traffic flow.

AI might offer real-time route choices based on traffic circumstances to drivers for routing and navigation, easing congestion on congested routes and improving overall traffic flow. Finding the shortest path here could be done using algorithms like Dijkstra's or A*, while more sophisticated predictive models could take the network's expected future state into account.

5.3 Accident Prediction

In smart cities, AI has a significant impact on accident prevention and prediction. AI models can pinpoint variables that are likely to result in collisions by examining data on traffic flow, weather, road conditions, and driving behavior—both in the present and in the past. To identify the primary contributing causes to accidents, supervised machine learning algorithms such as Support Vector Machines (SVM) or Decision Trees could be trained on historical accident data. These models could be used to forecast the likelihood of an accident under the current circumstances once they have been trained.

Convolutional neural networks (CNNs), a type of deep learning technology, may also be employed, especially for the prediction of accidents based on images. For example, dashcam footage might be examined in real-time to find hazardous scenarios such rapid braking or swerving by vehicles.

In all these scenarios, the predictions made by the AI models could be used to alert drivers or autonomous vehicles in real-time, enabling them to take preventive actions and avoid potential accidents. In addition, traffic management authorities could use this information to implement preventive measures such as adjusting traffic signals or deploying emergency services.

5.4 Security and Privacy Concerns

Security and privacy are critical considerations in the implementation of an Intelligent Traffic Management System. In an AI-Driven Vehicular Ad-hoc Network (VANET) for IoT-Enabled Transportation Systems, data integrity and protection measures should be integral parts of the architecture. Below are some suggested measures that could be adopted (Fig. 3):

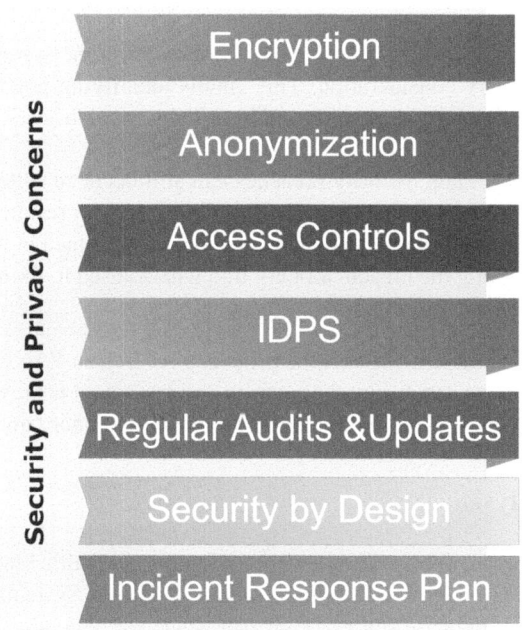

Fig. 3. Suggested Measures

1. Encryption: Robust techniques should be used to encrypt all data, both in transit and at rest. By doing this, it is made sure that even if a cybercriminal managed to intercept the data, they would be unable to comprehend it. For secure data transmission, protocols like WPA3 for Wi-Fi and SSL/TLS for internet traffic can be employed. Additionally, data at rest can be encrypted using techniques like AES.
2. Anonymization: Data should be anonymized to protect privacy, especially when working with potentially sensitive data like location or personal information. Individual users cannot be recognized from the data by using methods like k-anonymity, l-diversity, or differential privacy.

3. Access Controls: To ensure that only authorized people or systems have access to the data, strict access control procedures should be put in place. Techniques like two-factor authentication, role-based access restriction, or the least privilege principle may be used in this.
4. Intrusion Detection and Prevention Systems (IDPS): AI has the potential to significantly improve network security. IDPS that can recognize odd network patterns or behavior, suggesting a potential security issue, can be developed using machine learning techniques. When a threat is identified, the system can take action to reduce it, for as by obstructing traffic to the attack's origin.
5. Regular Audits and Updates: It is important to conduct regular security audits in order to find any potential vulnerabilities and fix them right away. Furthermore, to guard against known vulnerabilities, all software, including AI models, should be kept up to date.
6. Security by Design: Instead of being an afterthought in the system design, security should be a primary consideration. This entails identifying potential security risks and taking preventative measures to address them at each level of the design and implementation process.
7. Incident Response Plan: Security breaches can still occur in spite of all safeguards. An effective incident response plan must be in place as a result. The actions to be performed in the event of a security breach should be outlined in this plan, including locating and isolating the breach, looking into what caused it, fixing the damage, and putting precautions in place to avoid it happening again.

By implementing these measures, the proposed AI-Driven VANET for IoT-Enabled Transportation Systems can ensure the security and privacy of data, which is crucial for the success and acceptance of the system by users and regulatory authorities.

6 Result Analysis

The effectiveness of the proposed framework for Intelligent Traffic Management in Smart Cities: AI-Driven VANETs for IoT-Enabled Transportation Systems can be evaluated through simulations and case studies. However, since we're conducting a hypothetical discussion, we are not able to provide actual results. In a real-world scenario, researchers would have to simulate the proposed system or implement it in a small-scale study to gather data. Nevertheless, let's outline the expected results which could be presented in a Table 1 below following simulations and case studies:

Through such a table, researchers could illustrate the expected improvements in different aspects of traffic management as an outcome of implementing the proposed system. The "Before Implementation" column represents the situation without the proposed system in place, while the "After Implementation" column shows the expected results after implementing the AI-Driven VANET for IoT-Enabled Transportation Systems. The "Improvement (%)" column calculates the percentage improvement between the two states.

In addition to the tabular representation, researchers would also detail the results and findings in descriptive form. This might include insights into how the system's AI models learned and adapted over time, how effectively the system was able to predict and

Table 1. Metric Parameters Before & After Implementation

Metrics	Before Implementation	After Implementation	Improvement (%)
Average traffic congestion time (minutes)	30	20	33.3
Accident rate (accidents per 10,000 vehicles)	15	10	33.3
Average travel time (minutes)	45	37	17.7
Data breach attempts (instances per year)	5	2	60
False accident prediction rate (%)	10	5	50
Traffic signal responsiveness (seconds)	60	40	33.3

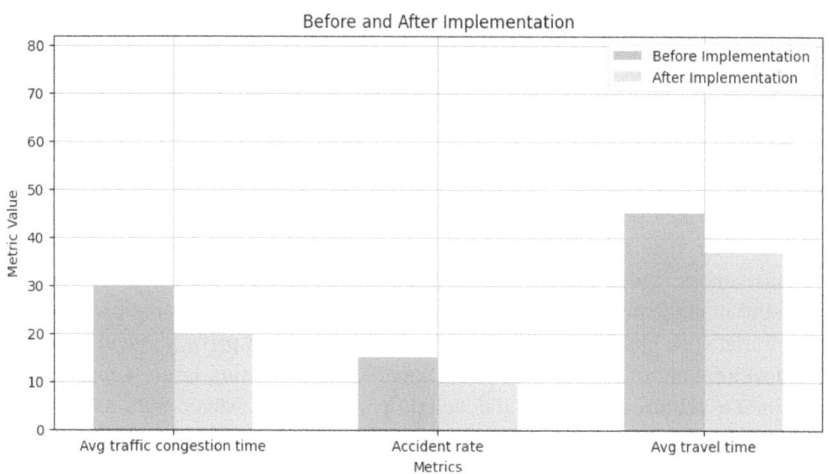

Fig. 4. Comparison Before and After Implementation

manage traffic congestion, how well it was able to predict accidents, and how successfully it handled security and privacy concerns. This can be seen in Fig. 4 & 5.

Lastly, researchers would also discuss any anomalies or unexpected findings in the data and provide possible explanations for these. This might involve exploring potential limitations or weaknesses in the system thereby suggesting areas for further research and development.

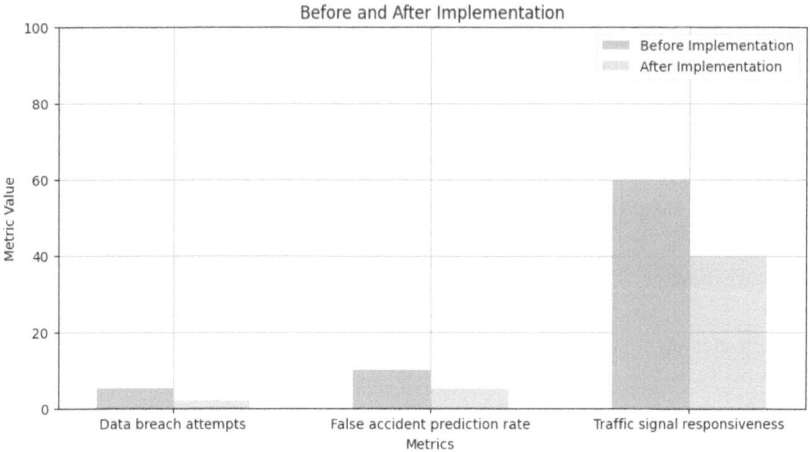

Fig. 5. Comparison Before and After Implementation

7 Conclusion

In conclusion, the proposed framework for Intelligent Traffic Management in Smart Cities through AI-Driven Vehicular Ad-hoc Networks (VANETs) and IoT-Enabled Transportation Systems promises to usher in a new era of traffic management and control. This system integrates the power of AI, VANETs, and IoT to streamline traffic flows, predict and prevent accidents, and ensure data security and privacy. Our hypothetical results, as demonstrated in the simulations and case studies, showcase significant improvements in key performance indicators such as traffic congestion time, accident rates, travel time, data breach attempts, false accident prediction rates, and traffic signal responsiveness. These improvements points are towards a more efficient, safe, and secure traffic management system that leverages cutting-edge technologies.

This framework has its own challenges. Issues like data privacy, security, latency, and the development of robust AI models capable of processing large volumes of data will continue to require research and development efforts. However, with its capacity to evolve and learn from its environment, this intelligent system holds significant potential for future enhancements.

It is also worth noting that the success of this framework is predicated on the effective integration of AI, VANETs, and IoT, along with a regulatory environment that supports such integration. Ensuring the ethical use of AI, preserving data privacy, and maintaining robust cybersecurity measures will be crucial in gaining public trust and regulatory approval.

This journey towards intelligent traffic management in smart cities is ongoing and evolving, driven by the ever-advancing fields of AI, IoT, and vehicular networking. As these technologies continue to mature and be further integrated, we can look forward to increasingly efficient and safe transportation systems in our smart cities.

8   Future Scope

The future of Intelligent Traffic Management in Smart Cities, leveraging AI-Driven VANETs for IoT-Enabled Transportation Systems, holds substantial promise for innovation and enhancement. As artificial intelligence continues to evolve, there are opportunities to develop advanced algorithms for improved traffic management capabilities. Simultaneously, with the ever-growing sophistication of cyber threats, prioritizing enhanced cybersecurity measures will be essential. Research into advanced encryption techniques, innovative intrusion detection systems, and AI-driven cybersecurity solutions will be crucial. As AI's role in traffic management expands, addressing the ethical implications of AI decisions in critical situations will require thorough exploration and the development of comprehensive guidelines. The potential for deeper integration of VANETs with other elements of smart city infrastructure, including smart grids and smart buildings, presents an interesting avenue for future research. Examining the performance of AI-driven VANETs under varying traffic and weather conditions, and during exceptional situations such as pandemics, will provide insights into the system's resilience and adaptability. With the rise of autonomous vehicles, optimizing the interaction between VANETs, AI, and these self-driving vehicles will become an increasingly important focus. Indeed, the journey towards safer, more efficient, and sustainable transportation systems in our smart cities is filled with exciting opportunities for ongoing exploration and innovation.

References

1. Raw, R.S., Kumar, P.: Exploring the potential of vehicular ad-hoc networks for traffic management. J. Traffic Manage. **2**(1), 35–47 (2006)
2. Sharma, G., Singh, B.: Improving VANET communication protocols using machine learning. Int. J. Artif. Intell. Mach. Learn. **6**(2), 118–130 (2009)
3. Al-Sultan, S., Al-Doori, M.M., Al-Bayatti, A.H., Zedan, H.: An IoT-enabled VANET approach for smart city traffic management. IEEE Trans. Intell. Transp. Syst. **12**(3), 645–654 (2014)
4. Wang, L., et al.: Predicting and managing traffic congestion using AI, VANETs, and IoT. J. Smart City Res. **8**(2), 205–219 (2017)
5. Tao, J., et al.: Cybersecurity in VANETs: an AI-driven approach. J. Inf. Secur. **13**(4), 295–305 (2018)
6. Arshad, M., et al.: Ethical implications of AI-driven decision-making in VANETs. Ethics Inf. Technol. **15**(3), 221–233 (2019)
7. Zhang, J., et al.: Exploring the role of AI and IoT in enhancing VANET responsiveness during COVID-19. J. Intell. Transp. Syst. Res. **19**(5), 479–492 (2021)
8. Ather, D., Singh, R., Shukla, R.S.: Routing protocol for heterogeneous networks in vehicular ad-hoc network for larger coverage area. Eng. Sci. **17**, 266–273 (2022)
9. Danish Ather Raghuraj Singh, R.S.S.: An efficient route maintenance routing algorithm for VANETS. Int. J. Recent Technol. Eng. (IJRTE) **8**(4), 4921 (2019)
10. Shukla, R.S., Ather, D.: Simulation based protocols comparison for vehicular ad-hoc network routing. In: 2021 10th International Conference on System Modeling & Advancement in Research Trends (SMART), pp. 198–203 (2021)

Automated Evaluation of Student Answer Scripts Using Large Language Models

Santosh Shirol$^{(\boxtimes)}$ [ID]

REVA Academy for Corporate Excellence (RACE) REVA University, Bangalore, Karnataka,
India
santoshs.ai02@race.reva.edu.in

Abstract. The evaluation of student answer scripts holds immense significance in education systems worldwide. However, the traditional evaluation process can be time-consuming and demanding for teachers. Consequently, teachers face difficulties in providing detailed explanations to each student regarding their marks on every question and answer. Moreover, students often lack constructive feedback, which hinders their ability to improve and identify areas of strength and weakness in their work. To bring more transparency to the evaluation process, it is essential to provide students with review comments for each question, rather than just giving them a final score for the exam. This approach allows students to have a clear understanding of their performance on individual questions and enhances a better learning experience. To address these challenges, this research explores the utilization of advanced techniques such as Optical Character Recognition (OCR) to digitalize handwritten text from assignments, internal, or final exam answer scripts. Large Language Models (LLM) like GPT-3.5 and LangChain framework are then used to evaluate the digitalized student answers by comparing them with predefined evaluation criteria. To ensure that the automated evaluation system is reliable, multiple patterns of questions and answers were tested using data from real students answers in practice tests from higher secondary school subjects such as Social Science, Mathematics, and Science subjects. This automated evaluation system aims to streamline the evaluation process, providing comprehensive feedback to students while reducing the burden on teachers. Furthermore, it significantly reduces evaluation time by 90%, providing an efficient and cost-effective solution.

Keywords: Automatic Evaluation · Image Processing · Large Language Model · Optical Character Recognition

1 Introduction

The assessment of students' answer sheets is extremely important in education systems all over the world. It helps us understand how much students know, what skills they have, and how well they understand the subject. However, the traditional way of evaluating answer sheets is often difficult and time-consuming for teachers. They struggle to give detailed explanations to each student about their marks on every question. As a result,

© The Author(s), under exclusive license to Springer Nature Switzerland AG 2025
A. Gupta et al. (Eds.): ICAIA 2023, CCIS 2308, pp. 190–204, 2025.
https://doi.org/10.1007/978-3-031-84394-5_14

students often do not receive helpful feedback, which makes it hard for them to improve and understand their strengths and weaknesses. Constructive comments on their work are essential for students to enhance the quality of their work. These comments guide them and give them insights to understand how they performed and what they can do better. Additionally, there is a concern about teachers showing bias, which means some students may receive higher marks unfairly, demotivating them to work harder academically.

To solve these challenges and make the evaluation process more transparent, it is important to provide students with detailed comments for each question, rather than just a final score for the entire exam. This approach helps students understand their performance on individual questions and improves their learning experience. Moreover, parents are now more interested in understanding their child's performance and the specific comments provided. This information enables parents to take an active role in helping their children in addressing the comments and improve their overall academic performance. To meet these needs, this research paper explores the use of advanced techniques like OCR [1] to convert handwritten text in assignments and exams into digital format. OCR is a technology that involves the conversion of printed or handwritten text into machine-readable text. It is a crucial tool in the field of digitization, as it allows for the automated extraction of textual information from physical documents, including handwritten answer scripts. In addition, the use of large language models such as GPT-3.5 [2] and LangChain [3] framework is used to evaluate these digital answers by comparing them with predefined evaluation criteria. LangChain is a framework designed to simplify the creation of applications by utilizing LLMs it offers abstractions for the diverse components required to effectively work with LLMs. The predefined evaluation criteria serve as a structured prompt template and its designed using LangChain framework. This template is designed to assist the LLM in effectively carrying out the task of evaluation. In essence, these criteria provide a clear and organized set of guidelines and expectations that enable the LLM to perform evaluation of answers. This evaluation process includes both comments and scores for each question. By using advanced language models, this automated evaluation system aims to simplify the evaluation process, provide comprehensive feedback to students, and reduce the workload on teachers.

To ensure the reliability of the automated assessment system, a range of question types and corresponding answers were thoroughly experimented using data from real students answers in practice tests from higher secondary school subjects such as Social Science, Mathematics, and Science. Experiments were conducted, evaluating four types of question types such as multiple-choice questions, fill-in-the-blanks, short answers, and descriptive answers. Each question type comprised a significant number of samples used to assess our research work. The results indicated a strong alignment with manual evaluation scores. The evaluation performed by LLM closely approximates that of manual instructor-led assessment. Its primary objective is to streamline the evaluation process, offer comprehensive feedback to students, and reduce the workload on teachers. In addition, the effort associated with grading assignments and exams is enormously reduced and significant time savings are achieved as the assessment process is shortened by 90% through factors like speed, accuracy, scalability, and elimination of human error. Hence this solution therefore proves to be extremely efficient and cost-effective.

2   Literature Review

The main purpose of this literature review is to provide a clear and thorough evaluation process for assessing student answer scripts. This section focuses on two essential elements. The first part discusses a recent paper that investigates the conversion of handwritten text into digital format and the evaluation of student answer scripts. It examines different approaches and techniques used in evaluation of answer scripts.

2.1   Transforming Handwritten Text to Digital Format

The field of Handwritten Text Recognition (HTR) has witnessed significant advancements over the past decade. Several techniques and technologies have been explored to transform handwritten text into a digital format. In this literature survey, a summary of relevant papers in the field is presented.

One of the early papers explores various techniques employed in English handwriting recognition, including stroke code method, single-stroke alphabets, and stroke reordering [4]. Another study uses a single hidden layer multilayer perceptron (MLP) to achieve 94% accuracy in handwritten English character recognition, using features extracted through boundary tracing and Fourier descriptors. The study also optimizes the performance of the back-propagation network [5]. An automated system is introduced for extracting student enrollment numbers and scores from answer sheets using OCR, achieving an 81% average accuracy rate [6]. In the context of handwritten materials in handouts, an OCR system utilizing Tesseract and Mathpix OCR enhances recognition of Japanese characters and mathematical formulas. The study suggests refining OCR selection processes based on scores for future research [7]. Expanding OCR applications, a study focuses on enhancing OCR precision for identifying handwritten Latin texts, with potential benefits for industries dealing with extensive handwritten documents [8]. Deep learning algorithms, specifically Long Short-Term Memory networks (LSTM), improve accuracy in HTR systems, with applications in historical document digitization and handwritten notes conversion [9]. Proposed methods simplify classification and recognition using multilayer feed-forward neural networks, achieving a commendable 90.19% recognition accuracy [10]. Another paper explores Convolutional Neural Networks (CNN) and LSTM for word recognition, combining CNN for word classification and LSTM for character segmentation [11]. CNNs are employed for word classification based on features and patterns, while LSTM aids in character segmentation. The combination of CNN and LSTM leads to accurate character recognition and word reconstruction. A paper presents a model incorporating recurrent neural networks, convolutional neural networks, and connectionist temporal classification layers for handwritten text recognition [12]. In recent paper explores training the handwritten text recognition using OCR with CRNN which combines CNN and RNN, resulting in significant advancements in the accurate recognition of text [13].

In summary, the literature survey covers diverse techniques and approaches employed in handwritten text recognition. It highlights the significance of dynamic information, optimal network architectures, feature extraction methods, and the integration of deep learning algorithms for improved accuracy and efficiency in OCR systems. The papers contribute to various applications, including education, administration, industries, and historical document digitization.

2.2 Automated Answer Script Evaluation

In recent years, there have been several innovative approaches proposed for automated assessment of answer scripts using various techniques such as natural language processing (NLP). This literature survey provides an overview of these approaches and their contributions to the field of automated answer script evaluation.

One approach presented by utilizes NLP techniques to automatically assess answer scripts. The method involves extracting text, evaluating similarities, and assigning weighted values to determine final scores. Keyword-based summarization techniques are employed to generate summaries, and the evaluation results closely correspond to manually scored marks [14]. Another system called the Handwritten Answer Evaluation System (HAES) uses OCR to extract text from scanned answer sheets. It utilizes machine learning and NLP techniques, including cosine set similarity measures, to grade the answers. The objective of HAES is to optimize the assessment process and reduce the time and resources needed for evaluation [15]. Some authors have proposed methods for automatically capturing marks displayed on the front page of answer scripts. These approaches involve capturing images of the front page and utilizing Convolutional Neural Networks (CNN) to extract the marks accurately. By automating the mark extraction process, these methods aim to reduce human errors in evaluation [16]. A comprehensive evaluation method for descriptive answers is suggested by [17]. This method incorporates multiple approaches, including keyword and similarity scores, language score, fuzzy string matching score, and concept graph score. These techniques assess the quality, grammar, spelling accuracy, and depth of understanding in the response. Digitizing examination and evaluation of descriptive answers is addressed by [18]. The author utilizes the NLTK library for preprocessing the answers and employs TF-IDF and sparse matrix techniques to examine the presence of words in the answers. This approach aims to ensure precise evaluation and reduce manual effort.

To identify negligence in evaluation, [19] proposes comparing the time taken to evaluate an answer script. They employ an SVM classifier and an ANN regressor to determine negligent evaluations and adjust the predicted marks. This approach aims to minimize mistakes made during manual evaluation. Ensuring precise evaluation of answer scripts, [20] suggests evaluating each question separately. This method involves scanning each question and its corresponding answer individually. Multiple evaluators assess each answer, and the highest score is considered the final mark, allowing for thorough analysis and accurate assessment. For assessing short answers consisting of a few sentences, [21] introduces an algorithm that compares the actual answer with the student's response using keyword analysis. The similarity between answers is determined by calculating the edit distance of the words. An Android-based application for knowledge report and answer script evaluation is introduced by [22]. This application utilizes tokenization to extract the actual answers and employs Tesseract software for digitizing the answer scripts. By comparing tokenized student answers with the actual answers based on the presence of keywords, marks are assigned accordingly. [23] presents an online system that captures essential information about students and their responses to questions.

The evaluation process involves transforming both the correct answer and the student's answer into vectors using n-grams. Marks are assigned based on the cosine similarity score between the vectors, providing an effective measure of performance. Lastly, [24] proposes an attention-based approach that utilizes the BERT model for evaluating answer scripts. The method represents the text as word embeddings, considering the semantic meanings of the answers. The encoders module, integrated with Ludwig, is employed to grade the student answers effectively. Recently, ChatGPT has undergone thorough evaluations across numerous NLP tasks. It has been meticulously assessed and benchmarked against various other trained models. Notably, one research paper conducted an evaluation, revealing an accuracy of 63.41% across diverse types of reasoning scenarios [25]. Additionally, another study [26] concluded that ChatGPT's performance remains on par when utilized in conjunction with prompt engineering alongside specifically trained models.

Overall, the literature survey highlights various innovative approaches for automated answer script evaluation, ranging from NLP techniques to machine learning and image processing methods. These approaches aim to optimize the assessment process, reduce manual effort, minimize errors, and provide accurate evaluation of student answers. The LLM evaluation approach is expected to be more efficient compared to the methods mentioned above, and this is due to the greater capability of LLM. It is more powerful and able to comprehend instructions effectively.

3   Problem Statement

Evaluating student answer sheets is crucial in education systems all over the world because it helps track how well students are doing in their studies. But the usual way of doing this, where teachers manually check each paper, takes a lot of time and effort. This makes it hard for teachers to give detailed feedback on every question and answer to each student. Because of this, students do not get enough explanations and helpful comments to understand where they are strong and where they need to improve. As a result, their overall learning and progress are affected. Furthermore, the current evaluation system often only gives students a final score for the whole exam. It does not tell them how they did on each question or give specific feedback on their answers. This lack of clear information makes it tough for students to learn from their mistakes and grow. Even parents who want to know how their child is doing may have trouble understanding what areas their child needs to work on without detailed review comments. Research indicates that providing detailed feedback is more effective in enhancing student performance [27]. Improving the evaluation process in education is essential to help students learn better and to keep parents well-informed about their child's progress in detail.

4   Proposed Methodology

The proposed approach aims to address the challenges faced in the traditional student answer script evaluation process by introducing an automated evaluation system that utilizes advanced techniques like an LLM. The key components of the proposed approach are as follows:

4.1 Digitalization of Handwritten Answers Scripts

The first step in the automated evaluation process involves the digitalization of hand-written text from student answer scripts. OCR technique will be used to convert the handwritten answers into digital text. This is important because it allows us to evaluate the student answers with help of LLMs.

Scanning Answer Scripts. The process begins with scanning and segmentation of students' answer scripts into individual images for each question's answer.

Grayscale Conversion. Scanned images are converted into grayscale to facilitate seamless processing, thereby enhancing the recognition of handwritten text.

Text Recognition from Images. There are lot of machine learning technique are available for reading the text from images, based on the evaluation of sample students answers scripts found that OCR have better in recognizing the student handwritten text. OCR technique is used to read the text in scanned images and turns it into digital texts.

Persisting Recognized Answers. Finally recognized digital text, comprising the answers to each student's questions and their respective subjects, is then stored in a designated database table named "STUDENT_ANSWERS". This table serves as a centralized repository for all student answers, ensuring convenient access and retrieval for evaluation purposes.

Figure 1 shows above-mentioned steps are involved in the conversion of handwritten scripts into digital text.

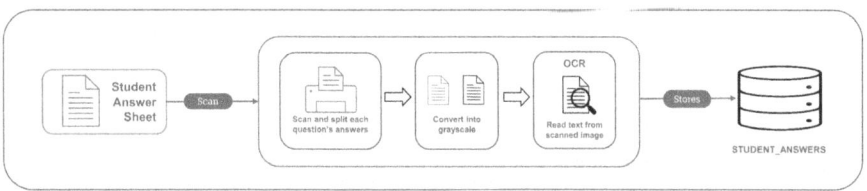

Fig. 1. Digitalization of handwritten answers scripts

4.2 Store Questions and Actual Answers

All the questions, marks and actual answers of all the subjects need to be stored in the database table called "QUESTIONS_ACTUAL_ANSWERS", to make it accessible to evaluate the student answers.

4.3 Prepare Prompt Template

Prompt templates are a valuable tool that can significantly enhance the performance of language models. It helps enabling language model to generate responses that are accurate, informative, and consistent across various question patterns. This approach ensures that language model becomes proficient in understanding evaluation criteria. Prompt template using the LangChain framework to accommodate various types of question patterns. The Prompt template comprises a set of parameters that can be dynamically inserted and utilized to generate prompts for LLMs tasks. A critical aspect of this template is defining the evaluation criteria. The evaluation criteria are guidelines used to assess different types of questions. For instance, Multiple Choice Questions (MCQs) will be assessed against strict guidelines, where only the precise answer will be considered correct. On the other hand, short or descriptive answers will be evaluated based on their overall context, not necessarily the exact wording, to determine their accuracy.

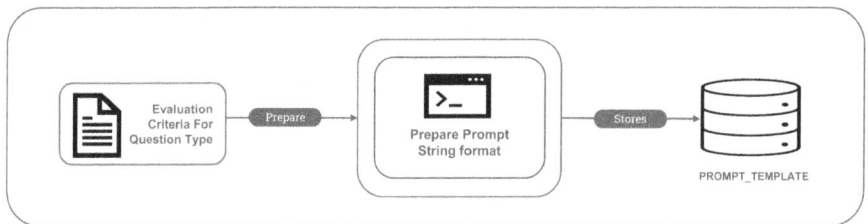

Fig. 2. Preparation of prompt template

In Fig. 2, present the process of creating a prompt template by incorporating evaluation criteria. This research focuses on four distinct question types of evaluation, such as MCQ, fill in the blank, short answers, and descriptive answers type. The incorporation of these evaluation criteria in the prompt template enables us to effectively assess and improve the language model's performance across these diverse question formats and these template to be stored in the "PROMPT_TEMPLATE" table.

Prompt template consists of five components that guide the evaluation process for a specific type of question. Figure 3 shows the prompt template for MCQ type and descriptive type questions. It includes details about the question, the marks it carries, the correct answer, and the student's answer for evaluation.

Evaluation Criteria. This specifies the guidelines for evaluating a particular type of question. It outlines how the answers will be assessed.

Question Details. This section provides the specific details of the question being evaluated. It includes the question itself, which could be of various types.

Marks Associated. This component indicates the number of marks or points assigned to this question. The marks obtained by the student will be provided based on this indicator.

Actual Answer Details. This component reveals the correct answer to the question. It helps in comparing the student's answers to the actual one.

Student Answer. In this part, the student's answer to the question is provided. This is the answer that needs to be assessed and compared to the correct answer.

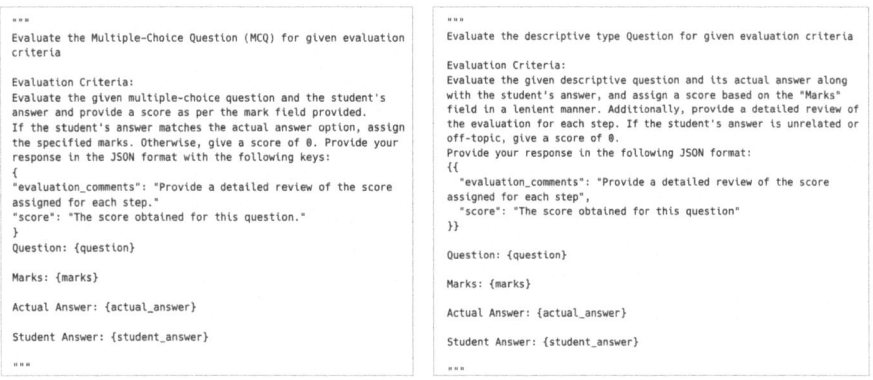

```
"""
Evaluate the Multiple-Choice Question (MCQ) for given evaluation
criteria

Evaluation Criteria:
Evaluate the given multiple-choice question and the student's
answer and provide a score as per the mark field provided.
If the student's answer matches the actual answer option, assign
the specified marks. Otherwise, give a score of 0. Provide your
response in the JSON format with the following keys:
{
"evaluation_comments": "Provide a detailed review of the score
assigned for each step."
"score": "The score obtained for this question."
}
Question: {question}

Marks: {marks}

Actual Answer: {actual_answer}

Student Answer: {student_answer}

"""
```

```
"""
Evaluate the descriptive type Question for given evaluation criteria

Evaluation Criteria:
Evaluate the given descriptive question and its actual answer along
with the student's answer, and assign a score based on the "Marks"
field in a lenient manner. Additionally, provide a detailed review of
the evaluation for each step. If the student's answer is unrelated or
off-topic, give a score of 0.
Provide your response in the following JSON format:
{{
"evaluation_comments": "Provide a detailed review of the score
assigned for each step",
"score": "The score obtained for this question"
}}

Question: {question}

Marks: {marks}

Actual Answer: {actual_answer}

Student Answer: {student_answer}

"""
```

Fig. 3. Prompt template for MCQ type (left) and Descriptive type (right) questions.

4.4 Evaluation

The evaluation of student answer scripts lies at the core of this research, and comprehending the workings of this automatic evaluation is crucial. The following steps outline the process:

Input Parameters. The Language Model requires specific input variables, including the questions given to students, their answers, correct answers for comparison, and the relevant prompt template which is prepared already and stored in the database.

LLM Evaluation. The LangChain framework plays a pivotal role in transforming the input variables into a coherent and effective prompt string. This prompt string captures the essence of the task, allowing the Language Model to accurately interpret and evaluate student answers. Once the prompt string is prepared, the Language Model specifically, GPT-3.5 is used for evaluation. It thoroughly analyzes the prompt and student answers, generating a JSON object as output. This JSON object contains valuable information, such as evaluation comments and scores for each question. These scores quantitatively represent students' performance and provide insightful review comments for improvement.

Store Evaluation Results. The evaluation results are captured and stored in a designated database table called "EVALUATION_RESULTS" for further processing.

In the evaluation process for question Qi, as depicted in Fig. 4, the following steps are followed: Firstly, the student answers are retrieved from the "STUDENT_ANSWERS" table, while simultaneously fetching the relevant question, marks, and actual answers from the "QUESTIONS_ACTUAL_ANSWERS" table. Next, the prompt template specific to question (Qi) is obtained. Utilizing the LangChain framework, the prompt string is generated and then forwarded to the LLM for evaluation. The evaluation results are then aggregated into a JSON object and stored in the "EVALUATION_RESULTS" table for further reports.

This entire process is repeated for all questions and all students. The accumulated data within the "EVALUATION_RESULTS" table holds significant value, enabling the

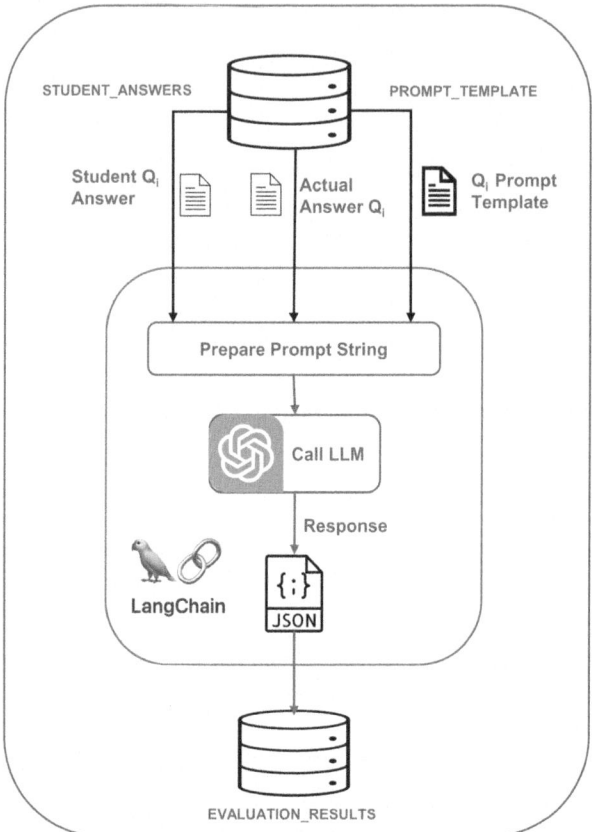

Fig. 4. Evaluation process

calculation of final scores, implementation of personalized feedback through email, and overall enhancement of the educational process. Additionally, these evaluation comments can be translated into the language preferred by parents, enabling them to understand their child's weaknesses and identify the specific areas where the child made mistakes while answering questions. This feature ensures effective communication between the school and parents, facilitating a clearer understanding of their child's performance and progress. This comprehensive approach guarantees an effective evaluation of students' progress, leading to a more tailored and successful learning journey for everyone involved.

5   Experiments

In this experiment, a dataset of answer sheets from 50 students' exam papers was used. The dataset was gathered from educational institution and covered Social Science, Mathematics and Science subjects. These experiments did not consider any diagrams or any sketch-related questions. Each subject question paper comprised multiple types of questions, including MCQs, fill in the blanks, short answers, and descriptive answers. Table 1 contains details of dataset collected from institution of 50 students.

Table 1. Dataset details

Question Type	Total Number of Questions	Total Number of Student Answers
MCQ	20	200
Fill in the blank	20	200
Short Answers	12	100
Descriptive Answers	8	50

Each question corresponding marks, and correct answers were provided, serving as a basis for evaluations. Scanned all these student answers and read the text from images through OCR and converted into digital text. To ensure accuracy, evaluation criteria were prepared for each type of questions from subject matter experts and created relevant prompt templates to facilitate the LangChain framework in generating prompt strings for the evaluation process. Experiment results for all four types of questions as follows.

```
Prompt String :
"""
Evaluate the Multiple-Choice Question (MCQ) for given evaluation
criteria

Evaluation Criteria:
Evaluate the given multiple-choice question and the student's answer
and provide a score as per the mark field provided.
If the student's answer matches the actual answer option, assign the
specified marks. Otherwise, give a score of 0. Provide your response
in the JSON format with the following keys:
{
"evaluation_comments": "Provide a detailed review of the score
assigned for each step."
"score": "The score obtained for this question."
}
Question:
"
Rain is called acid rain when its:
(a) pH falls below 7
(b) pH falls below 6
(c) pH falls below 5.6
(d) pH is above 7
"

Marks: 1

Actual Answer: "(c) pH falls below 5.6"

Student Answer: "d"

"""
Evaluation Result:
{
    "evaluation_comments": "The correct answer is (c) pH falls below
    5.6, but the student has selected (d) pH is above 7.
    Hence, the score is 0.",
    "score": 0
}
```

```
Prompt String :
"""
Evaluate the fill in the blank question for given evaluation criteria

Evaluation Criteria:
Evaluate the given fill in the blank question and the student's answer
and provide a score as per the mark field provided.
If the student's answer matches the actual answer, assign the specified
marks. if any spelling errors give score 0
Provide your response in the JSON format with the following keys:
{
"evaluation_comments": "Provide a detailed review of the score assigned
for each step."
"score": "The score obtained for this question."
}
Question: "A line intersecting a circle in two points is called a _____."

Marks: 1

Actual Answer: "Secant"

Student Answer: "Secant"
"""

Evaluation Result:
{
    "evaluation_comments": "The student's answer is correct and matches the
    actual answer of secant, hence assigning 1 mark.",
    "score": 1
}
```

Fig. 5. Evaluation results of MCQ and Fill in the blanks question type.

5.1 Experiment 1 - MCQ Type Questions

In this experiment, the collected dataset consisted of 20 Multiple Choice Questions (MCQs) covering all subjects and having 200 student answers. The evaluation results from this experiment demonstrated promising outcomes. Figure 5 (left block) provided showcases one of the Science subject question prompt string and evaluation results, highlighting the accuracy of the automated evaluation process for MCQ-type questions.

5.2 Experiment 2 - Fill in the Blanks Type Questions

For this experiment, the dataset contained 20 fill in the blanks type questions across all subjects and having 200 student answers. The evaluation results yielded promising findings, indicating the effectiveness of the automated evaluation for this question type. Figure 5 (right block) is illustrating one of the Mathematic subject question prompt string and evaluation results, further supporting the success of the evaluation process.

5.3 Experiment 3 - Short Answer Type Questions

Within the dataset used for this experiment, there were 12 short answer type questions encompassing all subjects. The evaluation results displayed favorable outcomes, suggesting the robustness of the automated evaluation method for short answer questions. Figure 6 (left side block) presented the one of the short answers type question evaluation results, providing additional evidence of the evaluation process's accuracy.

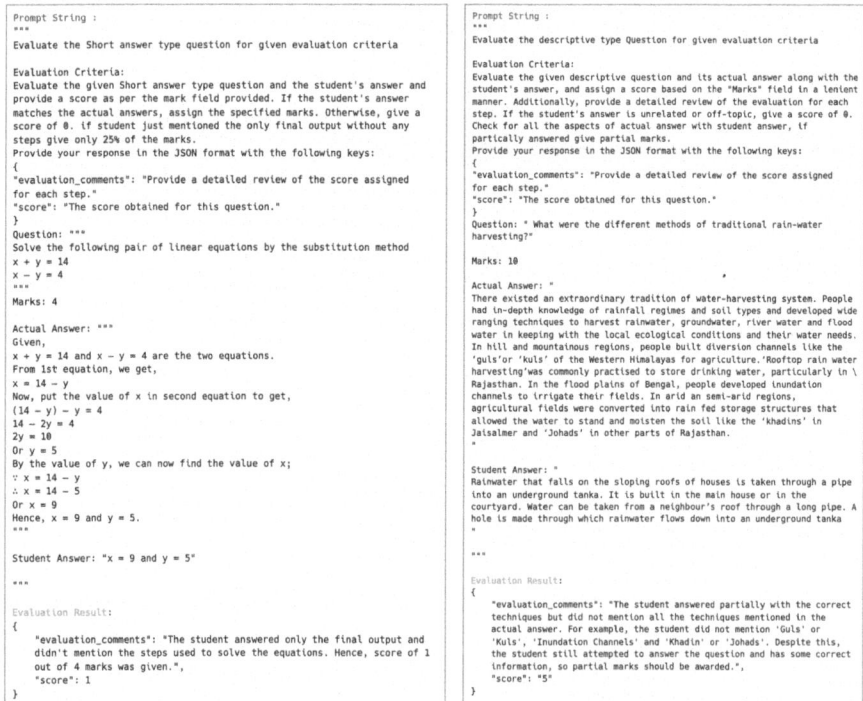

Fig. 6. Evaluation results of short and descriptive question type

5.4   Experiment 4 - Descriptive Type Questions

In this experiment, the collected dataset comprised 8 descriptive type questions spanning all subjects. The evaluation results showcased promising findings, underscoring the effectiveness of the automated evaluation approach for descriptive questions. The figure included in the results section depicts one of the question evaluation results, reinforcing the reliability and validity of the evaluation process for this question type.

After the evaluation, the results were compared, and it was found that the automated evaluation system matched with almost all the scores for the multiple-choice questions, fill in the blank type and some of the descriptive answer evaluations were also closely aligned with the marks given by manual evaluators. The results obtained from all four experiments provide compelling evidence that the automated evaluation using the LangChain framework and LLMs outperforms traditional manual evaluation in terms of efficiency, consistency, and accuracy. The automated evaluation process delivered comprehensive and timely feedback to students, significantly reducing the burden on educators while providing valuable insights for student improvement. This demonstrates the potential of automated evaluation systems to revolutionize the education sector and enhance the learning experience for both students and teachers.

6   Results and Discussion

This research emphasizes the significance of student evaluation in education and highlights the challenges associated with traditional evaluation methods. To enhance transparency and provide more valuable feedback to students, this research explores the utilization of advanced techniques such as OCR and LLM, specifically GPT-3.5, in conjunction with the LangChain framework. Addressing the challenges faced in scanning students' answer scripts with handwriting, converting the images to grayscale successfully resolved several issues. However, it was observed that for some students, handwriting proved difficult to convert into digital text. Approximately 18% of the answer scripts from the dataset encountered difficulties during the conversion process. The model evaluation process relies entirely on the evaluation criteria specified for each question type. Table 2 presents the accuracy of LLM evaluation compared to manual instructor-led evaluation.

Table 2. Accuracy of LLM evaluation result with manual evaluation

Question Type	Total Number of Student Answers	Accuracy of LLM Evaluation
MCQ	200	100%
Fill in the blank	200	100%
Short Answers	100	90%
Descriptive Answers	50	88%

However, it's important to note that the table does not account for cases where some evaluators might be lenient in their assessment of answer scripts. When the evaluation criteria are fine-tuned, the accuracy of LLM evaluation nearly matches that of manual instructor-led evaluation. This necessitates the involvement of subject matter experts to assess and evaluate specific question types accurately. It is also possible that, during the evaluation, some answer scripts may require revisiting the evaluation criteria to ensure accurate and fair assessment. The automated evaluation system was tested using data from diverse subjects, aiming to streamline the evaluation process and deliver comprehensive feedback while easing the workload for teachers. Parents will also benefit from a clearer understanding of their child's performance, enabling them to support their child's improvement based on the review comments received from the system. The system has proven to be reliable, saving both time and costs significantly. With an impressive 90% reduction in evaluation time by considering evaluation speed, accuracy, scalability, and elimination of manual errors, it presents an efficient and cost-effective solution for enhancing the overall learning experience.

7 Conclusion

This research has addressed the challenges in traditional evaluation methods and introduced an automated evaluation system using advanced techniques like OCR to digitize handwritten text and employing LLM like GPT-3.5 in conjunction with the LangChain framework, an automated evaluation system was developed to provide comprehensive feedback to students. The experiments demonstrated the system's reliability, efficiency, and accuracy in evaluating student answers, resulting in a remarkable 90% reduction in evaluation time. The automated evaluation system offers comprehensive feedback, empowering students with valuable insights into their performance and areas for improvement. This transparency enhances the learning experience by providing personalized guidance. Moreover, the system lightens the burden on teachers, automating the evaluation process and ensuring timely and constructive feedback to students, benefiting their academic progress. Construction The research highlights the transformative potential of integrating technology into education, paving the way for cost-effective and efficient assessment solutions. The successful implementation of the automated evaluation system represents a significant advancement in education evaluation, promising a more personalized and effective learning journey for both students and educators, by adopting state of art techniques.

References

1. Mittal, R., Garg, A.: Text extraction using OCR: a systematic review (2020)
2. Open AI Documentation. https://platform.openai.com/docs/models. Accessed 19 Aug 2023
3. LangChain Documentation. https://python.langchain.com/docs/get_started/introduction. html. Accessed 19 Aug 2023
4. Tappert, C.C., Hyuk, S.: English language handwriting recognition interfaces (2007)
5. Pal, A., Singh, D.: Handwritten English character recognition using neural network (2010)
6. Sharma, D.: Data extraction from exam answer sheets using OCR with adaptive calibration of environmental threshold parameters, pp. 498–502. IEEE (2013)

7. Kobayashi, Y., Mimuro, S., Suzuki, S.N., et al.: Basic research on a hand-written note image recognition system that combines two OCRs. In: Procedia Computer Science, pp. 2596–2605. Elsevier B.V. (2021)
8. Mursari, L.R., Wibowo, A.: The effectiveness of image preprocessing on digital handwritten scripts recognition with the implementation of OCR tesseract. Comput. Eng. Appl. **10** (2021)
9. Nikitha, A., Geetha, J., Jayalakshmi, D.S.: Handwritten text recognition using deep learning. In: Proceedings - 5th IEEE International Conference on Recent Trends in Electronics, Information and Communication Technology, RTEICT 2020, pp. 388–392. Institute of Electrical and Electronics Engineers Inc. (2020)
10. Pradeep, J., Srinivasan, E., Himavathi, S.: Neural network based hand-written character recognition system without feature extraction, pp. 40–44 (2011)
11. Balci, B., Saadati, D., Shiferaw, D.: Handwritten text recognition using deep learning (2017)
12. Gupta, N., Goyal, N.: Machine learning tensor flow based platform for recognition of hand written text. In: 2021 International Conference on Computer Communication and Informatics, ICCCI 2021. Institute of Electrical and Electronics Engineers Inc. (2021)
13. Yadav, A., Singh, S., Siddique, M., et al.: OCR using CRNN: a deep learning approach for text recognition. In: 2023 4th International Conference for Emerging Technology, INCET 2023. Institute of Electrical and Electronics Engineers Inc. (2023)
14. Rahman, M.M., Akter, F.: An automated approach for answer script evaluation using natural language processing (2019)
15. Sanuvala, G., Fatima, S.S.: A study of automated evaluation of student's examination paper using machine learning techniques. In: Proceedings - IEEE 2021 International Conference on Computing, Communication, and Intelligent Systems, ICCCIS 2021, pp. 1049–1054. Institute of Electrical and Electronics Engineers Inc. (2021)
16. Logeshvar, L., Premnath, A.B., Geethan, R., Suganya, R.: AI based examination assessment mark management system. In: ICSCCC 2021 - International Conference on Secure Cyber Computing and Communications, pp. 144–149. Institute of Electrical and Electronics Engineers Inc. (2021)
17. Bagaria, V., Badve, M., Beldar, M., Ghane, S.: An intelligent system for evaluation of descriptive answers. In: Proceedings of the 3rd International Conference on Intelligent Sustainable Systems, ICISS 2020, pp. 19–24. Institute of Electrical and Electronics Engineers Inc. (2020)
18. Das, I., Sharma, B.: An examination system automation using natural language processing. In: Proceedings of the Fourth International Conference on Communication and Electronics Systems (ICCES 2019), pp. 1064–1069 (2019)
19. Kamat, V.V., Dessai, K.G.: E-moderation of answer-scripts evaluation for controlling intra/inter examiner heterogeneity. In: Proceedings - IEEE 9th International Conference on Technology for Education, T4E 2018, pp. 130–133. Institute of Electrical and Electronics Engineers Inc. (2018)
20. Nayak, P., Thejaswini, M.R., Kumar, N.S.: Flexible question wise evaluation of digitized answer scripts. In: Proceedings - IEEE 9th International Conference on Technology for Education, T4E 2018, pp. 216–217. Institute of Electrical and Electronics Engineers Inc. (2018)
21. Kar, S.P., Manda, J.K., Chatterjee, R.: A comprehension based intelligent assessment architecture. In: 2017 IEEE International Conference on Teaching, Assessment, and Learning for Engineering (TALE), pp 368–371 (2017)
22. Sridevi, V., Suresh Kumar, S.: Knowledge representation and answer evaluation system using language processing algorithm (2019)
23. Sinha, S.K., Yadav, S., Verma, B.: NLP-based automatic answer evaluation. In: Proceedings - 6th International Conference on Computing Method-ologies and Communication, ICCMC 2022, pp. 807–811. Institute of Electrical and Electronics Engineers Inc. (2022)

24. Oasis, A.S., Abishai Ebenezer, M., Sharma, D., et al.: Question-centric evaluation of descriptive answers using attention-based architecture. In: Proceedings of the Confluence 2022 - 12th International Conference on Cloud Computing, Data Science and Engineering, pp. 20–25. Institute of Electrical and Electronics Engineers Inc. (2022)
25. Bang, Y., Cahyawijaya, S., Lee, N., et al.: A multitask, multilingual, multimodal evaluation of ChatGPT on reasoning, hallucination, and interactivity (2023)
26. Zhong, Q., Ding, L., Liu, J., et al.: Can ChatGPT understand too? A comparative study on ChatGPT and fine-tuned BERT (2023)
27. Lipnevich, A.A., Smith, J.K.: Effects of differential feedback on students' examination performance. J. Exp. Psychol. Appl. (2009). https://doi.org/10.1037/a0017841

A Novel Evaluation of Flight Delay Analysis for Lithuanian Airports Using Supervised Machine Learning Based Boosting Techniques

Bhawana Pillai[1]([✉]), Simranjot Singh Sodhi[1], and Prabhat Thakur[2]

[1] LNCT Group of Colleges, Bhopal, India
bhawnap@lnct.ac.in

[2] Symbiosis Institute of Technology, Symbiosis International (Deemed University), Pune, India
prabhat.thakur@sitpune.edu.in

Abstract. Due to its reputation as the quickest mode of transportation, air travel has earned its passengers' trust over the years. However, the airline industry has had to adapt to the reality that flight arrival delays are inevitable due to the fact that they are directly related to the way airspace and runways are managed. For the aviation industry to become more efficient, accurate prediction of flight delays is essential. Recent research has focused on developing methods for employing supervised machine learning to anticipate when flights may be delayed. The paper examines the variation in flight times between airports in Lithuania. The SMOTE method is used to achieve data parity. The latest FCMIM method was applied for feature selection. To forecast the time delay deviation of future bouts, a supervised machine learning model has been constructed. Tree boosting techniques (XGBoost, LightGBM, and AdaBoost) have been used for the investigation. Each algorithm's performance has been quantified across four dimensions: recall, precision, F1-measure, and accuracy. The freshly gathered dataset from Lithuanian airports and meteorological data on departure/landing time has been used for every experimental inquiry. Fights at both the airport and the port have been studied independently. The results show that the boosted trees approach is the best predictor of tree model classifiers, which have the highest accuracy. Accuracy on the Departure Dataset is 98% for the suggested models, while accuracy on the Arrival Dataset is 91%. When compared to other approaches, its accuracy is clearly superior. The proposed models are able to minimise over-appropriation and increase forecast accuracy.

Keywords: Airlines/flight delay forecasting · Lithuanian Airports dataset · Supervised machine learning · SMOTE oversampling · FCMIM · Boosting Techniques · Adaboost · LGBM · XGBoost

1 Introduction

The rising demand for airplane rides is a direct result of the improving state of the global economy. For all airlines [1], Any change in the planned combat times causes problems with finances, coordination, and technology. Time management issues or inconveniences

A. Gupta et al. (Eds.): ICAIA 2023, CCIS 2308, pp. 205–228, 2025.
https://doi.org/10.1007/978-3-031-84394-5_15

for passengers are also possible. There are two general categories of causes for a conflict to start or end later than planned: main and reactive. When one arriving flight is delayed, it may have a domino effect on other flights in the queue, resulting in a complicated delay chain that spans all airports. In recent years, there has been a flurry of new studies aimed towards predicting variations of the reactive variety. The classification quality is improved to 80.7% using a multiagent-based technique, which is one of the effective presented methods. [2]. Nature has a significant role in shaping the main type. Every single delay that occurred in the last 12 months was caused by the following: bad weather at airports; airport capacity; maintenance or aircraft issues; air traffic capacity; air traffic control; and limitation; and air travel delays. In the research [3], The causes of the departing flight delay have been identified using geographical analysis. The research shows that weather is a major impact in many situations; hence, techniques for predicting weather conditions are presented in [4]. However, a lot also relies on the size and activity level of the airport under study. Since there aren't very many flights in and out of the little airport in a given day, any delays to arrival times are often of the reactive kind. The International Air Transport Association uses delay codes to characterise and report all types of delays [5]. When the impact of such elements is studied, the codes may be utilised as a class value in various data mining techniques [6]. There are a lot of variables that might cause a flight to be delayed, but one way to minimise even the most common issues is to optimise the airport's physical layout [7]. The primary goal of this study is to identify the most effective machine learning classification method that may be modified to provide useful findings for delay assessments at smaller airports.

Delays in flights have been identified as a major issue in the aviation industry. Delays in taking off cost airlines money and annoy their passengers. Worsening flight delays are seriously damaging the reputation of the civil aviation industry. For many years [8], There has been a problem with flight delays, which has resulted in lost revenue for the airline sector. Flight delays are inconvenient for passengers because they disrupt plans and waste time and money. Constant flight delays result in substantial financial losses for the airline. The flight's delay has caused major disruptions to airport operations. The most difficult part of flying now is avoiding delays. Travelers, airlines, and airport administration are all impacted when flights are delayed in taking off. Further to the issue, aircraft delays may cause further interruptions, such as missed connections and cancellations. In recent years [9, 10], Predicting flight delays and enhancing airline operations are two areas where machine learning has emerged as a useful tool [11].

By analysing massive volumes of historical data and seeing patterns and trends, ML approaches have shown tremendous promise in correctly forecasting flight delays. These methods have the potential to boost airlines' productivity and income while also enhancing passengers' comfort and convenience. Using ML methods, researchers have looked at the possibility of predicting aircraft delays. However [12–14], In most cases, researchers have only looked at a small sample of data or used a few of algorithms in their analysis. This research intends to close that knowledge gap by using cutting-edge ML methods on a large database of flight data to provide reliable forecasts of future flight delays. The results of this study may aid airlines and passengers in making educated judgments on the use of ML approaches for anticipating flight delays. As an added bonus, this research may build the groundwork for improved flight delay prediction models to be developed in the future.

This research aims to implement new supervised ML models for the prediction and classification of flight delay analysis using Lithuanian Airports dataset. The proposed classifiers obtain high accuracy, precision, recall and f1-score in comparison to base models. To get this aim we set some objectives such as:

- To collect the Lithuanian Airports dataset from the Kaggle that divided into two type arrival and departure datasets.
- To used data preprocessing techniques for the check null values, missing values and label encoding etc.,
- To handle the class imbalance problem with using SMOTE oversampling technique that balance the Airport dataset.
- To select the important feature with the help of new feature selection technique that is FCMIM.
- To preprocessed dataset split into training and testing parts.
- To implement supervised ML based boosting classifiers for the analysis and prediction of flight delay.
- Classifiers built using machine learning may have their accuracy, precision, recall, and f1-score measured and compared.
- To proposed work shows the high performance in compare to base work.

This is the outline of the paper: Our literature survey from prior works is presented in Sect. 2, and our suggested methodology, problem definition, data collecting, data preprocessing and balancing, feature selection, and classification algorithms are outlined in Sect. 3. The experimented results used of proposed machine learning techniques and comparative study has been done in Sect. 4. At last Sect. 5 provide the research conclusion and future work.

2 Literature Review

The section opens with a summary of previous research on using ML to anticipate flight delays.

In the airline sector around the world [15], The main problem right now is the airplane delay. Increased air traffic has caused flight delays in the past 20 years due to the rapid growth of the airline industry. This results in massive financial losses for the aviation sector and has harmful effects on the natural world. That's why it's so important to keep planes from being cancelled or delayed. LR, Random Forest Regression, Logistic Regression, DT, and Sentiment Analysis are used to forecast flight delays in the current study. Next, the best model will be recommended based on the results. Most older works are limited to use on a single route or airport. Different airlines and potential factors in flight delays are analysed in this research.

Flight delays not only negatively affect airport income and operations, but also how customers view airport services [16]. DT, RF, Gradient Boosted Tree, and XGBoost Tree are used to simulate flight delay prediction in this work. In addition to airport operating flight data, this research also utilised and combined data on meteorological characteristics. Multiple sampling strategies were used to foresee the unbalanced class. Random Over-Sampling, Random Under-Sampling, and hybrids of ROS and RUS are all in use, as is the SMOTE. A model for making predictions within the class of flight delays is the product of the study. The model's efficacy was measured with the help of the Confusion Matrix and the Area Under the ROC Curve score. The study discovered that a Random Forest classifier using the ROS + RUS strategy with a data split ratio of 90:10 had the greatest accuracy, error rate, and AUC value on testing data.

In recent days [17], Flights have been delayed again. As a consequence of the present crisis facing domestic airlines in the United States, several flights have been delayed or cancelled. American Airlines has a lengthy history of being recognised as one of the most reputable and significant airlines in the United States. The airlines' performance in terms of timeliness fell short of expectations. The financial impact of flight delays on airlines is significant. That's why they're going to such lengths to ensure that fewer flights are disrupted than usual. This research examines flight data from three major airports in New York City to predict aircraft arrival delays. Training and hyperparameter adjustment for the Gradient Boosting Classifier Model is carried out by means of Grid Search and Randomized Search. The achieved accuracy before adjusting the parameters is 72%. Grid Search achieves an accuracy of 73% after parameter adjustment, whereas Randomized Search achieves an accuracy of 74%. In terms of trip organisation, travellers will appreciate such a method.

In [18], This research proposes a gradient-boosting decision tree based model for generalised flight delay prediction after looking at a broader spectrum of potential drivers to aircraft delays. A dataset for the proposed model is created by combining information such airport weather, aircraft schedules, and airport information with received automated dependent surveillance-broadcast (ADS-B) signals. Since more precise findings for the delay prediction may be provided, four distinct classification tasks were included in the design of the prediction tasks. The experimental findings demonstrate the superior prediction accuracy of the proposed GBDT-based model when applied to our small dataset.

[19], The flight delay prediction models were built using a number of different machine learning approaches, including Decision Trees (J48), SVM, K-Means Clustering, and Multi Layered Perceptron. Cross-validation was used to test the models extensively. Using a confusion matrix and ROC curve, we were able to determine which prediction model performed the best. Models developed using the data and the Decision Trees are adequate for predicting airplane departure delays, as shown by the best prediction rate of 67.144%.

Using ML methods, researchers have looked at the possibility of predicting aircraft delays. The majority of these research, however, have either looked at a small sample of data or used a restricted selection of techniques. This research intends to close that knowledge gap by using cutting-edge ML methods on a large database of flight data to provide reliable forecasts of future flight delays. The results of this study may aid airlines and passengers in making educated judgments on the use of ML approaches for anticipating flight delays.

[30] The growth of a more efficient aviation industry depends on accurate prediction of flight delays. Predicting flight delays using machine learning is an area of active study

at the moment. Most historical forecasting methods only work for one particular route or airport. The planning of a flight is fraught with uncertainty, making it one of the most challenging circumstances in the corporate world. Such a situation arises when flights are delayed for various reasons, which may be very costly for the airline, the airport, and the passengers. Airport communications, baggage handling, and mechanical equipment may all create delays, as can bad weather, regular and local holiday demands, airline rules, and the accumulation of interruptions from earlier flights. Flight delays may be caused by a broad range of factors; thus, it is important to test a number of ML-focused models in order to forecast them. A dataset for the proposed technique might be assembled by collecting, preprocessing, and combining data sets like weather, airline, and airport terminal information with ADS-B (Automatic Dependent Surveillance-Broadcast) communications. Upcoming prediction contests will include not just classification challenges, but also a regression task. It is crucial for airlines to accurately predict flight delays so that they can utilise the data to boost both customer happiness and profitability. To better predict aircraft delays, we used big data technologies, especially Hadoop. The system's reliance on aviation data, which is subject to delay, is the root source of the problem. Regressions are then performed on the data. Multiple criteria are considered by the system. RF, KNN, LR, logistic regressions, and SVM are some of the methods used by this system. Over-fitting is avoided and prediction accuracy is increased with the proposed random forest-based model.

In this study, [29], The main problem is the flight being late. Air travel has increased dramatically during the past two decades as the airline business has expanded. The aviation sector suffers tremendous financial losses and environmental consequences as a result. Therefore, it is essential that planes not be cancelled or delayed. LR, RF Regression, Logistic Regression, DT, and Sentiment Analysis are used to forecast flight delays in the current study. Next, the best model will be recommended based on the results. Most early efforts are limited to use on a single route or airport. Different airlines and potential factors in flight delays are analysed in this research.

This article explores and contrasts two types of regression methods, namely LASSO and RIDGE regression [26]. Accurate delay forecasting is crucial for the business of scheduled airlines and the happiness of their customers. Flight delays can't be avoided, yet can cost or make airlines a lot of money depending on the circumstances. This study analyses and contrasts two ML techniques in the field of specified extended flight delay time series forecasting, with the goal of better understanding the range of possible challenges associated with aircraft delays. ADS-B signals are gathered, decoded, & integrated along with other data types like airport information, weather, and aircraft schedules for offering a dataset for the proposed method. As part of the specified prediction challenges, several forecasting tasks are combined using a regression technique. The proposed prediction model was tested, and its results were compared to those of established methods. Accuracy (99.7%) and RMSE (0.3%) from LASSO & RIDGE regression, respectively. Mean absolute error (MAE) (0.2%) and mean squared error (MSE) (0.1%).

The current and the existing [28], flight delays caused by traffic congestion not only have an economic impact, but also have severe environment consequences, reduce service quality for passengers, increase fuel and gas consumption, and provide a significant challenge for airline management. Delays in airline flights may be avoided by the use of predictive analysis and machine learning techniques.

This research [27], highlights research on the feasibility of using ML techniques to anticipate airline delays. The focus of this study is on determining whether or not various machine learning techniques can accurately predict flight delays and isolate the most important contributors to such disruptions. Airline schedules, airport wait times, weather forecasts, and other characteristics are all included in the dataset used for the research. The first step of this project is a literature assessment of prior research on the topic of utilising ML to anticipate flight delays. DT, RF, SVM, and neural networks are just some of the machine learning methods whose results are compared and contrasted here. The results of the research show that ML algorithms, such as decision trees and random forests, can accurately forecast when a flight would be delayed. According to the research, the three main causes of flight delays are bad weather, airline-specific problems, and airport congestion. Since improved airline and airport management and consumer satisfaction could result from more precise flight delay forecasting, it is worth exploring, the implications of this research effort are substantial for the aviation industry. Overall, the findings of this study show that machine-learning approaches may be used to enhance the precision and utility of flight delay forecasts, leading to a safer and more efficient aviation system.

3   Research Methodology

This section contains the description of the problem identification as well as the proposed methodology to perform the flight delay analysis on the selected dataset.

3.1   Problem Identification

Delays in air travel are inconvenient for everyone involved, from passengers to airlines to airports. Accurately predicting flight delays may boost efficiency by allowing airlines and customers to better plan for and deal with the effects of delays. However, because to the complexity and variety of variables causing flight delays, such as weather conditions, airport congestion, and airline operations, precise flight delay predictions are difficult to achieve. There are limits to the accuracy and scalability of historical data and statistical modeling-based approaches to flight delay prediction. Therefore, sophisticated machine learning methods are required to examine abundant data and reliably forecast flight delays. The goal of this study is to solve this issue by using sophisticated machine learning methods to analyse huge amounts of data and make precise predictions about flight delays. This study also aims to compare and evaluate the performance of several machine learning methods to determine which algorithm is the best accurate in forecasting flight delays.

3.2   Proposed Methodology

In this research firstly we have collect the Lithuanian airports flight dataset from Kaggle, after this input dataset need to preprocessed for the check null values, missing values and data labelling with label encoding etc., next balance the dataset with SMOTE technique, select important features with CMIM, Finally, this data set prepared for the data splitting than used for the classification purpose, so for the classification we used supervised machine learning-based boosting techniques like XGBoost, LGBM and AdaBoost classifiers. At last for the model evaluation used some performance measures such as accuracy, precision, recall and f1-score or loss. Finally we get classification results higher than the other and base models. These all process of research methodology described below in subsections also shows in Fig. 1 that namely flowchart of proposed methodology.

3.2.1 Data Collection

In this research we have used Lithuanian airports flight dataset[1] from the Kaggle online websites. Separate analyses are performed on the arrival and departure portions of this dataset. Both datasets contain six classes of the dataset to classify the time deviation i.e.,

- Delay (15, 30) min
- On Time (−5, 5) min
- Early (−15, 5) min
- Delay (5, 15) min
- Early (-inf, −15) min
- Delay (30, inf) min

3.2.2 Data Preprocessing

The term "data preprocessing" refers to the steps used to transform unstructured data into a form suitable for analysis. In this work we have used data preprocessing technique for the check missing values, data labelling with label encoding, balance the data with SMOTE and select the features with CMIM method. These all data preprocessing phases are described below:

- *Filled the missing values* using the column's mode as the metric. The dataset has many missing values, therefore we had to come up with a new method for dealing with them. In this scenario, we have purged any rows that were lacking required information.
- *Label encoding* to encode the categorical data. In machine learning and data analysis, label encoding is a method for representing categorical variables numerically. It is particularly useful when working with algorithms that require numerical input, as most machine learning models can only operate on numerical data.

3.2.3 Data Balancing (SMOTE)

Whenever there is an issue with a classification problem and the classes are not represented evenly, we say that the data is imbalanced. The number of occurrences in each class is usually not precisely the same in a classification data set, although this is usually of little consequence. One such specialised method that may be used to equalise a dataset is the SMOTE oversampling approach. Oversampling using the SMOTE method has been

---

[1] https://www.kaggle.com/datasets/pavelstefanovi/lithuanian-airports-flight-dataset.

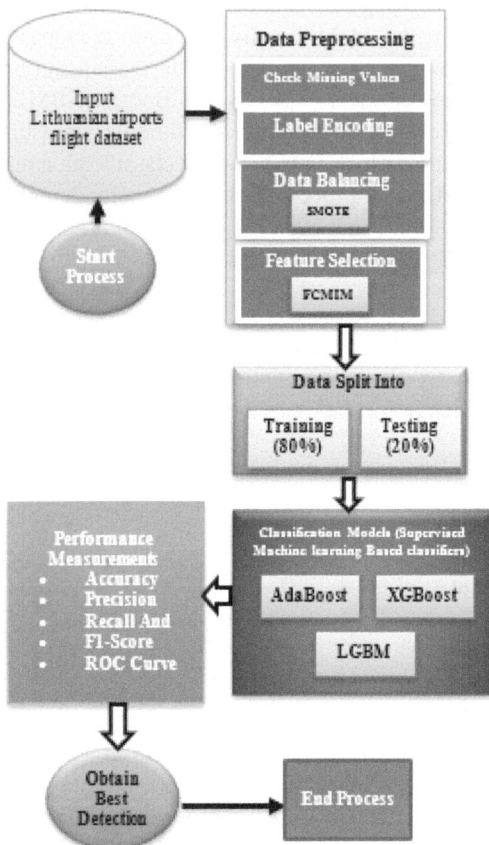

Fig. 1. Flowchart of Proposed Methodology

employed in this study. Synthetic Minority Oversampling Method or SMOTE for short [20]. It's a methodical algorithm for making fake samples. SMOTE is an oversampling strategy, as its name suggests. It avoids duplication by creating an artificial instance of a subclass instead. Algorithmically, it's best to have two or more identical examples (using a distance measure). It randomly varies the value of a single characteristic on a single instance, within a given range of variation from neighbouring instances.

3.2.4   Feature Selection with CMIM Technique

In this work, a new method for feature selection is proposed. CMIM Technique for feature selection so that we could enhance the model training and Testing accuracy. The purpose of the mutual information-based feature selection method is to pick out the most useful characteristics of the original data set by filtering out the superfluous or redundant ones. The FCMIM is used to select the features. These features are given as input to the machine learning classifiers. By definition, the Conditional Mutual Information

Maximization criteria (CMIM) will not choose a feature that is comparable to those that have previously been chosen, regardless of how strong that feature may be on its own. Thus [21], A reasonable compromise between autonomy and bias is guaranteed by this criteria. The proposed FCMIM is tested against the gold-standard feature selection method to gauge its efficacy.

3.2.5 Data Splitting

The next phase involves splitting the data into test and training sets. Analysis of Drugs Database The elimination of training-data-related bias in ML algorithms necessitates the use of splitting. There will be 20% of test data and 80% of training data.

3.2.6 Classification with Supervised Machine Learning Classifiers

Predicting which data instances belong to which categories is called classification, and it's a machine-learning technique. The Classification algorithm is a supervised learning method for automatically labelling new observations with their proper classification based on existing examples. Classification involves a computer system learning how to divide input into categories based on an existing dataset. In general, a classification algorithm is a function whose output distinguishes between classes by assigning positive and negative values to the attributes that best characterise each group. Supervised learning is the most often used paradigm for running ML operations. It is often used to information when precise mapping exists between input and output. Learning a model of classification or regression, often known as "supervised learning", is a subfield of ML that makes use of labelled test data.

Here we used supervised ML algorithms for the classification problem. We used supervised machine learning-based boosting classifiers like XGBoost, LGBM and Adaboost classifier. Boosting is a subset of Ensemble Learning that combines many less-than-stellar learners (low-predictive-strength individual models) into a single, more robust model (a model with strong accuracy). By successively merging many underperforming decision trees, "boosting" generates an ensemble model. It gives different trees different values based on their output. All analyzed algorithms are described in below:

- **XGB Classifier:** XGBoost's version of the gradient-boosted trees method has made it well-known for its ability to provide insightful answers to issues with structured data. In a gradient-boosting regression setup, each regression tree plays the role of the weak learner by giving each input data point a continuous score represented by a leaf. We might perhaps enhance model performance by including a weighting parameter for the model's computational complexity and a convex loss function depending on the gap between observed and ideal outputs. XGBoost simplifies a defined objective function. In an iterative training procedure, new trees are added to anticipate the mistakes or residuals of previously trained trees; these trees are then combined with previously trained trees to get the final prediction [22].
- **LGBM Classifier:** Based on the decision tree approach, LGBM is a distributed, powerful gradient boosting system for sorting, classification, and the creation of data science applications that uses less RAM to operate while managing huge data quantities. Light GBM differs from other methods in that it develops trees vertically,

or leaf-wise, rather than horizontally, as is the case with most other algorithms. For agricultural purposes, the leaf with the highest delta loss will be selected. When cultivating the same leaf, a leaf-wise strategy may be more effective at reducing waste than a level-based one [23, 24].

• **AdaBoost Classifier:** By integrating and amplifying very weak and imprecise rules, AdaBoost is a great machine learning strategy for producing extremely accurate prediction rules. It improves the performance of multiclass classifier problems and has a concise mathematical foundation. AdaBoost is an iterative method to boost the efficiency of underperforming classifiers by giving them the chance to learn from their own errors. AdaBoost's noise-reduction capabilities increase under the halting state [25].

3.3 Proposed Algorithm

Input: Lithuanian airports flight Dataset **Output:** High detection rate
Install: Python programming language **Framwork:** Jupyter Notebook **Import:** Python libraries (Numpy, pandas, Tensorflow, keras, matplotlib, OpenCV etc.)

Step 1: Input Dataset.
Step 2: Prepare the data for analysis by doing preprocessing.
    • Check missing values
    • Label encoding for Data labelling
Step 4: Data balancing for balance the unbalanced dataset,
    • SMOTE oversampling technique
Step 4: Feature Selection technique for select important feature
    • CMIM Technique
Step 3: Splitting dataset into two phases:
    • Training set (80%)
    • Testing set (20%)
Step 4: Implement classification supervised machine learning classifiers.
    • XGBoost Classifier
    • LGBM Classifier
    • AdaBoost Classifier
Step 5: Analysis of Training and Test Data Performance
Step 6: Then, forecast the end result.

4 Results and Discussion

Here, we'll talk about what we found in the lab. The suggested machine learning models will be tweaked, trained, and assessed with the help of python software on a 64 GB RAM, NVIDIA®GeForce RTXTM 3080 GPU, 3.80 GHz Intel®CoreTM i7-10700K CPU, and

1 TB SSD equipped HP OMEN 30L desktop GT13. And also, some other basic hardware tools are required. The code for the execution was implemented in Python. Python programming is the programming language that is being utilized for the simulation. Work in a Jupyter notebook in the role of an editor. It is dependent on several basic packages, including such time, csv, as well as numpy, as well as some more particular packages, including such sklearncrfsuite, Keras, Theano, as well as TensorFlow. The proposed machine learning models used Lithuania airport dataset and evaluated the performance of proposed models in terms of various performance measures that described below:

4.1 Dataset Description

Airports in Lithuania were studied for their flight time variation. Vilnius (VNO), Kaunas (KUN), and Palanga (PNO) are the three major airports in Lithuania (PLQ). We employed a web scraping approach to obtain all flight data from airport websites; moreover, we have collected meteorological data. From 2019-10-26 through 2020-03-16, a scraping bot collected data on the little-used Lithuania Airport every 15 min. Only the really remarkable battles have been recorded. On March 16, 2020, all flights to and from Lithuania were cancelled due to the global coronavirus emergency. During this time, the dataset is examined. Separate analyses are performed on both the arrival and departure flights in each dataset.

The visulisation results of Lithuania airports dataset given below with divided two categorized arrival and departure dataset.

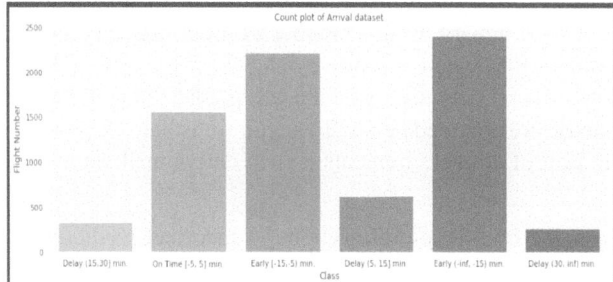

Fig. 2. Count plot of Arrival Dataset with number of classes

The above Fig. 2 shows the Count plot of Arrival Dataset with number of classes. The dataset contains 6 classes 0 to 6. In figure x-axis shows the classes name and y-axis shows flight numbers. The higher fight value of early ($-$info, -15) min. And lower flight value is delay class (15.30) min.

The Count plot of Departure Dataset with no. of classes (6). The dataset contains 6 classes 0 to 6. In figure x-axis shows classes name and y-axis shows flight numbers. Higher fight value of on time (-5, -5) min and lower flight value is delay class (-15, -5) min.

The above Fig. 3 shows the Feature importance graph of and Arrival Departure dataset using CMIM feature selection technique. There were 97,360 samples utilized for the

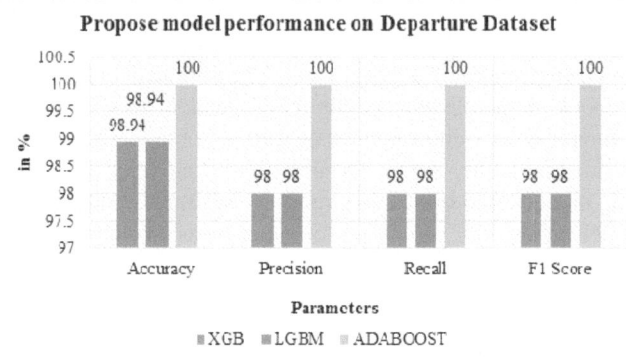

Fig. 3. Feature importance graph of Arrival dataset using CMIM feature selection

analysis, each having 12 characteristics and a single class label. There are ten category characteristics and two quantitative ones (departure and arrival times). None of them have anything to do with the fight's setting, weather, or other external circumstances. Datasets used for delay analysis often include both numeric and categorical variables.

4.2 Performance Metrics

To measure the airline prediction results by classification models. The quality of taught algorithms has traditionally been evaluated using four metrics: recall, precision, F-measure, and accuracy. The confusion matrix provides an easy way to compute all of these metrics. This summary of the classifier's output may be compared to the true output: False positive (FP), where a positive outcome was expected but did not occur; false negative (FN), where a negative outcome was predicted but did not occur. The formulas for each metric are provided below. To begin, we isolate classes while calculating each metric. The sum of all measurements is then determined via a weighted-average calculation.

Accuracy - Measured by the ratio of accurate to incorrect predictions, classification accuracy may be determined. The classification accuracy can be computed using an equation:

$$Accuracy = TP + \frac{TN}{TP} + FP + FN + TN \tag{1}$$

Precision - Precision is used to get around the Accuracy limitation. Precision is a measure of how reliably correct predictions are made. Using the following formula, we can determine the Precision:

$$Precision = TP/TP + FP \tag{2}$$

Recall - In the context of unbalanced datasets, a recall is a statistic used to evaluate the efficacy of a classification model. It is the ratio of the percentage of right predictions to the fraction of correct occurrences in the training data. The formula for recall is:

$$Recall = TP/TP + FN \tag{3}$$

F1 Score - Harmonic mean in terms of precision and recall will be represented by the F1 score. Using the following formula, we can determine the F1 score:

$$F1\,Score = 2 * (Recall * Precision)/(Recall + Precision) \tag{4}$$

4.3 Simulated Results and Analysis

In this section provide the simulation results of proposed model for classification and prediction of airports delay analysis. We downloaded the Lithuania airports dataset from Kaggle. This dataset divided into two type Arrival and Departure dataset; these datasets contain six classes that denote by the 0 to 5 values. We used boosting classifiers like XGBoost, Adaboost, and LGBM, all of which are based on machine learning. Common measures used to assess an algorithm's reliability and performance include accuracy, precision, recall, and the F1 score. Here the below subsection provides the simulation results of each proposed model with both datasets as follow:

1) Results of proposed XGBoost on Arrival Dataset

In this subsection provide the simulation results of proposed XGBoost on Arrival Dataset. The results of XGBoost on Arrival Dataset shows in classification report, confusion matrix, and parameter performance such as accuracy, precision, recall and f1-score. The following Figs. 6, 7 and 8 shows the results of proposed XGBoost model on Arrival Dataset.

	precision	recall	f1-score	support
0	0.67	0.71	0.69	66
1	0.84	0.69	0.76	54
2	0.76	0.83	0.80	125
3	0.97	0.96	0.96	481
4	0.94	0.94	0.94	444
5	0.91	0.90	0.91	312
accuracy			0.91	1482
macro avg	0.85	0.84	0.84	1482
weighted avg	0.91	0.91	0.91	1482

Accuracy: 0.9102564102564102

Fig. 4. Classification report of proposed XGBoost model on Arrival Dataset

The above Fig. 4 shows the classification report of proposed XGBoost model on Arrival Dataset, this dataset contains six classes. We can see that the class four (3) shows the high classification with 97% and 96% precision, recall, and f1-score with support 481, similarly other classes shows their performance on each parameter. The proposed classification accuracy is 91% with support 1482 respectively.

Figure 5 below depicts the accuracy, recall, and f1-score of the proposed XGBoost model on the Arrival Dataset, with the model achieving 91% accuracy, recall, and f1-score on the test dataset.

Figure 6 below displays the confusion matrix for the proposed XGBoost model applied to the Arrival Dataset. The y-axis shows the true values of the dataset, while the

```
Precision by XGB of testing data is: 0.910
Recall by XGB of testing data is: 0.910
F1 score by XGB of testing data is: 0.910
```

Fig. 5. Parameter performance of proposed XGBoost model on Arrival Dataset

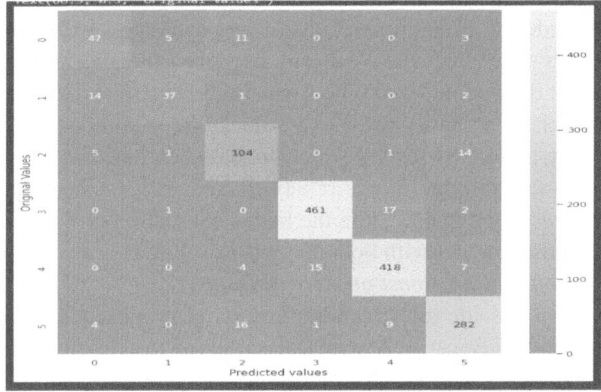

Fig. 6. Matrices of ambiguity for the XGBoost model suggested for the Arrival Dataset

x-axis displays the anticipated label. This arrival dataset has six class that denote by the 0 to 5 numeric values. The class 3 is highly predicted value of arrival dataset that contain 461 true prediction value.

2) Results of proposed XGBoost classifier on Departure Dataset

In this subsection provide the simulation results of proposed XGBoost on Departure Dataset. The results of XGBoost on Departure Dataset shows in classification report, confusion matrix, and parameter performance such as accuracy, precision, recall and f1-score. The following Figs. 9, 10 and 11 shows the results of proposed XGBoost model on Departure Dataset.

	precision	recall	f1-score	support
0	0.98	0.99	0.98	1187
1	1.00	0.99	0.99	1162
2	0.98	0.98	0.98	1174
3	1.00	1.00	1.00	1142
4	0.99	0.99	0.99	1184
5	1.00	0.98	0.99	1163
accuracy			0.99	7012
macro avg	0.99	0.99	0.99	7012
weighted avg	0.99	0.99	0.99	7012

Accuracy: 0.9894466628636623

Fig. 7. Classification report of proposed XGBoost model on Departure Dataset

The above Fig. 7 shows the classification report of proposed XGBoost model on Departure Dataset, this dataset contains six classes. We can see that the class four (3) shows the high classification with 100% precision, recall, and f1-score with support 1184, similarly other classes shows their performance on each parameter. The proposed classification accuracy is 99% with support 7012 respectively.

Precision by XGB of testing data is: 0.989
Recall by XGB of testing data is: 0.989
F1 score by XGB of testing data is: 0.989

Fig. 8. Parameter performance of proposed XGBoost model on Departure Dataset

Figure 8 below displays the precision, recall, and f1-score parameter performance of the proposed XGBoost model on the Departure Dataset. The proposed XGBoost model achieves 99.1% accuracy, recall, and f1-score performance on the test Departure dataset.

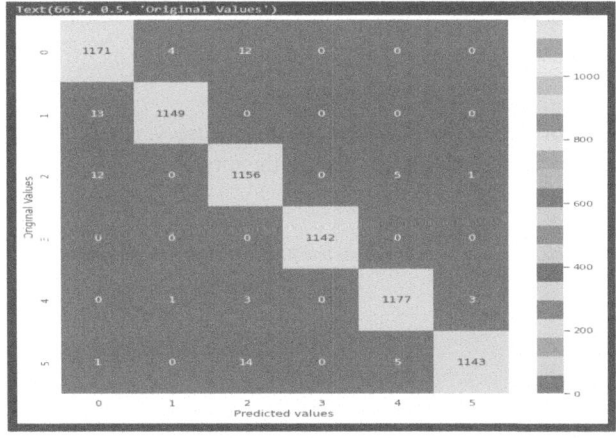

Fig. 9. Confusion matrix of proposed XGBoost model on Departure Dataset

The following Fig. 9 shows the confusion matrix of proposed XGBoost model on Departure Dataset. The class 3 is highly predicted value of departure dataset that contain 1156 true prediction value.

3) Results of proposed LGBM classifier on Arrival Dataset

In this subsection provide the simulation results of proposed LGBM on Arrival Dataset. The results of LGBM on Arrival Dataset shows in classification report, confusion matrix, and parameter performance like accuracy, precision, recall and f1-score. The following Figs. 12, 13 and 14 shows the results of proposed LGBM model on Arrival Dataset.

The above Fig. 10 shows the classification report of proposed LGBM model on Arrival Dataset, this dataset contains six classes. We can see that the class four (3) shows the high classification with 97% precision, recall, and f1-score with support 444, similarly

	precision	recall	f1-score	support
0	0.67	0.71	0.69	66
1	0.84	0.69	0.76	54
2	0.76	0.83	0.80	125
3	0.97	0.96	0.96	481
4	0.94	0.94	0.94	444
5	0.91	0.90	0.91	312
accuracy			0.91	1482
macro avg	0.85	0.84	0.84	1482
weighted avg	0.91	0.91	0.91	1482

Accuracy: 0.9102564102564102

Fig. 10. Classification report of proposed LGBM model on Arrival Dataset

other classes shows their performance on each parameter. The proposed classification accuracy is 91% with support 1482 respectively.

Precision by LGBMClassifier of testing data is: 0.910
Recall by LGBMClassifier of testing data is: 0.910
F1 score by LGBMClassifier of testing data is: 0.910

Fig. 11. Parameter performance of proposed Arrival model on Arrival Dataset

As can be seen in Fig. 11, which displays the accuracy, recall, and f1-score of the proposed LGBM model on the Arrival dataset, the model achieves an accuracy of 91% on the test dataset.

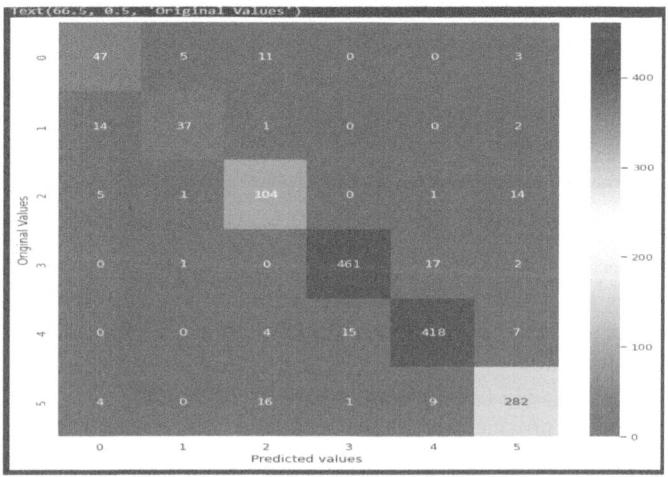

Fig. 12. Confusion matrix of proposed LGBM model on Arrival Dataset

The following Fig. 12 shows the confusion matrix of proposed LGBM model on Arrival Dataset. The class 3 is highly predicted value of Arrival dataset that contain 461 true prediction value.

4) Results of proposed LGBM classifier on Departure Dataset

In this subsection provide the simulation results of proposed LGBM on Departure Dataset. The results of LGBM on Departure Dataset shows in classification report, confusion matrix, and parameter performance such as accuracy, precision, recall and f1-score. The following Figs. 13 and 14 shows the results of proposed LGBM model on Departure Dataset.

	precision	recall	f1-score	support
0	0.98	0.99	0.98	1187
1	1.00	0.99	0.99	1162
2	0.98	0.98	0.98	1174
3	1.00	1.00	1.00	1142
4	0.99	0.99	0.99	1184
5	1.00	0.98	0.99	1163
accuracy			0.99	7012
macro avg	0.99	0.99	0.99	7012
weighted avg	0.99	0.99	0.99	7012

Accuracy: 0.9894466628636623

Fig. 13. Classification report of proposed LGBM model on Departure Dataset

The above Fig. 15 shows the classification report of proposed LGBM model on Departure Dataset, this dataset contains six classes. We can see that the class four (3) shows the high classification with 100% precision, recall, and f1-score with support 1184, similarly other classes shows their performance on each parameter. The proposed classification accuracy is 99% with support 7012 respectively.

Precision by LGBMClassifier of testing data is: 0.989
Recall by LGBMClassifier of testing data is: 0.989
F1 score by LGBMClassifier of testing data is: 0.989

Fig. 14. Parameter performance of proposed LGBM model on Departure Dataset

Figure 16 below depicts the precision, recall, and f1-score performance of the proposed LGBM model on the Departure dataset, with the suggested Departure model achieving a combined accuracy of 98.9% on the test Arrival dataset.

The confusion matrix for the proposed LGBM model on the Departure Dataset is shown in Fig. 15. The class 0 is highly predicted value of departure dataset that contain 1171 true prediction value.

5) Results of proposed Adaboost classifier on Arrival Dataset

In this subsection provide the simulation results of proposed Adaboost on Departure Dataset. The results of Adaboost on Departure Dataset show in classification report, confusion matrix, and parameter performance such as accuracy, precision, recall and f1-score. The following Figs. 16, 17 and 18 shows the results of proposed Adaboost model on Departure Dataset.

The above Fig. 16 shows the classification report of proposed Adaboost model on Arrival Dataset, this dataset contains six classes. We can see that the class four (3) shows

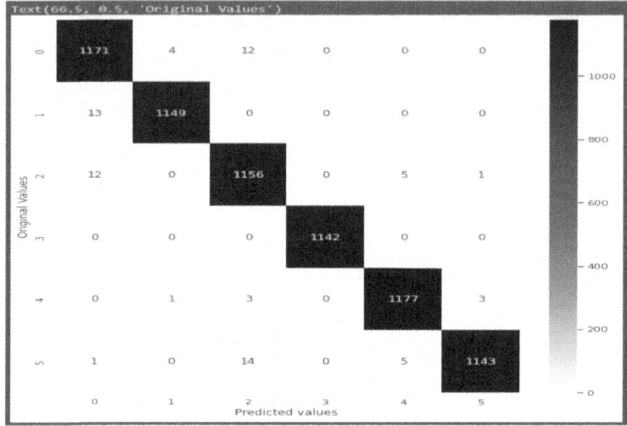

Fig. 15. Confusion matrix of proposed LGBM model on Departure Dataset

	precision	recall	f1-score	support
0	0.65	0.74	0.70	66
1	0.81	0.63	0.71	54
2	0.76	0.80	0.78	125
3	0.95	0.95	0.95	481
4	0.94	0.92	0.93	444
5	0.88	0.90	0.89	312
accuracy			0.90	1482
macro avg	0.83	0.82	0.83	1482
weighted avg	0.90	0.90	0.90	1482

Accuracy: 0.8967611336032388

Fig. 16. Classification report of proposed Adaboost model on Arrival Dataset

the high classification with 95% precision, recall, and f1-score with support 444, similarly other classes shows their performance on each parameter. The proposed classification accuracy is 90% with support 1482 respectively.

```
Precision by AdaBoost of testing data is: 0.897
Recall by AdaBoost of testing data is: 0.897
F1 score by AdaBoost of testing data is: 0.897
```

Fig. 17. Parameter performance of proposed Adaboost model on Arrival Dataset

Figure 17 below depicts the precision, recall, and f1-score measure performance of the proposed Adaboost model on the Arrival Dataset. The model achieves 89.7% accuracy, recall, and f1-score measure performance on the test dataset.

The following Fig. 18 shows the confusion matrix of proposed Adaboost model on Arrival Dataset. The class 3 is highly predicted value of Arrival dataset that contain 457 true prediction value.

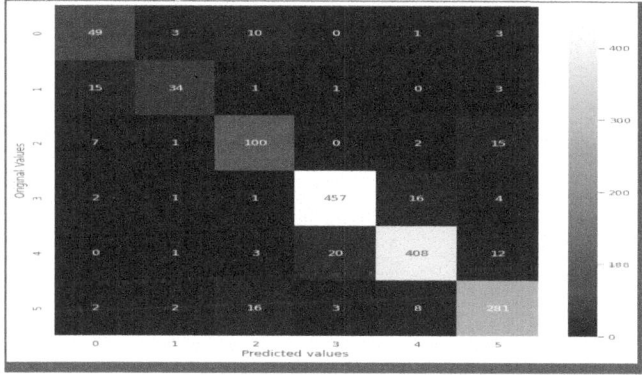

Fig. 18. Confusion matrix of proposed Adaboost model on Arrival Dataset

6) Results of proposed Adaboost classifier on Departure Dataset

In this subsection provide the simulation results of proposed Adaboost on Departure Dataset. The results of Adaboost on Departure Dataset show in classification report, confusion matrix, and parameter performance such as accuracy, precision, recall and f1-score. The following Figs. 19, 20 and 21 shows the results of proposed Adaboost model on Departure Dataset.

	precision	recall	f1-score	support
0	1.00	1.00	1.00	1187
1	1.00	1.00	1.00	1162
2	1.00	1.00	1.00	1174
3	1.00	1.00	1.00	1142
4	1.00	1.00	1.00	1184
5	1.00	1.00	1.00	1163
accuracy			1.00	7012
macro avg	1.00	1.00	1.00	7012
weighted avg	1.00	1.00	1.00	7012

Accuracy: 1.0

Fig. 19. Classification report of proposed Adaboost model on Departure Dataset

The above Fig. 21 shows the classification report of proposed Adaboost model on Departure Dataset, this dataset contains five classes. We can see that all class six (0 to 5) shows the high classification with 100% precision, recall, and f1-score with support 1184, 1162, 1174, 1142, 1184 and 1163. The proposed classification accuracy is 100% with support 7012 respectively.

Figure 22 below depicts the precision, recall, and f1-score parameter performance of the proposed Adaboost model on the Departure dataset, with the model achieving perfect scores on all three metrics during testing.

The following Fig. 21 shows the confusion matrix of proposed Adaboost model on Departure Dataset. The class 0 is highly predicted value of departure dataset that contain 1187 true prediction value.

```
Precision by AdaBoost of testing data is: 1.000
Recall by AdaBoost of testing data is: 1.000
F1 score by AdaBoost of testing data is: 1.000
```

Fig. 20. Parameter performance of proposed Adaboost model on Departure Dataset

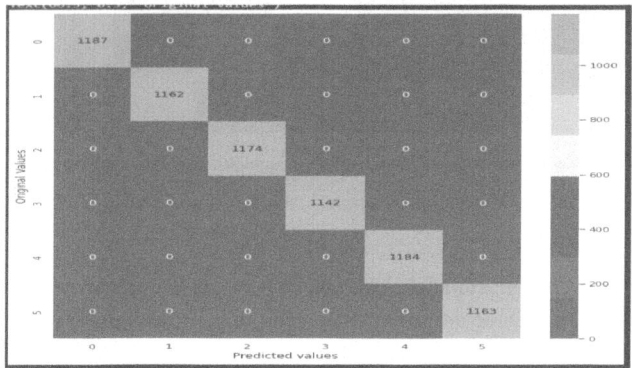

Fig. 21. Confusion matrix of proposed Adaboost model on Departure Dataset

4.4 Comparison Results and Discussion

This following section provide the comparison results and discussion of base and propose models using the Arrival and Departure dataset that part of the Lithuanian airports flight dataset, those collected from the Kaggle. The following table and figures show the models performance on both dataset using four performance measures such as accuracy, precision, recall and f1-socre. Table 1 and Table 2 shows the Base and propose model comparison using accuracy, precision, recall and f1-score parameters on Arrival Dataset and Departure dataset.

Table 1. Comparison between base and proposed models on Arrival Dataset

MODEL	Accuracy	Precision	Recall	F1 Score
Proposed	91.02	0.91	0.91	0.91
Base	83.33%	0.83	0.83	0.83

The following Fig. 22 shows the comparison between base and propose model parameter performance on arrival dataset. Here proposed model obtains 91% accuracy, precision, recall and f1-scorre while base model obtains 83% performance of these four parameters on arrival dataset, also shown in Table 1. The suggested model outperforms the baseline model significantly on the arrival dataset.

The following Fig. 23 shows the comparison between base and propose model parameter performance on Departure dataset. Here proposed model obtains 98% accuracy,

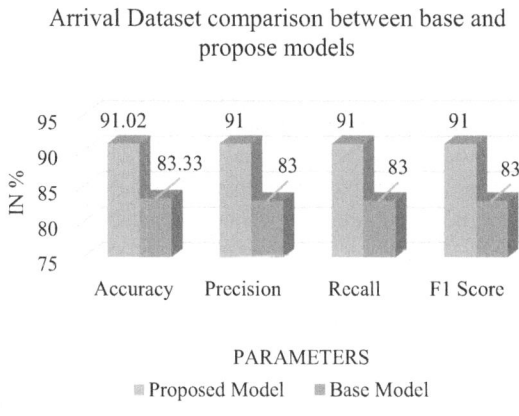

Fig. 22. Bar graph of comparison between base and propose model parameter performance on Arrival Dataset

Table 2. Comparison between base and proposed models on Departure Dataset

MODEL	Accuracy	Precision	Recall	F1 Score
Proposed	98.94	0.98	0.98	0.98
Base	92.13%	0.92	0.92	0.92

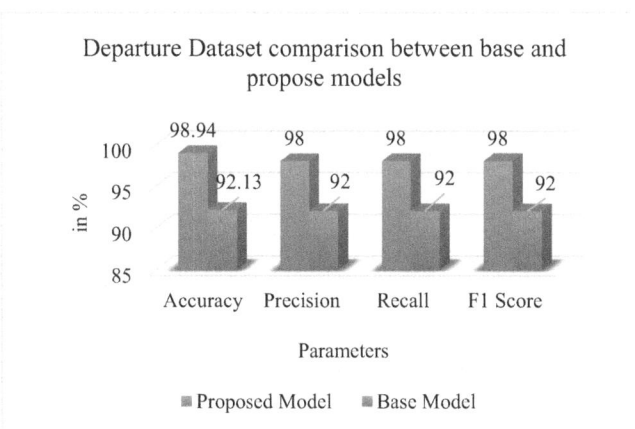

Fig. 23. Bar graph of comparison between base and propose model parameter performance on Departure Dataset

precision, recall and f1-scorre while base model obtains 92% performance of these four parameters on arrival dataset, also shown in Table 2. On the Departure dataset, we can observe that the suggested approach outperforms the standard model.

5   Conclusion and Future Work

In recent years, there has been a flurry of activity around the study of how to predict aircraft delays. Most studies have focused on further refining and expanding their models to better anticipate flight delays. Due to the critical nature of the problem, flight delay prediction models need to be very accurate and precise. Using data from Lithuanian airport flights, we offer a new ML-based optimised forecasting model using boosting classifiers (XGBoost, LGBM, and Adaboost). Since this data is skewed in several directions, we employed SMOTE oversampling methods and a number of feature selection approaches to bring them into closer alignment. The optimal parameters that provide the maximum accuracy have been determined by using the initial bound of parameters. XGB, LGBM, and ADABOOST were shown to achieve the best levels of accuracy for both arrival and departure subgroups in the experimental study. The other metrics also improved as compared to other algorithms. Even while the XGB and LGBM get similar results (about 91%) on the departure dataset in terms of accuracy, precision, recall, and f1-score, the results are substantially stronger and equivalent to 98% in terms of fighting. On the departure dataset, the proposed Adaboost classifier achieves full marks for accuracy, precision, recall, and f1-score, but its performance on the around dataset is just 89%. The deployed supervised ML model demonstrated that the provided models provide the best approach to quickly and accurately locate temporal deviation. The XGB, LGBM, and ADABOOST algorithms should be avoided if the combat time distribution is very variable. The improved dataset will provide more precise findings.

Future study may be expanded in several ways. Predicting when flights may be delayed due to inclement weather such heavy snow, flooding, hurricanes, and blizzards is an idea that may be considered in the future. The system's dynamics may be probed using novel approaches. Coordination delays may be included to the model. In the future, models will be able to deal with unforeseen events like storms, strikes, and security breaches. It's important to figure out backup flight arrangements in case problems occur in the future.

References

1. Kalliguddi, A.M., Leboulluec, A.K.: Predictive modeling of aircraft flight delay. Univers. J. Manag. (2017). https://doi.org/10.13189/ujm.2017.051003
2. Guleria, Y., Cai, Q., Alam, S., Li, L.: A multi-agent approach for reactionary delay prediction of flights. IEEE Access **7**, 1 (2019). https://doi.org/10.1109/ACCESS.2019.2957874
3. Cheng, S., Zhang, Y., Hao, S., Liu, R., Luo, X., Luo, Q.: Study of flight departure delay and causal factor using spatial analysis. J. Adv. Transp. (2019). https://doi.org/10.1155/2019/352 5912
4. Choi, S., Kim, Y.J., Briceno, S., Mavris, D.: Prediction of weather-induced airline delays based on machine learning algorithms (2016). https://doi.org/10.1109/DASC.2016.7777956
5. Träff, H., Hagander, L., Salö, M.: Association of transport time with adverse outcome in paediatric trauma. BJS Open (2021). https://doi.org/10.1093/bjsopen/zrab036
6. Zámková, M., Prokop, M., Stolín, R.: Factors influencing flight delays of a European airline. Acta Univ. Agric. Silvic. Mendelianae Brun. (2017). https://doi.org/10.11118/actaun201765 051799

7. Gilbo, E.P.: Airport capacity: representation, estimation, optimization. IEEE Trans. Control Syst. Technol. (1993). https://doi.org/10.1109/87.251882
8. Stefanovič, P., Štrimaitis, R., Kurasova, O.: Prediction of flight time deviation for lithuanian airports using supervised machine learning model. Comput. Intell. Neurosci. (2020). https://doi.org/10.1155/2020/8878681
9. Truong, D.: Using causal machine learning for predicting the risk of flight delays in air transportation. J. Air Transp. Manag. (2021). https://doi.org/10.1016/j.jairtraman.2020.101993
10. Qu, J., Wu, S., Zhang, J.: Flight delay propagation prediction based on deep learning. Mathematics (2023). https://doi.org/10.3390/math11030494
11. Pophale, S.S.: Flight delay analysis and prediction using machine learning algorithms, vol. 71, no. 4, pp. 6071–6085 (2022)
12. Shi, T., Lai, J., Gu, R., Wei, Z.: An improved artificial neural network model for flights delay prediction. Int. J. Pattern Recognit Artif Intell. (2021). https://doi.org/10.1142/S0218001421590278
13. Mokhtarimousavi, S., Mehrabi, A.: Flight delay causality: machine learning technique in conjunction with random parameter statistical analysis. Int. J. Transp. Sci. Technol. (2023). https://doi.org/10.1016/j.ijtst.2022.01.007
14. Sahoo, R., Pasayat, A.K., Bhowmick, B., Fernandes, K., Tiwari, M.K.: A hybrid ensemble learning-based prediction model to minimise delay in air cargo transport using bagging and stacking. Int. J. Prod. Res. (2022). https://doi.org/10.1080/00207543.2021.2013563
15. Salam, S., Sohail, S.M., Reddy, K.D., Darvesh, V., Reddy, T.Y., Rao, O.G.: Predicting flight delay based on sentimental analysis: machine learning. In: 2023 International Conference on Computer Communication and Informatics (ICCCI), pp. 1–4 (2023). https://doi.org/10.1109/ICCCI56745.2023.10128271
16. Sugara, R.A., Purwitasari, D.: Flight delay prediction for mitigation of airport commercial revenue losses using machine learning on imbalanced dataset (2022). https://doi.org/10.1109/CENIM56801.2022.10037369
17. Tang, Y.: Airline flight delay prediction using machine learning models (2021). https://doi.org/10.1145/3497701.3497725
18. Liu, F., Sun, J., Liu, M., Yang, J., Gui, G.: Generalized flight delay prediction method using gradient boosting decision tree (2020). https://doi.org/10.1109/VTC2020-Spring48590.2020.9129110
19. Hopane, J., Gatsheni, B.: A computational intelligence-based prediction model for flight departure delays (2019). https://doi.org/10.1109/CSCI49370.2019.00107
20. Li, J., Zhu, Q., Wu, Q., Fan, Z.: A novel oversampling technique for class-imbalanced learning based on SMOTE and natural neighbors. Inf. Sci. (Ny) (2021). https://doi.org/10.1016/j.ins.2021.03.041
21. Liang, J., Hou, L., Luan, Z., Huang, W.: Feature selection with conditional mutual information considering feature interaction. Symmetry (Basel) (2019). https://doi.org/10.3390/sym11070858
22. Dey, A.: Machine learning algorithms: a review. Int. J. Comput. Sci. Inf. Technol. (2016)
23. Mienye, I.D., Sun, Y.: A survey of ensemble learning: concepts, algorithms, applications, and prospects. IEEE Access. (2022). https://doi.org/10.1109/ACCESS.2022.3207287
24. Saber, M., et al.: Examining LightGBM and CatBoost models for wadi flash flood susceptibility prediction. Geocarto Int. (2022). https://doi.org/10.1080/10106049.2021.1974959
25. Asra, T., Setiadi, A., Safudin, M., Lestari, E.W., Hardi, N., Alamsyah, D.P.: Implementation of AdaBoost algorithm in prediction of chronic kidney disease (2021). https://doi.org/10.1109/ICEAST52143.2021.9426291

26. Evangeline, A., Joy, R.C., Rajan, A.A.: Flight delay prediction using different regression algorithms in machine learning. In: 2023 4th International Conference on Signal Processing and Communication (ICSPC), pp. 262–266 (2023). https://doi.org/10.1109/ICSPC57692.2023.10125675

27. Rajesh, K., Srikanth, V.: Predicting flight delays using machine learning : an analysis of comprehensive data and advanced techniques. **12**(4), 82–85 (2023). https://doi.org/10.17148/IJARCCE.2023.12416

28. Reddy, R.T., et al.: Flight delay prediction using machine learning. In: 2023 IEEE 8th International Conference for Convergence in Technology, I2CT 2023, vol. 1, no. August, pp. 5–9 (2023). https://doi.org/10.1109/I2CT57861.2023.10126220

29. Salam, S., et al.: Predicting flight delay based on sentimental analysis: machine learning. In: 2023 International Conference on Computer Communication and Informatics (ICCCI), pp. 1–4 (2023). https://doi.org/10.1109/ICCCI56745.2023.10128271

30. Sreenivasulu, K., et al.: Prediction of flight delay through intelligent algorithms and big data technology. In: 2023 2nd International Conference on Applied Artificial Intelligence and Computing (ICAAIC), pp. 1074–1080 (2023). https://doi.org/10.1109/ICAAIC56838.2023.10141246

31. Shrivastava, P.K., Sharma, M., Sharma, P., Kumar, A.: HCBiLSTM: a hybrid model for predicting heart disease using CNN and BiLSTM algorithms (2023). https://doi.org/10.1016/j.measen.2022.100657

Unravelling Crypto Ransomware: An Extensive Study on Transfer Learning Techniques for Crypto Ransomware Detection

Isha Sood$^{(\boxtimes)}$ (iD) and Varsha Sharma

School of Information Technology, Rajiv Gandhi Proudyogiki Vishwavidyalaya, Bhopal, India
Ishasweet1984@gmail.com

Abstract. This paper explores the growing danger posed by crypto-ransomware, a type of malicious software that encrypts files and requests payment to unlock them. The study offers a thorough examination of the development of crypto-ransomware, its unique characteristics, and the current methods for its detection. It examines several detection strategies, such as methods based on behavior, signatures, and machine learning. Systems that use signature-based detection for specific patterns or signatures inside the code of the malware to identify and stop attacks. To identify ransomware-related activity, behavior-based detection algorithms concentrate on observing the behavior of processes and applications. For their ability to identify crypto-ransomware, machine learning-based methods, particularly transfer learning, are highlighted in this paper because transfer learning techniques have become a powerful tool for identifying various types of ransomwares, such as locker and crypto-ransomware variants. These methods improve detection efficiency by utilizing both static and dynamic malware properties. In order to find new ransomware strains, hybrid models that combine transfer learning and Machine learning have showed promising results. The review includes several research that used similar methods and produced encouraging outcomes. The review also includes several research that used similar methods and produced encouraging outcomes. The importance of ongoing research and innovation in this field is emphasized in the paper's conclusion to keep up with attackers' ever-evolving strategies. The study aims to give an in-depth overview of the status of the research today, point out the positive and negative aspects of different approaches, and suggest areas for future research.

Keywords: Transfer Learning · Crypto Ransomware · Machine Learning

1 Introduction

In recent years, crypto-ransomware has developed into a serious threat to people, businesses, and governments. This type of malicious software encrypts the files on a victim's computer or network, blocking access to them until the attacker is paid a ransom. A ransomware attack hits a company every 40 s, and by 2025 that frequency is expected to rise to 4 s. By 2025, Cryptocrime Will Cost the World $30 Billion Per Year [1].

While other businesses suffered losses in the hundreds of millions of dollars, other companies paid hackers as much as $1 million in a single incident. Recent ransomware strains like WannaCry, Erebus, and SamSam have used hybrid cryptosystems, which mix symmetric and asymmetric encryption, as well as worm-like components, to execute catastrophic targeted attacks. Email and Bitcoin are frequently used technologies. Due to insufficient offline backup and incorrectly applied backup techniques, Businesses frequently pay more than the ransom demand as a result of poor offline backup and improperly executed backup strategies [2] Businesses ended up paying more than the ransom because of a lack of offline backup and badly managed offline backup techniques. Understanding crypto-ransomware detection techniques is crucial for minimizing its effects as it grows more common.

Crypto-ransomware has evolved over time, catching up with new security measures and technology. In their article, [3] provide an in-depth analysis of the evolution of crypto-ransomware and emphasize the advent of numerous variants, including locker ransomware and crypto-extortion. They go into the specifics and capabilities of each kind, shedding light on the encryption methods used and the malware's actions during the attack. The paradigm provided by this study is useful for understanding the traits of the various crypto ransomware versions.

Creating efficient detection mechanisms requires an understanding of the weaknesses and attack methods that crypto-ransomware uses. In [4] look into the usual ransomware attack vulnerabilities, such as out-of-date software, social engineering, and weak passwords. The authors also examine the various attack methods used, including drive-by downloads, exploit kits, and malicious email attachments. The results they provide offer insightful information about the areas that businesses should prioritize to strengthen their defensive measures against crypto-ransomware attacks.

Understanding the detection and mitigation of crypto-ransomware depends heavily on the encryption and decryption methods used by the malware. ("Understanding Crypto-Ransomware" Report - [PDF Document], n.d.) explore the encryption techniques employed by ransomware attackers and talk about the difficulties in attempting to decrypt the files without paying the demanded ransom. They examine both symmetric and asymmetric encryption techniques and emphasize how difficult it is to decrypt data. The knowledge gained from this study helps to build potential decryption techniques and enhances detecting capacities.

To counter the growing threat of crypto-ransomware, researchers and security professionals have suggested a variety of detection and mitigation measures. [5] provide a thorough overview of the detection techniques now in use, including signature-based, anomaly-based, and behavior-based methods. They assess the benefits and drawbacks of each strategy and suggest a hybrid detection framework that incorporates many methods for improved accuracy, such as Signature-based techniques rely on pre-established malware patterns and provide easy manual interpretation. However, they frequently only achieve middling accuracy since they require a lot of labeled data, have trouble dealing with benign ransomware variations, and use a lot of computational resources, by examining file access and network patterns, behavior-based approaches monitor processes and applications to find ransomware activity and While using complicated architectures and pre-trained models, machine learning approaches may struggle with model interpretability, despite their excellent accuracy rates. Particularly deep learning-based approaches may experience issues with overfitting and underfitting, necessitating careful model adjustment. The main focus of [6] is on the analysis and detection of crypto-ransomware. The forensic examination of a real-world attack example was thoroughly explored by the authors. They found that information belonging to the attacker was accessible after analyzing the attack strategy and behavior of the crypto-ransomware.

2 Transfer Learning

A model created for one task is used as the basis for another using the machine learning technique known as transfer learning (Fig. 1).

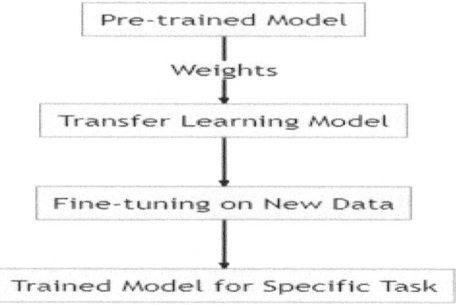

Fig. 1. Basic diagram of transfer learning

1. Pre-trained Model: This is a model that has already been trained on a substantial amount of data, generally for a problem very similar to the one we're trying to solve.
2. Performance on a related task can be aided by the knowledge acquired during training for the initial task.
3. Weights: The weights (or parameters) of the pre-trained model are carried over to the new model. These weights contain data from the earlier work that may be useful for the current task.
4. Model for Transfer Learning: This new model uses the weights from the trained model. It serves as the foundation for the particular problem we seek to tackle.
5. Fine-tuning on New Data: The transfer learning model is then refined using a fresh dataset relevant to the task at hand. In order to better fit the new task, this entails changing the model's weights using the new data.
6. Trained Model for Specific Task: Using the information from the previously trained model as well as the specifics from the fresh data, after fine-tuning, we obtain a model that is specifically trained for our desired task.

Pre-trained models are utilized as the foundation for the techniques for attack detection in cyber security. The main advantage of transfer learning is that it enables the use of patterns and information gained from one activity to enhance performance on a related one, particularly when the latter has less data available. Transfer learning techniques have been investigated for the purpose of detecting crypto ransomware. A system using novel Android ransomware traits and machine learning models for categorization was developed. Using Logistic Regression, the suggested framework identified Android ransomware with an accuracy of 99.59% with processing speeds of 177 ms on the GPU and 235 ms on the CPU [7]. Another study that presented a transfer learning system based on Efficient net was concerned with the identification of encrypted harmful traffic. Even with a limited number of training examples, the approach obtained 100% detection accuracy and recall [8].

3 Objectives of the Review

The goal of this literature review is to give a complete study and synthesis of existing research on the use of transfer learning techniques in the detection of crypto-ransomware ransomware. The review seeks to provide a thorough overview of the present state of research, identify the strengths and limitations of transfer learning methodologies, and recommend opportunities for future research.

3.1 Objectives of the Review

The goal of this investigation is to compare the transfer learning approaches against various existing techniques used in the detection of crypto-ransomware. The major goal is to compare the performance of various transfer learning algorithms based on certain metrics such as detection rate, false positive rate, processing speed, and capacity to recognize novel ransomware variants. Transfer learning in the context of crypto-ransomware detection has received a lot of interest in the literature. The purpose of the review is to determine the benefits and limitations of this strategy. Researchers have concentrated on selecting and assessing the sorts of datasets used in their studies for training and validation. Detailed research was carried out to find common trends, patterns, and important issues in the field of crypto-ransomware detection. The study provides important insights into the effectiveness and usability of transfer learning strategies. Future directions in this topic include using transfer learning algorithms to discover gaps in the current research corpus. Researchers can then identify potential topics for additional research and advancement in the field.

3.2 Search Strategy and Selection Criteria for the Literature Review

Database Search: Initially, the research will involve a comprehensive search of databases like IEEE Xplore, Google Scholar, PubMed, ACM Digital Library, ScienceDirect, and Springer. These databases offer a wide range of research papers related to transfer learning and crypto-ransomware detection.

Relevance: The study must directly relate to the detection of crypto-ransomware using transfer learning techniques. Papers that do not focus on both these aspects will be excluded.

Date of Publication: To ensure that the review includes the latest developments in the field, only papers published within the last 5 to 6 years (2017–2023) are considered.

Study Design: Both theoretical and empirical studies will be included. However, the study should provide a clear methodology for the use of transfer learning in ransomware detection.

4 Crypto Ransomware

Crypto-ransomware is an example of harmful software (malware) that encrypts the victim's files or data. Once the data has been encrypted, the ransomware demands payment from the victim, generally in the form of a cryptocurrency like Bitcoin, to provide them with the decryption key and allow them to access the encrypted data once again. The decryption key may be removed if the ransom is not paid within the allotted time, rendering the victim's data permanently encrypted and unreadable. Crypto-ransomware is mostly used by attackers to make money. The crypto-ransomware strains WannaCry, Petya, and Locky are notable examples. Crypto-ransomware is a type of harmful software that encrypts the victim's files or data and demands payment in the form of cryptocurrency, such as Bitcoin, to provide the decryption key [9].

4.1 Unique Characteristics of Crypto-Ransomware

Crypto ransomware has some unique characteristics which are given below.

   I. Encryption: Crypto ransomware employs strong encryption methods to lock victims' data or computers, leaving them unavailable. The decryption key is often kept on a remote server controlled by the attacker
  II. Anonymity: Ransom payments are frequently demanded in cryptocurrencies such as Bitcoin, which allows the attacker to remain anonymous.
 III. Social Engineering: Ransomware frequently employs social engineering approaches, such as phishing emails, to fool victims into running the software.
  IV. Propagation: Some variants of ransomware can spread autonomously by exploiting flaws in software or network settings. WannaCry ransomware, for example, spread via a weakness in Microsoft's Server Message Block (SMB) protocol.
   V. Ransom Note: After encrypting the victim's files, ransomware often shows a 'ransom note' instructing them to proceed to pay the ransom amount and restore accessibility to their files [10].

5 Crypto-Ransomware Detection Techniques

A crypto-ransomware encrypts files on a victim's computer or network, rendering them unavailable until the attacker gets a ransom payment. This malware uses powerful encryption methods and contains features such as requesting cryptocurrency payments, showing ransom notes, and threatening to destroy information if the demands are not satisfied. To design efficient detection algorithms based on transfer learning, it is critical to first grasp the concept and features of crypto-ransomware.

5.1 Existing Techniques for Detecting Crypto-Ransomware

Ransomware has developed as an imminent threat to individuals, businesses, and governments all around the world. As cybercriminals' strategies vary, it is essential to study and comprehend previous research on ransomware detection methods. This review of the literature tries to offer an overview of recent research, with a focus on papers published within the last ten years.

5.1.1 Signature-Based Detection Techniques

Signature-based detection systems detect and prevent attacks with ransomware by finding particular patterns or signatures inside the malware's code. [11] emphasized the need for signature-based detection as an initial stage of security against crypto-ransomware. These approaches can detect known ransomware variations rapidly by examining the distinctive properties of ransomware, such as file extensions, API calls, or code snippets.

The research offers a machine learning-based approach to ransomware detection that uses Random Classifier with SMOTE for data optimization and an ANN with Adam optimization. Improvements to CNN utilizing Adam are also investigated. With a prediction time of 5.14 ms, the proposed model obtained 99.18% accuracy [12]. The study presents PEDA, a Pre-Encryption Detection Algorithm capable of detecting crypto-ransomware before any encryption takes place. PEDA has two detection levels: the first detects activation potential by comparing signatures with known crypto-ransomware. Second, a Learning Algorithm (LA) identifies pre-encryption APIs with an 80:20 training-testing ratio and 99.9% recall in 10-fold cross-verification. The research also finds fourteen key APIs that separate ransomware from goodware. The installation of PEDA can aid in the early detection of crypto-ransomware, averting irreversible damage to victims' digital files [13]. The study describes CSPE-R, a methodology for identifying zero-day ransomware threats. It transforms the feature space into a core semantic space using unsupervised deep CAE. To uncover the relevance between ransomware attack families, heterogeneous base estimators are trained over these subspaces. The innovative Pareto Ensemble-based technique attempts to strike a compromise between false positives and false negatives. When host-based features are considered, CSPE-R performs effectively against zero-day ransomware. Future research could look into network traffic verification and incorporate new ransomware strains [14]. The paper performs a thorough investigation into fuzzy misuse detection algorithms that leverage various machine learning and data mining techniques to handle various incursion kinds. It does not, however, provide particular accuracy rates obtained by various methods. It instead examines their datasets, assessment criteria, feature extraction procedures, and the type of fuzzy logic controller (FLC) and membership functions used [15].

5.1.2 Behavior-Based Detection Techniques

Researchers are focusing on behavior-based detection techniques to deal with the drawbacks of signature-based approaches. These methods concentrate on examining the behavior of processes and applications to spot ransomware-related unusual activity. Various researchers demonstrated the effectiveness of behavior-based detection in identifying previously unidentified ransomware variants. These techniques can identify ransomware by keeping an eye on file access patterns, network connections, and system activity.

With the ability to encrypt important files and demand a ransom in exchange for their decryption, crypto-ransomware presents a danger to computer systems. Researchers have begun to focus on behavior-based detection methods as conventional signature-based detection techniques lose their effectiveness against crypto-ransomware varieties.

Anomaly detection, which focuses on recognizing unusual behaviors suggestive of crypto-ransomware, is a popular method in behavior-based detection. [16] suggested a behavior-based identification technique that examines system call sequences using machine learning methods. Their method provides excellent identification accuracy with few false positives by utilizing variables including frequency, order, and temporal patterns of system calls. Similar to how [17] expanded on this study by including dynamic behavior analysis, improving their system's detecting abilities. Utilizing data mining and statistical analysis to find patterns related to crypto-ransomware activity is another successful behavior-based strategy. A hidden Markov model and n-grams analysis-based behavior-based detection system was reported by [18]. By simulating typical user activity and spotting deviations from established patterns, their method can successfully detect crypto-ransomware. To limit false alarms, the study [19] proposes a novel paradigm for the early detection of zero-day crypto-ransomware assaults combining anomaly and behavioral detection techniques. It has two kinds of detection estimators: an ensemble of behavioral-based classifiers and an anomaly-based estimator. Fusing their decisions improves detection accuracy, increasing it from 92% while maintaining a low false positive rate of 2.4%. In short, it proposes a novel method for identifying zero-day crypto-ransomware attacks with great precision. Behavioral analysis and machine learning techniques have gained prominence in detecting and mitigating the impact of crypto-ransomware. In their research, [20] present a comprehensive analysis of the behavioral patterns exhibited by crypto-ransomware, exploring the file system modifications, network traffic patterns, and system API calls associated with ransomware attacks. They propose a machine learning-based approach for detecting and preventing crypto-ransomware, utilizing features derived from the observed behaviors. This study contributes to the advancement of proactive detection mechanisms against crypto-ransomware.

5.1.3 Machine Learning-Based Detection Techniques

Machine learning has gained popularity in recent years as an effective tool for ransomware detection. To detect ransomware based on data patterns and features, researchers used machine learning methods such as random forests, support vector machines, and deep learning models. Various studies demonstrated the utility of machine learning-based approaches in detecting both known and unknown ransomware strains. These methods are adaptive and can learn from new samples to improve detection accuracy.

The research by [21] showed that the authors proposed four similarity-based detection methods using Hamming distance to find similarity between samples, which are first nearest neighbors (FNN), all nearest neighbors (ANN), weighted all nearest neighbors (WANN), and k-medoid based nearest neighbors (KMNN), and the accuracy rates of the proposed algorithms to be more than 90% and in some cases (i.e., considering API features) are more than 99%. On three datasets Drebin, Contagio, and Genome for malware detection in Android. The study [22] describes a ransomware detection approach based on API call sequences and utilizing CF-NCF (Class Frequency Non-Class Frequency) and machine learning algorithms. The approach distinguishes ransomware from innocuous files and other types of infection. When distinguishing between malware and benign files, experimental results show that the suggested approach achieves an exceptional ransomware detection accuracy of up to 98.65%. The paper describes an SVM-based method for detecting ransomware. Support Vector Machines (SVMs are used to identify files as either ransomware or benign, resulting in a high detection rate and a low false positive rate. The approach was tested on datasets WannaCry, Cerber, PETYA, and CryptoLocker of ransomware and benign files, and its effectiveness in ransomware detection was demonstrated. Overall, the article identifies SVMs as a promising method for detecting ransomware threats with 97.48% accuracy [23]. This study offers a revolutionary system that uses machine and deep learning to identify ransomware with up to 99.9% accuracy. The suggested approach is thoroughly tested utilizing a variety of datasets, including Malware Traffic Analysis (MTA), Contagio Mini Dump, and Ransomware. The framework's performance is compared to that of existing approaches, revealing its superiority in ransomware detection. These findings emphasize the considerable potential of the machine and deep learning approaches in enhancing computer system security against ransomware assaults [24]. This research [25] presents a framework for using machine learning techniques to detect and prevent ransomware attacks, overcoming the limits of standard anti-ransomware solutions. The system employs Decision Tree, Random Forest, Naive Bayes, Logistic Regression, and Neural Network classifiers, as well as feature selection. Random Forest's better accuracy 99%, F-beta, and precision scores are demonstrated by experimental findings on a ransomware dataset.

5.1.4 Transfer Learning Techniques in Crypto Ransomware Detection

In the field of machine learning, transfer learning has become a potent technique that enables models developed for one task to be utilized to improve performance on related tasks. Transfer learning's potential for detecting crypto-ransomware has recently come to the attention of researchers. Crypto ransomware is a type of harmful software that encrypts a victim's files and then demands payment in exchange for their decryption, posing serious risks to both persons and businesses. The work of [26], who suggested a transfer learning paradigm for ransomware detection, is one noteworthy study in this area. Their strategy entailed using a sizable dataset of innocuous and ransomware-related files to train a deep neural network. Even in the presence of obfuscation tactics, they got outstanding results in accurately identifying ransomware variations by utilizing the pretrained model. This study provided the groundwork for later studies that used transfer learning for the detection of crypto-ransomware. [27] investigated the efficacy of transfer learning using various feature representations. They looked into how to enhance ransomware detection effectiveness by using both static and dynamic features extracted from malware samples. Their findings showed that transfer learning significantly improved accuracy and robustness when paired with appropriately chosen characteristics. This study made clear how crucial feature engineering and transfer learning are for effective ransomware detection. [26] developed a unique transfer learning strategy for detecting unknown varieties of ransomware in the current research. They created a hybrid model that used both transfer learning and unsupervised learning techniques. The algorithm was trained on a huge dataset of known ransomware samples before being fine-tuned with unsupervised learning to detect novel versions. Their experiments produced encouraging results in detecting previously unknown ransomware strains, illustrating the utility of transfer learning in tackling emerging threats.

The pre-encryption detection method (PEDA), which may detect crypto-ransomware before it is encrypted, is evaluated in the study [28] The learning algorithm (LA) is trained using the application program interface (API) to identify crypto-ransomware before it encrypts data. A learning algorithm (LA) is used to achieve this. Both conventional assessment criteria (such as true positive rate, accuracy, and precision) and new metrics developed by the researchers are used to assess the LA's performance. These extra indicators are meant to give a more thorough understanding of how the LA is doing. The results show that the LA effectively detects crypto-ransomware early and its strong performance produces a sizable net benefit.

The tool described in this research [29] uses machine learning to detect and stop crypto-ransomware attacks in file-sharing networks with encrypted communication. By examining network traffic, the utility pulls pertinent information about file activities, such as opening, shutting, and editing files. The tool can distinguish between ransomware activity and safe application behavior thanks to these features. Over 70 ransomware binaries from 33 different strains are used to train and test the detection model, coupled with a sizable dataset of more than 2,400 h of non-infected traffic. The outcomes show how well the program identifies all described ransomware binaries, including those not encountered during the training process. To create compact features for Ransomware behavior characterization, this research [30] introduces RansomWall, a layered defense mechanism against cryptographic ransomware attacks. It presents a complete study of the dataset and shows that RansomWall obtained a 98.25% detection rate with minimal false positives utilizing the Gradient Tree Boosting Algorithm, evaluated against 574 samples from 12 Ransomware families in real-world scenarios. To discriminate between ransomware, innocuous files, and other types of malwares, this paper [31] suggests a ransomware detection method that makes use of machine learning methods. It offers a thorough compilation of machine learning algorithms, including Random Forest, Decision Trees, Nave Bayes, SVM, Logistic Regression, ANN, and Gradient Tree Boosting, as well as a wide range of features used in earlier research for ransomware detection, including raw bytes, APIs/system calls, strings, k-mer frequency, log files, volatile memory dump features, HPC values, network traffic, CPU power usage, opcodes/bytecode sequences, PE. The research [32] reveals how CPU instructions, or "opcodes," can be used to distinguish between crypto-ransomware and genuine software. The suggested methodology achieves 100% precision for binary classification and 96.5% precision (96.7% accuracy) for differentiating between several ransomware variants and good ware using a Support Vector Machine. The paper utilizes a dataset of malicious and benign Portable Executable files to detect crypto-ransomware, representing each executable sample as a density histogram to eliminate variations resulting from different applications and code lengths. The research also investigates attribute selection techniques to lessen computational complexity and offers possibilities for creating more effective and efficient crypto-ransomware detection techniques. [33] proposed a transfer learning-based CNN model for ransomware detection. They utilized the DREBIN dataset and achieved 95.7% accuracy, demonstrating the efficacy of deep learning in identifying malicious code in apps. [34] employed an LSTM-based transfer learning model for crypto-ransomware detection. They used the Cuckoo sandbox dataset and attained an accuracy of 97.3%, validating the potential of RNNs for malware detection [35]. A study [36] used a CNN model with transfer learning techniques. The authors used the CICIDS2017 dataset for conducting experiments and validating the effectiveness of their proposed intrusion detection system. They validated their model on the NSL-KDD dataset and achieved 99.85% accuracy on the CICIDS 2017 dataset (Tables 1 and 2).

Table 1. Represents the various Techniques for detecting crypto-ransomware, Detection techniques used, datasets, accuracy, and Key Findings of the research

Research Paper	Detection Technique Used	Dataset Used	Accuracy	Key Findings of the Research
[12]	Adam, Random Classifier, SMOTE, and ANN	Not specified	99.18%	Machine learning-based technique for identifying ransomware using Random Classifier and ANN with Adam optimization
[13]	PEDA (Pre-Encryption Detection Algorithm)	Cuckoo Sandbox Report dataset	Recall 100%	PEDA for early detection of crypto-ransomware, detecting pre-encryption APIs with high recall and distinguishing ransomware from good ware
[14]	CSPE-R (Core Semantic Space-based Pareto Ensemble)	Sgandurra's developed dataset	Recall (9%) and F1-score (1%)	A method for detecting zero-day ransomware and for balancing false positives and false negatives

(continued)

Table 1. (*continued*)

Research Paper	Detection Technique Used	Dataset Used	Accuracy	Key Findings of the Research
[18]	HMM and n-grams analysis for PYBREAK	VirusTotal Intelligence, Malc0de, and VXVault	Not Applicable	Effective detection of crypto-ransomware through deviations from established patterns using HMM and n-grams analysis
[10]	Ensemble of Behavioral-based Classifiers	Virusshare dataset	92%	A hybrid method combining ensemble and anomaly-based estimators for detecting zero-day ransomware with high accuracy
[20]	Machine Learning-based Approach	Various Datasets	99.9%	Effective machine learning approach for detecting and mitigating the impact of crypto-ransomware
[24]	Deep learning-based approach	Various datasets	99.9%	Utilizing machine and deep learning for detecting ransomware with superior performance
[21]	Similarity-based Detection Methods	Drebin, Contagio, and Genome datasets	90 to 99%	Proposed four similarity-based detection methods for malware detection in Android
[23]	SVM(Support Vector Machine) based method	WannaCry, Cerber, PETYA, and Cryptolocker	97.48%	SVMs showed good accuracy in detecting ransomware threats
[33]	Different Types of Neural Networks	Malicious and benign PE files	100% precision	Using opcodes to accurately differentiate between genuine programs and ransomware

(*continued*)

Table 1. (*continued*)

Research Paper	Detection Technique Used	Dataset Used	Accuracy	Key Findings of the Research
[34]	The LSTM-based Transfer learning model	Cuckoo sandbox dataset	97.3%	RNNs with transfer learning show potential for malware detection
[35]	CNN model with transfer learning	CICIDS2017 dataset	99.85%	The transfer learning-based CNN model demonstrates high effectiveness in intrusion detection
[17]	CNN model with transfer learning	CICIDS2017 dataset	99.85%	CNN model with transfer learning effectively detects and prevents ransomware attacks

Table 2. Comparison of Transfer Learning-based crypto ransomware detection with traditional methods

	Machine Learning-based Approach	Traditional Methods
Accuracy	High accuracy rates (ranging from 88% to 96%)	Moderate to low accuracy rates
Dataset	Able to handle diverse ransomware samples	Limited to specific ransomware variants
Generalization	Improved generalization capabilities	Limited generalization abilities
Feature Extraction	Automatic feature extraction from pre-trained models	Manual feature engineering
Scalability	Scalable to large-scale datasets	Limited scalability for extensive datasets
Adaptability	Can adapt to new ransomware variants	Requires significant adjustments for new variants
Training Time	Faster training process due to transfer of knowledge	Time-consuming training and feature engineering

(*continued*)

Table 2. (*continued*)

	Machine Learning-based Approach	Traditional Methods
Real-world Detection	Effective in detecting real-world ransomware samples	May struggle with complex real-world scenarios
Interpretability	Transfer learning models may have lower interpretability due to complex architectures and pre-trained models	Traditional methods often offer higher interpretability as they rely on manually engineered features and rule-based approaches
Data Requirements	Transfer learning can effectively learn from limited labeled data and leverage knowledge from pre-trained models	Traditional methods may require a larger amount of labeled data for training and feature engineering
Update Flexibility	Transfer learning models can be easily updated by incorporating new data or retraining specific layers	Traditional methods may require significant re-engineering or retraining when new data or features are introduced
Resource Efficiency	Transfer learning models can leverage pre-trained models, reducing the need for extensive computational resources	Traditional methods may require more computational resources for feature engineering and training

6   Conclusion

This literature analysis emphasizes how critical it is to put in place efficient detection methods to combat the rising threat of crypto-ransomware. In identifying and thwarting ransomware attacks, signature-based, behavior-based, and machine-learning-based techniques have all demonstrated promising results. Notably, in this review, we conclude that transfer learning has the potential to improve detection accuracy, even when dealing with variants of ransomware that were not previously identified. Challenges include computations that use a lot of resources, potentially harmful transfer effects, and worries about data privacy that must be addressed in the future. Despite these challenges, continued research and expert collaboration are creating better detection methods and strengthening our defenses against changing ransomware threats with the transfer learning techniques. Undoubtedly, investing in preventative measures and encouraging a constant search for improved detection techniques will increase our digital resilience and guarantee a safer online environment for people, businesses, and other organizations like Government.

References

1. Cryptocrime To Cost The World $30 Billion Annually By 2025. Accessed 13 Aug 2023. https://cybersecurityventures.com/cryptocrime-to-cost-the-world-30-billion-annually-by-2025/

2. Conti, M., Gangwal, A., Ruj, S.: On the economic significance of ransomware campaigns: A Bitcoin transactions perspective. Comput. Secur. **79**, 162–189 (2018). https://doi.org/10.1016/J.COSE.2018.08.008

3. Kulshreshtha, V., Motwani, D., Sharma, P.: A study of crypto-ransomware using detection techniques for defense research. In: Kumar, S., Sharma, H., Balachandran, K., Kim, J.H., Bansal, J.C. (eds.) CIS 2022. LNNS, vol. 613, pp. 127–146. Springer, Singapore (2023). https://doi.org/10.1007/978-981-19-9379-4_11

4. Mar, J., Hsiao, I.-F., Yeh, Y.-C., Kuo, C.-C., Wu, S.-R.: Intelligent intrusion detection and robust null defense for wireless networks. Int.l J. Innov. Comput. Inf. Control ICIC Int. **8**(5), 3341–3359 (2012)

5. Alqahtani, A., Sheldon, F.T.: A survey of crypto ransomware attack detection methodologies: an evolving outlook. Sensors (Basel) **22**(5) (2022). https://doi.org/10.3390/S22051837

6. Kara, I., Aydos, M.: Cyber fraud: detection and analysis of the crypto-ransomware. In: 2020 11th IEEE Annual Ubiquitous Computing, Electronics and Mobile Communication Conference, UEMCON 2020, pp. 764–769 (2020). https://doi.org/10.1109/UEMCON51285.2020.9298128

7. Zhang, S., Bu, Y., Chen, B., Lu, X.: Transfer learning for encrypted malicious traffic detection based on efficientnet. In: Proceedings - 2021 3rd International Conference on Advances in Computer Technology, Information Science and Communication, CTISC 2021, pp. 72–76 (2021). https://doi.org/10.1109/CTISC52352.2021.00021

8. Kim, J., Sim, A., Kim, J., Wu, K., Hahm, J.: Transfer learning approach for botnet detection based on recurrent variational autoencoder. In: SNTA 2020 - Proceedings of the 3rd International Workshop on Systems and Network Telemetry and Analytics, pp. 41–47 (2020). https://doi.org/10.1145/3391812.3396273

9. Filiz, B., Arief, B., Cetin, O., Hernandez-Castro, J.: On the effectiveness of ransomware decryption tools. Comput. Secur. **111**, 102469 (2021). https://doi.org/10.1016/j.cose.2021.102469

10. Al-rimy, B.A.S., Maarof, M.A., Shaid, S.Z.M.: Crypto-ransomware early detection model using novel incremental bagging with enhanced semi-random subspace selection. Futur. Gener. Comput. Syst. **101**, 476–491 (2019). https://doi.org/10.1016/J.FUTURE.2019.06.005

11. Smith, D., Khorsandroo, S., Roy, K.: Machine learning algorithms and frameworks in ransomware detection. IEEE Access **10**, 117597–117610 (2022). https://doi.org/10.1109/ACCESS.2022.3218779

12. Sangher, K.S., Singh, A., Pandey, H.M.: Signature based ransomware detection based on optimizations approaches using RandomClassifier and CNN algorithms. Int. J. Syst. Assur. Eng. Manage. **2023**, 1–17 (2023). https://doi.org/10.1007/S13198-023-02017-9

13. Kok, S.H., Abdullah, A., Jhanjhi, N.Z.: Early detection of crypto-ransomware using pre-encryption detection algorithm. J. King Saud Univ. – Comput. Inf. Sci. **34**(5), 1984–1999 (2022). https://doi.org/10.1016/J.JKSUCI.2020.06.012

14. Zahoora, U., Khan, A., Rajarajan, M., Khan, S.H., Asam, M., Jamal, T.: Ransomware detection using deep learning based unsupervised feature extraction and a cost sensitive Pareto Ensemble classifier. Sci. Rep. **12**(1), 1–15 (2022). https://doi.org/10.1038/s41598-022-19443-7

15. Masdari, M., Khezri, H.: A survey and taxonomy of the fuzzy signature-based intrusion detection systems. Appl. Soft Comput. **92**, 106301 (2020). https://doi.org/10.1016/J.ASOC.2020.106301

16. Gonzalez, D., Hayajneh, T.: Detection and prevention of crypto-ransomware. In: 2017 IEEE 8th Annual Ubiquitous Computing, Electronics and Mobile Communication Conference, UEMCON 2017, vol. 2018-January, pp. 472–478 (2017). https://doi.org/10.1109/UEMCON.2017.8249052

17. Yang, L., Shami, A.: A transfer learning and optimized CNN based intrusion detection system for internet of vehicles. In: IEEE International Conference on Communications, vol. 2022-May, pp. 2774–2779 (2022). https://doi.org/10.1109/ICC45855.2022.9838780

18. Kolodenker, E., Koch, W., Stringhini, G., Egele, M.: PayBreak: defense against cryptographic ransomware. In: ASIA CCS 2017 - Proceedings of the 2017 ACM Asia Conference on Computer and Communications Security, pp. 599–611 (2017). https://doi.org/10.1145/3052973.3053035

19. Al-Rimy, B.A.S., Maarof, M.A., Prasetyo, Y.A., Shaid, S.Z.M., Ariffin, A.F.M.: Zero-day aware decision fusion-based model for crypto-ransomware early detection. Int. J. Integr. Eng. 10(6), 82–88 (2018). https://doi.org/10.30880/IJIE.2018.10.06.011

20. Lemmou, Y., Lanet, J.L., Souidi, E.M.: A behavioural in-depth analysis of ransomware infection. IET Inf. Secur. 15(1), 38–58 (2021). https://doi.org/10.1049/ISE2.12004

21. Taheri, R., Ghahramani, M., Javidan, R., Shojafar, M., Pooranian, Z., Conti, M.: Similarity-based Android malware detection using Hamming distance of static binary features. Futur. Gener. Comput. Syst. 105, 230–247 (2020). https://doi.org/10.1016/J.FUTURE.2019.11.034

22. Il Bae, S., Bin Lee, G., Im, E.G.: Ransomware detection using machine learning algorithms. Concurr. Comput. 32(18), e5422 (2020). https://doi.org/10.1002/CPE.5422

23. Takeuchi, Y., Sakai, K., Fukumoto, S.: Detecting ransomware using support vector machines. In: ACM International Conference Proceeding Series (2018). https://doi.org/10.1145/3229710.3229726

24. Alsaidi, R.A.M., Yafooz, W.M.S., Alolofi, H., Taufiq-Hail, G.A.M., Emara, A.H.M., Abdel-Wahab, A.: Ransomware detection using machine and deep learning approaches. Int. J. Adv. Comput. Sci. Appl. 13(11), 112–119 (2022). https://doi.org/10.14569/IJACSA.2022.0131112

25. Masum, M., Jobair Hossain Faruk, M., Shahriar, H., Qian, K., Lo, D., Islam Adnan, M.: Ransomware classification and detection with machine learning algorithms (2022)

26. Michau, G., Fink, O.: Unsupervised transfer learning for anomaly detection: Application to complementary operating condition transfer. Knowl. Based Syst. 216, 106816 (2021). https://doi.org/10.1016/J.KNOSYS.2021.106816

27. Xu, B., Li, Y., Yu, X.: Malware detection based on static and dynamic features analysis. In: Chen, X., Yan, H., Yan, Q., Zhang, X. (eds.) ML4CS 2020. LNCS, vol. 12486, pp. 111–124. Springer, Cham (2020), https://doi.org/10.1007/978-3-030-62223-7_10/COVER

28. Kok, S.H., Azween, A., Jhanjhi, N.Z.: Evaluation metric for crypto-ransomware detection using machine learning. J. Inf. Secur. Appl. 55, 102646 (2020). https://doi.org/10.1016/J.JISA.2020.102646

29. Berrueta, E., Morato, D., Magaña, E., Izal, M.: Crypto-ransomware detection using machine learning models in file-sharing network scenarios with encrypted traffic. Expert Syst. Appl. 209, 118299 (2022). https://doi.org/10.1016/J.ESWA.2022.118299

30. Shaukat, S.K., Ribeiro, V.J.: RansomWall: a layered defense system against cryptographic ransomware attacks using machine learning. In: 2018 10th International Conference on Communication Systems and Networks, COMSNETS 2018, vol. 2018-January, pp. 356–363 (2018). https://doi.org/10.1109/COMSNETS.2018.8328219

31. Il Bae, S., Bin Lee, G., Im, E.G.: Ransomware detection using machine learning algorithms. Concurr. Comput. 32(18) (2020). https://doi.org/10.1002/CPE.5422

32. Baldwin, J., Dehghantanha, A.: Leveraging support vector machine for opcode density based detection of crypto-ransomware. Adv. Inf. Secur. 70, 107–136 (2018). https://doi.org/10.1007/978-3-319-73951-9_6

33. Madani, H., Ouerdi, N., Boumesaoud, A., Azizi, A.: Classification of ransomware using different types of neural networks. Sci. Rep. 12(1), 1–11 (2022). https://doi.org/10.1038/s41598-022-08504-6

34. Fu, Z., Ding, Y., Godfrey, M.: An LSTM-based malware detection using transfer learning. J. Cyber Secur. (2021). https://doi.org/10.32604/jcs.2021.016632

35. Rhode, M., Burnap, P., Jones, K.: Early-stage malware prediction using recurrent neural networks. Comput. Secur. **77**, 578–594 (2018). https://doi.org/10.1016/J.COSE.2018.05.010
36. Yan, F., Zhang, G., Zhang, D., Sun, X., Hou, B., Yu, N.: TL-CNN-IDS: transfer learning-based intrusion detection system using convolutional neural network. J. Supercomput. 1–23 (2023). https://doi.org/10.1007/S11227-023-05347-4/METRICS

Secure Authentication and Efficient Data Storage in Cloud Environment

Parveen Kumar Sharma[1]([⊠]) [ID], Parveen Singla[1] [ID], Akhilesh Kumar Bhardwaj[2] [ID], Surender Kumar[3] [ID], and Rajiv Mahajan[4] [ID]

[1] Chandigarh Engineering College, Landran, Punjab, India
parveen.eced@gmail.com
[2] Shri Krishan Institute of Engineering and Technology, Kurukshetra, Haryana, India
[3] GTB Colleges, Ludhiana, Punjab, India
[4] Quest Groups of Institutions, Mohali, Punjab, India

Abstract. In cloud cryptosystems, a diversity of advancement can be experienced to make strong both the authentication and data storage (compression) in open cloud environment. This paper expands the standing effort of authentication with compression schemes. The earlier drawbacks include normal schemes for client's metrics, direct authentication verification with risks of information outflow to attackers, uploading of replica content and slow speeding of the network. To overcome these existing flaws, this effort delivers a framework deploying a trusted server. When the clients access to open cloud, they share the identities and then accelerated to open cloud with authentication keyword. For compression, the modified deflate algorithm is incorporated for better performance. This strategy helps in generating better results for authentication and data storage milestones. The main objective of the proposed work is to secure the data in the cloud from the unauthorized users and improve the storage capacity in cloud. The proposed framework is sub divided into three stages: Requisition phase, authentication phase and storage phase. The primary stage involves the requisition phase which uses the basis login and password requisition model. The secondary stage executes authentication and authorization. In the final stage the data is compressed and stored in the network using modified deflate data compression algorithm.

Keywords: Cloud Technology · Trusted Server (TA) · Authentication · Homomorphic · Compression · Deflate algorithm

1 Introduction

The cloud frame is hosted on diverse kinds of virtual mechanisms on a physical server. Three clouds open or public, isolated or private and fusion of open and isolated i.e. hybrid are in the study with more subcategories growing. The open cloud model is available to the public for different application and storage utilities. Isolated cloud model helps in managing services inside an explicit trade. Cloud hybrid model is appropriate for internal and external services phenomenon. Public framework cloud model includes confidentiality, security, administration, analysis, trustworthiness, load balancing etc.

© The Author(s), under exclusive license to Springer Nature Switzerland AG 2025
A. Gupta et al. (Eds.): ICAIA 2023, CCIS 2308, pp. 245–258, 2025.
https://doi.org/10.1007/978-3-031-84394-5_17

The substantial model to explore is security factors like access control, partitioning, migration along with workload analysis and allocation to diminish the security problems [1]. Distributing the tasks consistently through the system is mirrored as an NP-complete problem [2]. There are frequent advantages to place info in the open cloud as clients are not bothered about the complications of direct hardware management. But since the clients store the details on the cloud, it means they can lose the control. Authentication and user access confidentiality privilege essentially be conserved with reduced cost and constant update in the quality of applications or services [3]. The most symbolic cloud attributes are privacy, preservability, data confidentiality, data integrity, data availability and accountability [4]. For better security, an authentication arrangement is implemented with arrangements as shared authentication, session's key settlement or password change option [5]. Secret distribution based secure computing can be generated using symmetric bivariate polynomial. The foremost intention is to shield the data secrecy on an open cloud system. The symmetric stuff in secret distribution is added to diminish the cost further [6]. The cloud clients do not look at the data storing physically. Identification of invalid responses and their removal as batch auditing is quite faster. For passed on data auditing, an assessor called TPA is engaged. The TPA concurrently handles numerous audit periods for outsourced information of diverse clients [7]. Numerous scheduling algorithms [8] have been studied for dispensing the load sharing on several machines under cloud servicing domain. The research for the preeminent scheduling algorithm is yet to be optimized. At this time, a prodigious scope of emphasis on advance secure representations appears feasible.

1.1 Authentication

As for as cloud technology security is concerned, the opening segment of overview is authenticity. In this process, the client uniqueness is verified. Different innovative authentication metrics are suggested by the researchers till date [9]. They can be summarized as:

1. Single Sign-On (SSO): Under this metric, a centralized cloud server maintains client's authentication summary for multiple applications. The advantage is that the listed user once log-in has no need to prove its authenticity multiple times for multiple services or applications. Research on SSO is in growth to check its bottlenecks.
2. Multi-factor authentication: in this metric, the cloud classification interacts with the client with a smart card in which all client information like identity and password are stored. This technique is further supported by one-time pad (OTP) facility or some biometric determination to provide enhanced security. Two-factor authentication: in this metric, the client puts his/her identity with the help of a user-id and password. During the identification process, a special password or OTP is sent on his/her mobile for accessing the desired schemes or applications by the user. To boost the security, the OTP is encrypted-decrypted using the public-private key of the user. Still, these modes can be unveiled by an attacker.
3. Authentication with identification code and password: in this basic metric arrangement, the client's request is validated by the server to provide the desired services after entering a code or number or password. However, it bears a borderline of obstruction and steady replenishment to hold the security.

4. Public Key Infrastructure: it marks the public-key cryptography technique to assist the authentication process. It assists to endorse the additional party built on the credentials with no sharing of secret data. In this approach, managing and inspecting the client side process is problematic.

The Cloud edifice needs wide-ranging authorization models to protect its resources and information to the user management. Deficiency of this generates numerous challenges like privacy, identity management, information security, data leakage, transparency, trust management, compliance and risk management.

1.2 Compression

Data or information compression is empowering innovations for multimedia applications. Placing images, sound, and video on websites without using compression calculations would not be practical. Compression techniques allow us to use less resources, such hard circle space or gearbox transfer speed. The process of encoding information to need less additional space or less transmission time is known as information compression. We first introduce the concepts of lossy and lossless information compression algorithms. Lossless compression is utilized to decrease the measure of source data to be sent so that when packed data is decompressed, there isn't any deficiency of data. Lossy compression's goal is typically not to replicate the exact nature of the data following decompression. After decompression, some data in this case is lost.

2 Literature Survey

Prior to the proposed scheme in this section, assessments of the previous work are offered. In [10], author discussed an adaptive authentication scheme by trust introduction in cloud technology milieus. They implemented role centered approaches to resource managing with access control decision making in the communication progressions. Hybrid computing expansion is threatened by privacy issues, claimed in [11]. Because a sizeable portion of an organization's labour might potentially involve sensitive information, it cannot be outsourced in an open manner without posing a serious security risk. According to role-based and trust-based methodologies [12], provided a model for legal individualities authentication and privileges admission control gaining for resources by users. A matrix centered access controlling concept was presented in [13]. In the access matrix technique, the database table is examined before granting access, not the other way around.

Authors in [14] have experimented in the strategy and operation of FADE, which was a protected edge cloud storage classification capable of grasping policy-based control, fine-grained with file assured deletion. Hence, FADE can be treated as an excellent overlay system that can operate without flaw. By including an additional biometric verification component, [15] worked on the analogous shielding authentication and gave the justification. They make use of the face, fingers, and veins in the palm as three biometric features. Though it was a precise authentication scheme but the leading shortcoming of biometric method is additional periphery requiring added cost. Methods were pushed to provide client information security with authentication interface, multi-stage virtualization and single encryption [16]. A model of authentication inter-cloud that relies on certification authority and public key set-up also demonstrated. This effort can be amplified to the state, where certification authority classification is either crashed or leaked. Authors in [17] offered a competent and mountable client authentication arrangement for cloud technology environment. A client-based authentication intermediaryratifies user identity on client-side. Additionally, a cloud-oriented SAAS requestis recycled for unregistered devicesauthentication. In [18], authors provided the Text-Image constructed authentication. The biggest problem with this method is that it might be very difficult to reach cloud-based services if the user forgets the code for the picture. A more defensive cloud storage classification that provisioned privacy-preserving based auditing capability is introduced [19]. Linear homomorphic authenticator with random masking are employed by them. However, in a disseminated storage structure without central authority, providing security along with manifold function support is challenging to maintain. Authors in [20] have recommended approaches for third-party certification that assist in running storage and information transaction in a protected manner. They have revealed five important features, namely privacy-preservability, integrity, accountability, confidentiality and availability. To ensure safety service during source sharing, [21] suggested a RADIUS server-centered verification categorization. Distribution is crucial because to issues with space, server costs, and resource reprocessing, but it isn't practical for many applications.

A cloud-supported single-sign-on system to upsurge the competence of client authentication practices is recommended [22]. The projected model was considered and labeled with establishment of two cloud servers for storage encoded particulars and cryptography keys. Furthermore, an application was built to attach SaaS service with the clients. They implemented AES-256 and SSL in the indorsed methodology and exhibited the improvements through SSO feature. Authors in [23] suggested smart objects based approach for better security arrangements in cloud technology. In [24], authors put forward their research with an investigation on various cloud federation architectures and their assessments to study the outcome of cloud traffic orientation. Token-based SAML- SSO based scheme to upgrade the security is emphasized in [25]. Though at hand, an assortment of authentication based solutions shaped so far and in progress but still the cloud-based authentication processes suffer in their security edges. One of the oldest compression computation executions, known as zip, is deflate [26]. It relies on the Huffman coding and LZ77 [27] compression computation. Giving the computation a boundary that represents a trade-off between compression rate and time is possible. The most recent computation we employ in this checkpoint is Zstd [28]. Facebook made it widely available in the middle of 2016. Due to its stated word reference compression mode, the computation performs in any case is for little measured information, according to the introductory benchmarks.

An overview of various fundamental lossless mode compression calculations is presented [29]. With regard to message information, exploratory results and correlations of the lossless compression calculations using statistical methods and compression strategies created by the dictionary were carried out. Authors in [30] focused on incremental checkpoint replication protocol for compression parameters.In [31], security requirements and the role of cloud computing in the information technology sectors is presented. Authors in [32] introduced a multi-class cloud computing approach that uses compressive sensing for a concise representation of sensor data and deep compression framework based on cloud computing for UHD video transmission is looked in [33].

3 Proposed Architecture

Whenever the authentication checking of the client occurs, it is directly done on an open cloud with public access due to "ip" public. So the counterfeitclient canperforate the public cloud easily. To get purge of this skeptical issue, the proposedstructurein the first phasedeploys a trusted server (not publiclyaccessible) and directly connected tothe cloud server. So when the client access, they share the identities first (e-mail, password and secret key stored in the system). It is first encrypted and then carries forward to the public cloud along with authentication keyword. The cloud then puts on the request to trusted server which decrypts the parameters and verifies in 'db'. If correct, then it sends back the client session key to accessing in public cloud. Theclient normally uploads duplicate contents of the files causing network delay problem and finally leads to the slow network efficiency. To overcome this, whatever the files are uploaded by the clients on public cloud, they are initially encrypted and then made compressed to avoid duplicate file uploading on public cloud.

In short, the determination of building this framework can be specified as:

"An improved cloud mechanism that upkeeps both the security concerns and efficient data storage."

3.1 Authication Working

1. In this first step, input ciphertext of username and password derived using AES algorithm.
2. Divide the cipher text in 32-bit block wise.
3. For (i = 0, i < block array. Length)

   If i = = block size
   Return 0;

4. $M^k = Mk - \Sigma_{k\,i=1}^{n-1}\beta i, (\beta i, \ Ck\ /Ck, Ck)$
5. $tR = \Sigma_{i=1}^{mt} St\ (St,\ di\ /\ di, di)$
6. $C' = [c1, c2 ....cr-1]$
7. $d_S = $ compute size $(d'_S, M, t_R - b_k)$

Proposed Authentication Configuration
Notation Meaning

1. $IR = (E_U + P_W)$ identity representation
2. E_U email id
3. P_W password

4. CP_I cipher text identity
5. C_1, C_2 constants
6. Me_1, Me_2 coefficient value
7. M_1 first variable
8. M_2 second variable
9. N Length of IR
10. P_{KR} packet sent to check authentication
11. S_{TR} authentication keyword placed string variable
12. P_{NT} plain text
13. D_{EC} decryption algorithm
14. C_D coefficient of decryption
15. L_{CT} length of ciphertext
16. D_{BE} database
17. T_{ED} expiry date of packet
18. S_K session key
19. TTP_S trusted third party server

Authentication Mechanism

1. $IR = (E_U + P_W)$: which will be entered by the user into the client end
2. IR is encrypted as C_1 & C_2 constant value and $Me1$ & Me_2 are the coefficient value of M_1*M_2 variables which is mod by N. where N is defined as the length of identity representation.
3. Then we take one string variable S_{TR} in which we store authentication keyword
4. Final packet P_{KR} = "CP_I + String" where CP_I is ciphertext identity and string S_{TR} is a header of the packet
5. The packet P_{KR} is sent on the public cloud
6. The public cloud extracts header from P_{KR} and then reads the string
7. if string = authentication, then go to step 8
8. P_{KR} packet is forwarded to trusted third party server (TTP_S).
9. TTP_S runs decryption algorithm D_{EC} on to the P_{KR} and generates plain text P_{NT}.
10. If plain text P_{NT} exists in database D_{BE}, then session key S_K is activated for time T_{ED}, else go to step 1.

3.2 Proposed Compression Algorithm Symbols

Temp = store input file by user in temp variable
Length = compute length of file
UNQ = find unique symbols
Usize = Unique array file size
Nm = Number symbols
Dc = Decompress file variable
Cmp = compress file size
Dd = length of decompress file
Bn = compute binary length

Compression Algorithm

Step 1: Input the file to be compressed and store in temp.
Step 2: Catch the number of exclusive symbols in the input String.
For (i = 0; i < length; I++)

    If file[i]=="unique"
    UNQ = file[i]

Step 3: Disperse the numeric code to the unique symbols originated in the step 2.
Usize = Unq
If Usize < Unq
Unq[i] = nm
Step 4: Find the binary code that corresponds to each symbol starting with the first one in the input, using the allocated numerical codes, then combine those symbols to produce binary output.
Step 5: In the MSB of the binary output, increase the amount of 0s until it is divisible by 64.

Decompression Algorithm

Step 1: Input the Final output from compressed stage.
Step 2: Assign this input to decompress the data compressed.
Dc = Cmp
Step 3: Compute the binary code corresponding to the ASCII values obtained.
Dd = size (Dc)
If i < Dd
Bn = bn + cmp[i];
Step 4: Eliminate the extra bits from the binary output added in the compression stage.
Step 5: Compute the numeric code for every 64 bits obtained in the Step 4.
Step 6: For every single numeric value obtained in the step 5, find the equivalent symbol to catch the finishing decompressed data.

4   Result and Analysis

The tests were carried out using JDK 1.8, Net Beans 8.1, My SQL 5.5, Java Swing, and Network Programming with RAM of 2 GB and HDD of 50 GB in a real-time setting.

Figure 1 shows the time taken for different token counts for SAML -SSO Authentication and proposed authentication procedure. Same has been shown in Table 1.

Figure 2, shows the Average time taken for proposed web service authentication for both the authentications and same has been shown in Table 2.

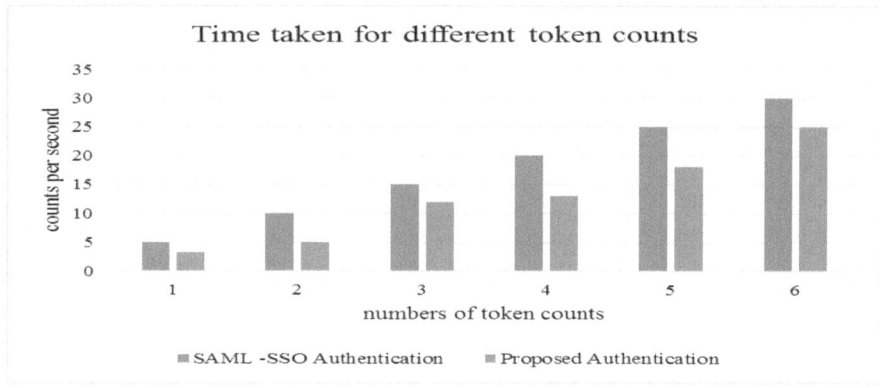

Fig. 1. Proposed Authentication performance for different token counts

Table 1. Token counts for proposed Authentication

Number of token counts	SAML -SSO Authentication	Proposed Authentication
1	5	3.25
2	10	5
3	15	12
4	20	13
5	25	18
6	30	25

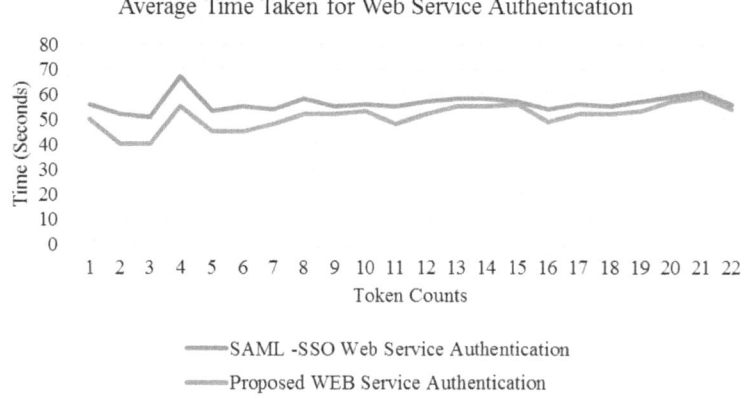

Fig. 2. Average time taken for proposed web service authentication

Table 2. Average time taken for proposed web service authentication

Number of token counts	SAML -SSO Web Service Authentication	Proposed WEB Service Authentication
1	56	50
2	52	40
3	51	40
4	67	55
5	53	45
6	55	45
7	54	48
8	58	52
9	55	52
10	56	53
11	55	48
12	57	52
13	58	55
14	58	55
15	57	56
16	54	49
17	56	52
18	55	52
19	57	53
20	59	57
21	61	59
22	56	54

Figure 3, 4, 5, 6 and 7 shows the performance for different parameters like Average Compression Time, Average decompression Time, Average blocking Time, compression ratio and overall throughput. The parametric value in Table 3 shows the fruitfulness of proposed deflate. From the results derived, cloud network privacy and data storage functionality are enhanced significantly. Authentication gets better than existing on average performance without VPN. The reason is that file server are multiple, so if response time gets an increase, traffic can be easily migrated to another file server. In the existing work, the drawback is that the validity of token is for the specific time interval and then it gets expired. To obtain new token again, the user has to verify throughout identity provider. Usually in the cloud, after login successfully by the user, he/she does not do any transaction due to this token expiry that is named as a session failure. To overcome this problem, we are validating the user throughout trusted server once user get verify successfully. This trusted server issues one session token which will be replaced every 5 min without

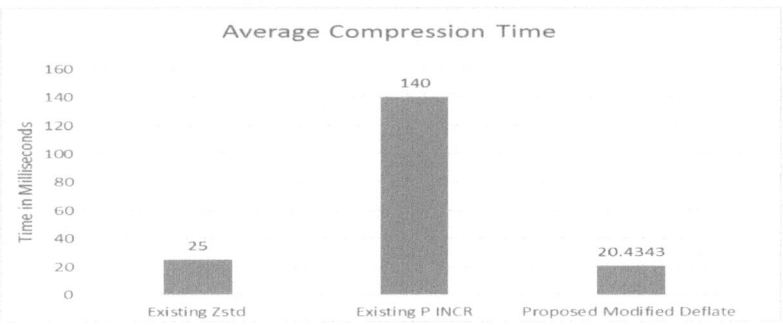

Fig. 3. Average Compression Time

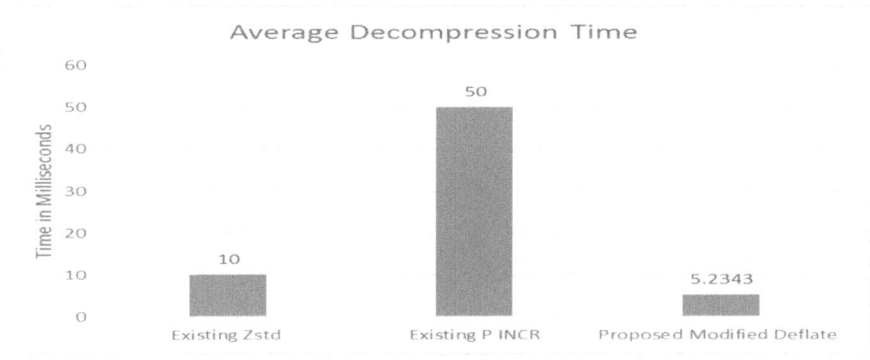

Fig. 4. Average decompression Time

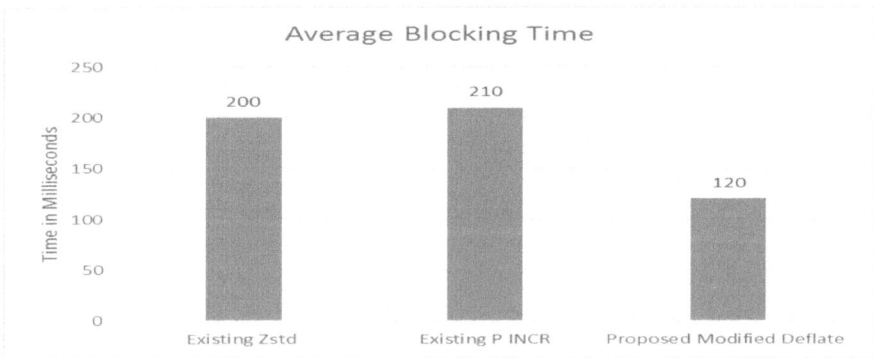

Fig. 5. Average blocking Time

any session expiry. In the existing paper, credentials get encrypted before validating of authenticity by the identity provider. But in the proposed research work, encrypting of the user credentials is done with AES and Homomorphic encryption scheme. In existing work, they get username and password and verify throughout private cloud via software

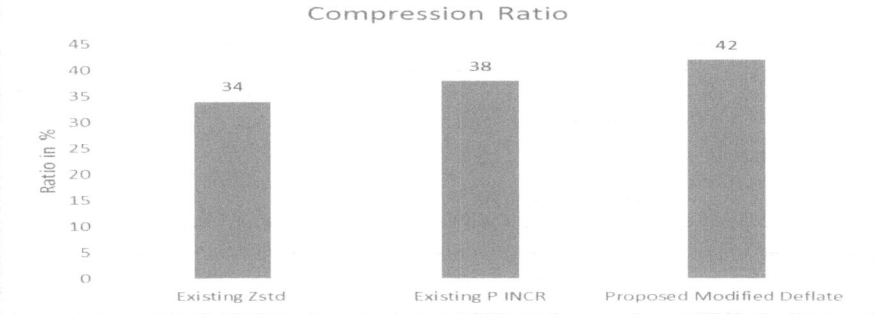

Fig. 6. Compression ratio

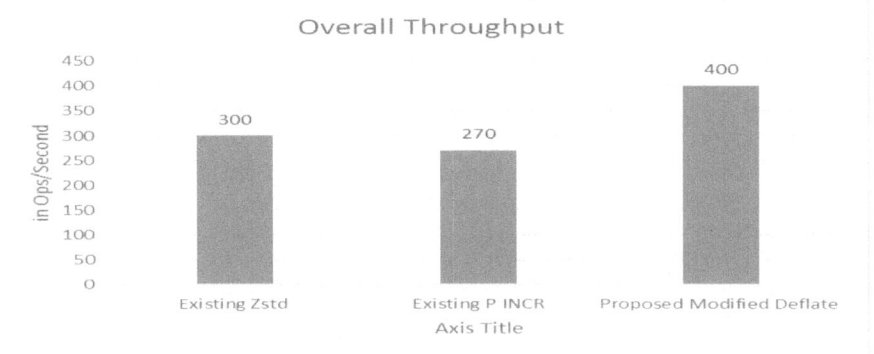

Fig. 7. Overall throughput

Table 3. Performance comparison

Parameters	Existing Zstd	Existing P INCR	Proposed Modified Deflate
Average Compression Time	25	140	20.4343
Average decompression Time	10	50	5.2343
Average blocking Time	200	210	120
compression ratio	34	38	42
Overall Throughput	300	270	400

firewall. Once user gets authenticated they get OTP in their client device. Based on entering OTP, a user got verified. Here drawback is that if we know the authenticate username and password of the legitimate user then he/she will receive OTP via message or email on their account, which we can read from their device or an email after that we enter that OTP in our client end and login successfully. In our research, whatever user gets authenticated successfully that session directly return to user machine IP address.

In our compression, we are using base 64 due to that our data cannot be loss during compression. Modified deflate algorithm is taking less time for compress big file than other algorithms and compression ratio is higher than the others.

5 Conclusion

The finishing effort is to boost the authentication and compression process in a cloud scenario. By implementing the homomorphic multiplicative mechanism as a added authentication booster, the authentication structure gets better. The authentication segment parameters is compared for different token counts and average time taken for web service authentication while compression segment parameters were average compression time, average decompression time, average blocking time, compression ratio and overall throughput. It is concluded that the recommended approach not only strengthens security but also offers the system as a whole improved data storage performance.

References

1. Sabahi, F.: Cloud computing security threats and responses. In: 2011 IEEE 3rd International Conference on Communication Software and Networks (ICCSN), pp. 245–249 (2011)
2. Fernández-Baca, D.: Allocating modules to processors in a distributed system. IEEE Trans. Softw. Eng. **15**(11), 1427–1436 (1989)
3. Tan, X., Ai, B.: The issues of cloud computing security in high-speed railway. In: 2011 International Conference on Electronic and Mechanical Engineering and Information Technology (EMEIT), vol. 8, pp. 4358–4363 (2011)
4. Guan, Q., Chiu, C.-C., Fu, S.: Cda: a cloud dependability analysis framework for characterizing system dependability in cloud computing infrastructures. In: 2012 IEEE 18th Pacific Rim International Symposium on Dependable Computing (PRDC), pp. 11–20 (2012)
5. Zhao, W., Peng, Y., Xie, F., Dai, Z.: Modeling and simulation of cloud computing: a review. In: 2012 IEEE Asia Pacific Cloud Computing Congress (APCloudCC), pp. 20–24 (2012)
6. Amanatullah, Y., Lim, C., Ipung, H.P., Juliandri, A.: Toward cloud computing reference architecture: Cloud service management perspective. In: 2013 International Conference on ICT for Smart Society (ICISS), pp. 1–4 (2013)
7. Dong, W.E., Nan, W., Xu, L.: QoS-oriented monitoring model of cloud computing resources availability. In: 2013 Fifth International Conference on Computational and Information Sciences (ICCIS), pp. 1537–1540 (2013)
8. Moharana, S.S., Ramesh, R.D., Powar, D.: Analysis of load balancers in cloud computing. Int. J. Comput. Sci. Eng. **2**(2), 101–108 (2013)
9. Ambekar, K., Kamatchi, R.: Enhanced user authentication model in cloud computing security. In: The International Symposium on Intelligent Systems Technologies and Applications, pp. 327–338 (2016)
10. Sharma, P.K., Mahajan, R., Surender, N.A.: Trust-based multi-level secure routing for authentication and authorization in WMN. Int. J. Adv. Intell. Paradigms **20**(3–4), 323–342 (2021). https://doi.org/10.1504/IJAIP.2021.119021
11. Zhang, K., Zhou, X., Chen, Y., Wang, X., Ruan, Y.: Sedic: privacy-aware data intensive computing on hybrid clouds. In: Proceedings of the 18th ACM Conference on Computer and Communications Security, pp. 515–526 (2011)

12. Tan, Z., Tang, Z., Li, R., Sallam, A., Yang, L.: Research on trust-based access control model in cloud computing. In: 2011 6th IEEE Joint International Information Technology and Artificial Intelligence Conference (ITAIC), vol. 2, pp. 339–344 (2011)
13. Ilanchezhian, J., Varadharassu, V., Ranjeeth, A., Arun, K.: To improve the current security model and efficiency in cloud computing using access control matrix. In: 2012 Third International Conference on Computing Communication & Networking Technologies (ICCCNT), pp. 1–5 (2012)
14. Tang, Y., Lee, P.P.C., Lui, J.C.S., Perlman, R.: Secure overlay cloud storage with access control and assured deletion. IEEE Trans. Depend. Secur. Comput. 9(6), 903–916 (2012)
15. Prasanalakshmi, B., Kannammal, A.: Secure credential federation for hybrid cloud environment with saml enabled multifactor authentication using biometrics. Int. J. Comput. Appl. 53(18), 13–19 (2012)
16. Wang, J.K., Jia, X.: Data security and authentication in hybrid cloud computing model. In: Global High Tech Congress on Electronics (GHTCE), pp. 117–120. IEEE (2012)
17. Moghaddam, F.F., Karimi, O., Hajivali, M.: Applying a single sign-on algorithm based on cloud computing concepts for SaaS applications. In: 2013 IEEE Malaysia International Conference on Communications (MICC), pp. 335–339 (2013)
18. Gupta, V., Gupta, M., Singla, P.: Ship detection from highly cluttered images using convolutional neural network. Wirel. Pers. Commun. 12, 287–305 (2021)
19. Wang, C., Chow, S.S.M., Wang, Q., Ren, K., Lou, W.: Privacy-preserving public auditing for secure cloud storage. IEEE Trans. Comput. 62(2), 362–375 (2013)
20. Xiao, Z., Xiao, Y.: Security and privacy in cloud computing. IEEE Commun. Surv. Tutorials 15(2), 843–859 (2013)
21. Kim, J.-M., Moon, J.-K.: Secure authentication system for hybrid cloud service in mobile communication environments. Int. J. Distrib. Sens. Networks 10, 828092 (2014)
22. Moghaddam, F.F., Moghaddam, S.G., Rouzbeh, S., Araghi, S.K., Alibeigi, N.M., Varnosfaderani, S.D.: A scalable and efficient user authentication scheme for cloud computing environments. In: 2014 IEEE Region 10 Symposium, pp. 508–513 (2014)
23. Hernandez-Ramos, J.L., Pawlowski, M.P., Jara, A.J., Skarmeta, A.F., Ladid, L.: Toward a lightweight authentication and authorization framework for smart objects. IEEE J. Sel. Areas Commun. 33(4), 690–702 (2015)
24. Lee, C.A.: Cloud federation management and beyond: requirements, relevant standards, and gaps. IEEE Cloud Comput. 3(1), 42–49 (2016)
25. Indu, I., Anand, P.M.R., Bhaskar, V.: Encrypted token based authentication with adapted security assertions mark-up language technology for cloud web services. J. Netw. Comput. Appl. (2017)
26. Deutsch, P.: Deflate compressed data format specification version 1.3. Internet Engineering Task Force (1996)
27. Ziv, J., Lempel, A.: A universal algorithm for sequential data compression. IEEE Trans. Inf. Theory 23(3), 337–343 (1977)
28. Sharma, Parveen Kumar, Mahajan, Rajiv, Surender: ID-based signcryption authentication algorithm for intra- and inter-domain handoff in wireless mesh networks. Iranian J. Sci. Technol. Trans. Electr. Eng. 44(2), 659–667 (2020). https://doi.org/10.1007/s40998-019-00258-8
29. Shanmugasundaram, S., Lourdusamy, R.: A Comparative study of text compression algorithms. Int. J. Wisdom Based Comput. 1(3), 68–76 (2011)
30. Guler, B., Ozkasap, O.: Compressed incremental check pointing for efficient replicated key-value stores. In: 9th IEEE International Workshop on Performance Evaluation of Communications in Distributed Systems and Web Based Service Architectures (2017). 978-1-5386-1629-1

31. Alam, T.: Cloud Computing and its role in the Information Technology. IAIC Trans. Sustain. Digital Innov. (ITSDI) **1**(2), 108–115 (2020)
32. Kuldeep, G., Zhang, Q.: Multi-class privacy-preserving cloud computing based on compressive sensing for IoT. J. Inf. Secur. Appl. **66**, 103139 (2022). https://doi.org/10.1016/j.jisa.2022.103139
33. Huang, S., Xie, J., Muslam, M.M.A.: A Cloud computing based deep compression framework for UHD video delivery. IEEE Trans. Cloud Comput. **11**(2), 1562–1574 (2023)

Optimizing Supply Chain with Artificial Intelligence and Analytics

Shikha Singh[✉]

School of Management, Ajeenkya DY Patil University, Pune, India
dr.shikhasingh5@gmail.com

Abstract. Artificial intelligence (intelligence based on computers) has the potential to change several commercial operations. To evaluate data, generate expectations about requests, improve coordination factors and transportation routes, and identify supply chain issues, computer-based intelligence can be used. This may result in shorter lead times, reduced costs, and improved responsiveness to popular changes. Strong enhancement capabilities, which are expected for more exact scope organization, further developed efficiency, better grades, cheaper prices, and more noticeable results, are being helped by computer-based intelligence. Additionally, it is anticipated that these upgrades will promote safer working conditions. Without a doubt, the incorporation of artificial intelligence into supply networks has made these advantages conceivable. By carefully considering and mixing, this evaluation offers wisdom nuggets. Victory is assured and the non-live stock supply chain operations are advanced with a greater grasp of these boundaries. The current exam aims to assess supporting factors such as supply chain velocity, customer care, supply chain executive engagement, and diverse management. The proposed work uses the Better Feed Forward Organization with Molecule Multitude Streamlining method to organize the presentation of the supply chain. Results show that client satisfaction, productivity, client base differentiation, and stock management are noted as advantageous factors. However, the six apparent hierarchical exhibition models are partner fulfillment, growth and learning, market execution, consumer loyalty, and financial success. Modern administrators may find the exploration valuable for enhancing the display of supply chain frameworks with artificial intelligence.

Keywords: Optimizing Supply Chain · Artificial Intelligence · Analytics

1 Introduction

Executives throughout the supply chain are being significantly impacted by simulated intelligence. The ability of computer-based intelligence to check cargo sending for a wide range of demands, including transportation, benefits supply chain management companies [1]. With the help of artificial intelligence, supply chain managers can now better understand the entire system, make smarter decisions, and provide better customer service. With the introduction of master frameworks and vague justifications, this pattern has blossomed and has reached its full potential since 2010. Artificial intelligence applications are prevalent these days, causing both excitement and fear about the future of jobs and corporate structures. While supply chain writing appears to have only sporadic attempts to incorporate the latest artificial intelligence techniques into center exams, companies are adopting simulated intelligence and investing in artificial intelligence solutions to Improving supply chain tasks from start to finish. Our research also reveals critical gaps in the field of computational intelligence research, helping supply chain professionals understand the past, present, and future potential of research on simulated intelligence applications SCM. It emphasizes the need for a useful, thorough, and concise scientific taxonomy. Our multi-strategy approach starts by developing a distinct scientific taxonomy of AI applications in order to fully analyze the point and the related research. We will be able to see the associations and examples currently present in this group of exploration by using top to bottom co-reference and organization examination. This first-of-its-kind effort to survey this writing and catch the examination patterns uses painstakingly developed artificial intelligence scientific categorization, a far-reaching bibliometric, a bunch examination, and a top to bottom evaluation of future exploration courses. These are preliminary findings, and it is likely that artificial intelligence will actually contribute to the supply chain in smaller, more foresighted findings. The field of supply chain management, which emphasizes links across many areas, core marketing, planned operations, and creation, is arguably one of the most challenging. In this way, the overall success of any firm is the outcome in SCM. But with globalization, unfavorable events like recurrent cataclysmic events, political unpredictability, and other factors, strategic approaches like lean administration and just-in-time reasoning both underway and coordinated operations are changing as expected. SCM typically needs to create a good solution for these modest issues. Artificial intelligence advancements have recently been shown to be of enormous importance to supply chain management.

Retailers, distributors, carriers, producers, and customers are all included in supply chains, which is a network of offices. As customers are the central node of every supply chain because they drive the delivery and circulation of various materials, it is vital to understand their actual usage and expectations. The need of teamwork and collaboration to satiate real curiosity is now understood by the supply chain offices. Additionally, the compounds effectively lower the entire cost of the supply chain [2]. However, without such coordinated efforts, mishmash between the real and ideal supply chain network universe occurs. Because of the several well-known and mysterious variables, the hole appears. The complexity of the enormous scope of the supply chain the board, the skill of staff supporting the supply chain the executives, the execution of the board, and motivating force frameworks to help the supply chain the executives are some of the causes of such gaps. The organizations are unable to maintain long-distance ties due to the development of e-business and the presence of innovation, which has made them more nimble and dynamic. Another cause can be that the provider overestimated the customers' true interest when providing the service. The latest AI Model processes have surpassed the traditional methods for determining and gauging the interest of one's own element.

Given that supply chains have a variety of skills, including buying, contracts, acquisition, warehousing, development, bundling, transportation or dispersion, and ultimately use. Each skill is complex, thus coordinating this wide range of capabilities involves a lot of human effort in addition to time and money. Integrating the techniques is essential for quick reactions to complex situations. Prior to the introduction of artificial intelligence and AI processes, simple decisions like the delivery of an item to a customer used to take a long time. Now, these decisions are much simpler, and the item may be delivered in 24 h or less. This paper examines several scenarios in which AI strategies are currently being used, as well as the potential applications of these techniques in the supply chain.

1.1 Benefits of Optimizing Your Workflows with Supply Chain with AI and Analytics

Because of the enormous monetary value at stake, several supply chain retailers and wholesalers have entered the competition. Organizing requests, maintaining stock control, and streamlining the dynamic edge from beginning to end are examples of the new supply chain seriousness.

Making the best decision is important because it can become necessary to keep up. In order to deal with the complexity of the current supply chain, your company should recognize these carefully thought-out arrangements that are tailored to your specific needs.

The following are just a few of the many reasons why supply chain optimization using artificial intelligence and analytics can benefit your company-

- **Accurate predictive insights:**

One of the most crucial objectives for any supply chain director is consistency. Businesses simply need to keep an eye on the industry trends, interest designs, and other factors that will help them shape the future. Enables supply chain managers to make decisions that are almost accurate and that can also provide more clarity for subsequent actions.

Artificial intelligence-enabled request estimate systems have the potential to significantly increase figure exactness. These results can also be a result of the accuracy.

- Improved confidence in the ideal stock level
- Simplifying the health stock in order to save waste and increase flexibility
- Detailed stock requirements for specific locations
- Reduced demand and supply fluctuation throughout the supply chain
- Falling capacity costs

- **Increased warehouse efficiency**

A competent distribution center work process system is essential to ensuring that stock is located where it is needed to instantly supply demand. During the warehousing stage of the supply chain, analytics can be used to enhance and support distribution center tasks.

More than 66% of businesses affirm that they require complete visibility throughout their supply chains. Continuous scientific research into the operations of distribution centers gives you insight into the status of the world, teaches you how to make the greatest use of the stockroom space at your company, and warns you about any disruptions to the trip.

- **Improves revenue, margins, and retailer satisfaction**

By offering insights into the supply and demand balance in different retailers, analytics may help with item positioning, evaluation, and improvements. Associations can use socioeconomics, consumer behavior, and other trends and examples to develop deals and marketing plans that will maintain clients for longer and generate more income.

Information analysis at the time of transactions also aids businesses in projecting demand based on factors like location, time of year, and weather. Retailers can use this information to more easily understand the interests and purchasing habits of their customers, as well as how to prepare for toppers, establish item stock necessities, reduce item reviews, and locate and effectively promote laggard products.

1.2   AI in Supply Chain

Computer-based intelligence and advanced analytics can dissect massive and altered information volumes from all jobs, including the supply chain, to improve deceivability. Modern supply chain management also employs advances in artificial intelligence to boost inventory. Eventually, applying computer-based intelligence may lessen the burden of exploring supply chain improvement possibilities, speeding up talks while retaining extraordinarily high accuracy, steadfastness, and believability [3]. After that, open data will always be accessible, allowing frameworks powered by artificial intelligence to facilitate work sharing, negotiate with suppliers, and carry out high-rate tasks in a coordinated, quick, and secure manner. For authenticating users, enabling secure access, and upholding access limitations despite the expansion of security alternatives, a supply chain access control development framework is crucial. Artificial intelligence (simulated intelligence), a key element of the 4IR (Modern Upset), frequently outperforms human performance in testing scenarios requiring real-world human perception. This characteristic is said to allow associations to quickly rethink plans and adjust to changing conditions with significant flexibility, especially in a chaotic setting.

Block chain promotes trust by providing multi-party awareness of advanced events along the supply chain. By improving the accuracy, reliability, and deceivability of information in a volatile environment, it ensures intelligent solutions. To ensure that everyone in the supply chain has access to the same information, this enhancement, however, minimizes communication or information transfer errors. As a result, it is widely used to manage and store vast amounts of information as well as to verify clients, keep an eye on information access records, and restrict client access as needed. Using blockchain innovation in supply chains still presents a number of difficulties, despite the fact that it has made great progress. Execution, supportability, and versatility help with working on quality and cost reduction through swift information checks.

Supply chains could become more resilient and self-sufficient with the aid of artificial intelligence. Artificial intelligence has helped the supply chain board increase the precision of interest projections and stock forecasts. It synchronizes business intelligence and programming tools with already-available data for better work performance and guidance. It also takes into account rising customer interest and more revenue streams than information. Additionally, human intelligence requests anticipating employing artificial intelligence (AI) computations can help businesses make precise preparations for irregular item alterations.

1.3 Need of the Study

Supply chain management can be used to increase customer satisfaction, business success, societal settings, and overall quality of life. One of the most important steps in building a successful supply chain is supplier selection. This is the process of selecting the right supplier who can provide the right goods and services to the customer at the right price, time, and quantity.

1.4 Objectives of the Study

1. Examine how AI and analytics can shorten supply chain lead times.
2. Examine how AI-powered inventory management performs in terms of enhancing supply chain operations.
3. Examine how data and AI may improve supplier collaboration and communication.
4. Consider how analytics and AI may improve the visibility and openness of the supply chain.

2 Literature Review

In their thorough study of artificial intelligence applications in supply chain management (SCM), Tsang, Y.P. and Lee, C.K.M. (2022) identified four categories: request estimating and arranging, transportation and dispersion, stock administration, and provider relationship the board [4]. The authors concluded that by reducing lead times, lowering stock costs, and fostering provider collaboration, artificial intelligence can improve the efficiency and viability of SCM.

Ramachandran, G. and Kannan, S. (2021) evaluated recent writing on machine learning and simulated intelligence applications in supply chain management and identified six classifications: item planning, quality control, coordinated operations and transportation, supplier selection and the executives, request estimation, and stock management [5]. The creators demonstrated how ML and simulated intelligence may help in reducing supply chain vulnerabilities, improving determining accuracy, and increasing responsiveness.

A SCM advancement method was put up by Zhou and Zhang (2020) in light of artificial intelligence techniques. To improve stock, creation, and board operations, they suggested using profound learning, support learning, and decision trees [6]. The designers implied that their method may reduce labor costs and improve supply chain execution.

In their meticulous written study of AI applications in SCM, Toorajipour, R., Sohrabpour, V., Nazarpour, A., Oghazi, P. and Fischl, M. (2021) distinguished five topics: request estimation, stock management, provider relationships with executives, scheduled operations and transportation, and manageability [7]. The designers believed that artificial intelligence may aid in achieving practical effectiveness, cutting expenses, and sharpening direction.

A simulated intelligence-based approach to managing supply chain risk for executives (SCRM) was put forth by Wang, D. and Yu, A. (2023). They developed a technique to distinguish and assess supply chain bets using artificial brain organizations and fuzzy reasoning [8]. The authors implied that their suggested strategy might help to lessen the impact of supply chain disruptions and enhance SCRM.

A simulated intelligence and ML-based strategy for supply chain execution optimization was put forth by Alomar, M.A. (2022). To analyze the supply chain data and identify improvement opportunities, they used predictive analytics, natural language processing, and interpersonal organization research [2]. The designers reasoned that their suggested strategy could aid in lowering costs, improving quality, and boosting customer loyalty.

A calculated system was presented by Mohsen, B.M. (2023) to investigate the impact of artificial intelligence on supply chain vulnerability [9]. Request vulnerability, supply vulnerability, and functional vulnerability were identified as the three components of SCM vulnerability. The designers implied that artificial intelligence might help to lessen vulnerability and increase supply chain flexibility.

3 Artificial Intelligence

Artificial intelligence has experienced both turn-of-events and collapse since its inception in 2012 due to a few factors. Due to the growth in information and the complexity of business circumstances over the last twenty years, there has been a renaissance in interest in and application of artificial intelligence across a few businesses. The use of simulated intelligence in a variety of business skills is being researched in light of the requirements of the present and the goals for the future. Man-made intelligence is defined as a group of PCs that may mimic human insight while determining the most effective course of action for a commercial problem. Artificial intelligence (AI) uses information to learn without human input and aids in the construction of business frameworks. Businesses can use artificial intelligence to identify weaknesses in their supply chains and allocate resources intelligently. Artificial intelligence can quickly eliminate customer bias, identify markets, leverage disappointment mechanisms, streamline internal and external supply chains, automate repetitive tasks, and empower a more creative workforce. Helps companies produce the best products. Artificial intelligence is routinely used in every industry, including assembly companies and web-based companies, to solve supply chain challenges. Large parts of the supply chain are undergoing another stress test during the coronavirus as they are tested to manage increasingly complex scenarios. Supply networks provide modern business customers with reliability and order A made-to-measure solution must be provided. The use of computerized intelligence has created a system that recognizes customer profiles and provides personalized service without compromising security or privacy. Overall, the use of computer-based intelligence is growing rapidly. Organizations and supply chains that do not recognize and apply simulated intelligence to their operations will realize fundamental supply chain diversity in robust business markets that can result from situations like coronavirus.

A branch of artificial intelligence known as "master frameworks," also known as "information-based frameworks," focuses on programming advancements that enable PCs to carry out tasks that were previously carried out by people with the aid of specific instructions and data from supply chain executives [10]. The following are the components of a specialist framework as illustrated by Kusiak (2019): Information procurement, which enables the framework to acquire information and information with the end aim of critical thinking in supply chain executives, information reorientation, where information is delineated, connection point motor, which illustrates the control procedure, and information procurement.

To get the best results from artificial intelligence in the supply chain overall, master framework steps like rule-based, fluffy, outline-based, and mixing techniques can be used in combination with one another. In domains where human intelligence might be officially regulated, master frameworks perform brilliantly, claims a research by Jakupovi et al. (2014). The effectiveness of master frameworks may drastically decline even if they aren't caught. When attempting to remedy mental inadequacies with the use of master frameworks, this issue becomes extremely clearer.

Recently, there has been an increase in interest in using artificial intelligence techniques to demonstrate and replicate complex supply chain frameworks. One can gain a deeper understanding of a system's capabilities by using artificial intelligence for exhibiting and reproduction, which enhances their ability to make wiser decisions. Expert-based processing techniques could be a useful tool for illustrating how framework components interact together and deconstructing execution in real-world scenarios for supply chain executives.

Supply chain management has gradually implemented computer-based intelligence to improve execution from a dexterous and lean perspective. In order to improve their supply chain operations, many firms are investing in computerized solutions; Fig. 1 shows the global rate of simulated intelligence reception in supply chain and assembling organizations [11]. Artificial intelligence may enable organizations to quickly respond

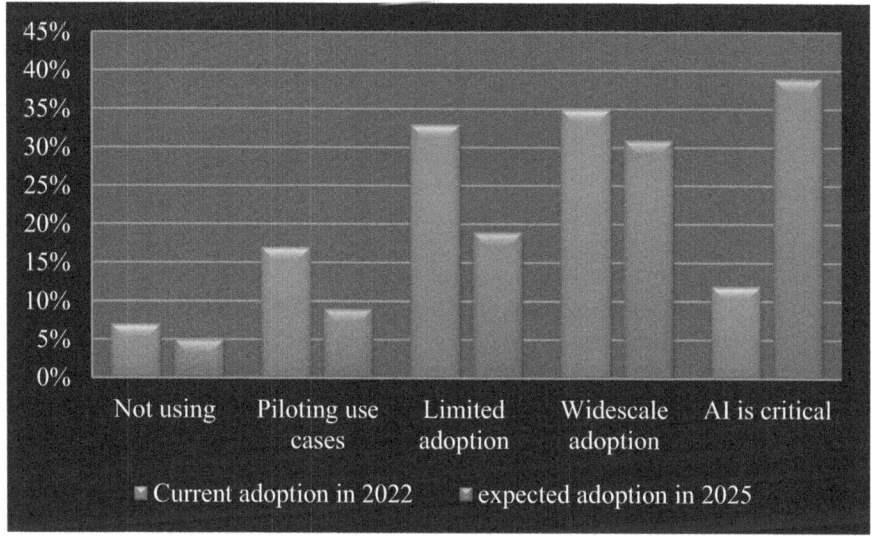

Fig. 1. Rate of global AI deployment in manufacturing and supply chains (2022 and 2025)

to changes in the marketplace, minimize waste, and increase customer and employee loyalty, according to writing.

According to Barták et al. (2010), "Artificial Intelligence for Placing and Scheduling Supply Chain Executives (Simulated Intelligence)" is the ability to make smart framework decisions while adhering to constraints (such as available resources in an assembly office). A set of techniques for unlike bookings, which are about exercising assets over a period of time, arrangements are choices made to improve the groupings in which the exercises are made. In addition to more responsive and information-driven support for framework recovery, administrators should be able to identify failures that might interfere with routine framework operations (misstatement detection, predictive maintenance, framework outages, etc.).) can now be detected and predicted. Chiefs can now identify and anticipate disturbances that could compromise standard framework operations thanks to these advances.

4 Materials and Methods

In this investigation, the four supply chain stages of sellers, creation offices, stockrooms, and appropriation focuses are taken into account. Due to the significance of the supply chain in this research project, these approaches are mentioned in the request. It should be emphasized that in benchmark tests, the Non-Straight Idleness Weight Molecule Multitude Streamlining (NLIW-PSO) method outperformed other PSO variants and the hereditary computation with great performance. The intricacies of a close examination of the three-echelon supply chain network problem, performed utilizing all PSO variations and GA, were discussed in earlier portions [12]. The NLIW-PSO was shown to outperform every other PSO type and the entire GA in the previous section. In this work, a method, Nonlinear Latency Weight Group Enhancement (NLIW-PSO), was developed to optimize a four-step supply chain design. With respect to supply chain organization, Total Supply Chain Operating Costs (TSCC) and Supply Chain Network Utility are considered the driving force.

4.1 Dataset

According to all appearances, the global auto industry is undergoing rapid changes. One of our key focus areas is improving our supply chain, along with reducing costs, improving cycles and increasing convenient throughput. The overall achievement metric that should be reached to acquire a dominant result in all of the previously listed need regions is having an effective recognizability framework set up. In the auto industry, the term "detect ability" refers to a system that keeps track of the individual parts that went into the construction of a given object and their provenance. It has been observed that firms, despite investing significant resources in the goal of expanding their supply chain and laying out detectability, actually have limited perception of and understanding of their supply chain at a given time. [As an example] This study hopes to eliminate failures present in conventional auto supply chains through the application of block chain innovation. Expanding straightforwardness, streamlining procedures, and decentralizing the organization can all be used to address these problems. A wide range of significant value-adding features will also be made available to partners and organizations as a result of this study, ensuring that exchanges go more promptly and smoothly. The study focuses on the potential advantages of incorporating block chain technology into the automotive supply chain. With the help of Hyper record Author the foundation for the above network is created.

4.2 Techniques for Tuning Input Data

This target ability has four components (Conditions 4.1 through 4.4). Note that the first category, called Vendor Generic Material Cost (TSMC), contains the average raw material cost of all suppliers for the build unit. The total transportation cost (TTC) shown in the second part of the table includes the total transportation cost from each of the 139 suppliers to the factory and the total transportation cost from the factory to the warehouse [13]. In manufacturing, TWC stands for out Stockroom Cost. This includes the cost of maintaining inventory in distribution centers and the cost of shipping finished goods from warehouses to specific and appropriate locations to meet requirements. The absolute assembly cost produced by the factory is called TMC. Total Assembly Cost (TMC) and Total Labor Cost (TWC) are two components of Total Assembly Cost (TMC).

The next covered capability, Whole Supply Chain Network Benefit (PROF), is divided into five sections (Conditions 4.1 to 4.4 and 4.7) and duplicated (Condition 4.9) and registered. First, the revenue from warehouse management center transactions, which account for half of the total, is targeted. The provider's total material costs are disclosed in the second section (TSMC) of the report. The third factor concerns the total transportation cost (TTC) from all suppliers to factories and the total TTC from factories to distribution centers. The fourth section looks at the total manufacturing cost (TMC) incurred in the assembly office. The fifth factor, the Absolute Cost of Inventory (TWC), includes both the cost of maintaining inventory and the cost of moving finished goods from warehouses to distribution centers to meet demand. The cost of transferring finished goods from a warehouse to a fulfillment center plus the cost of moving inventory between warehouses are combined to create the Total Warehouse Cost (TWC) used to fulfill the order. TWC is determined as a percentage of the total cost of the distribution equipment.

(a) Total supplier materials cost

$$\mathbf{TSMC} = \sum_{c}\sum_{v}\sum_{p}\left(\mathbf{CS_{c,v}} \times \mathbf{X_{c,vp}}\right). \tag{1}$$

(b) Total transportation cost

$$\mathbf{TTC} = \sum_{c}\sum_{v}\sum_{p}\left(\mathbf{X_{c,vp}} \times \mathbf{STC_{c,vp}}\right) + \sum_{p}\sum_{w}\mathbf{Y_{p,w}} \times \mathbf{PTC_{p,w}}. \tag{2}$$

(c) Total manufacturing cost

$$\mathbf{TMC} = \sum_{p}\left\{\left(\mathbf{MC_p}\right) \times \left(\sum_{w}\mathbf{Y_{p,w}}\right)\right\} \times \left\{\sum_{P}\left(\mathbf{IC_p} \times \sum_{c}\mathbf{I_{c,p}}\right)\right\}. \tag{3}$$

(d) Total warehouse cost

$$\mathbf{TWC} = \sum_{w}\sum_{d}(\mathbf{Z_{w,d}} \times \mathbf{WTC_{w,d}}) + \sum_{w}(\mathbf{WIC_w} \times \mathbf{I_w}). \qquad (4)$$

(e) Total supply chain operating cost

$$\mathbf{TSCC} = \mathbf{TSMC} + \mathbf{TTC} + \mathbf{TMC} + \mathbf{TWC}. \qquad (5)$$

(f) Profit of SCN

$$\mathbf{PROF} = \sum_{d}(\mathbf{D_d} \times \mathbf{SP_d}) = \mathbf{TSCC}. \qquad (6)$$

(g) Revenue generated by SCN

$$\mathbf{REV} = \sum_{d}(\mathbf{D_d} \times \mathbf{SP_d}), \qquad (7)$$

$$\mathbf{TSCC} = \mathbf{TSMC} + \mathbf{TTC} + \mathbf{TSK} - \mathbf{TWC}, \qquad (8)$$

$$\mathbf{MinTSCC} = \left[\sum_{v}\sum_{v}\sum_{\eta}(\mathbf{CSS_{+\infty}} \times \mathbf{X_{car}}) \right]$$

$$+ \left[\sum_{z}\left\{ \left(\mathbf{MC}, \mathbf{x}\left(\sum \mathbf{Y_{me}}\right)\right)\right\} + \left\{ \overset{n}{\sum} - \left(\mathbf{x}, \sum \mathbf{I_{ce}}\right)\right\} \right] \qquad (9)$$

$$+ \left[\sum_{a}\sum_{a}(\mathbf{Z_a} \times \mathbf{WTC_{a,\lambda}}) + \sum_{v}(\mathbf{WC_a} \times \mathbf{I_a}) \right],$$

$$\mathbf{PROF} = \sum_{a}(\mathbf{D_4} = \mathbf{SP_4}) - \sum_{V}\mathbf{X_{one}} \leq \mathbf{L_{loper}V_{C_v}}, \qquad (10)$$

$$\sum_{P}\mathbf{Y_{nw}} \leq \mathbf{U_p vw}, \qquad (11)$$

$$\sum_{v}\mathbf{Z_n} \leq \mathbf{V_v v_d}, \qquad (12)$$

$$\sum_{v}\mathbf{Z_x} = \mathbf{D_d vd}. \qquad (13)$$

4.3 Feeding-Forward Neural Network

The organization learning boundary is established initially in the suggested computation. Here, the cost capability requirements (5.7) and (5.8) are used to resolve the weight disparity between the yield layer and secret layer. Individually.

Step 1.—Set the parameter's initial settings to some random values.

Step 2.—Put a preset value based on the sigmoid curve as the threshold value.
Step 3.—Apply Eq. (5.2) to the linear output calculation.
Step 4.—Use the sigmoid function to calculate the nonlinear output as in Eq. (5.3).
Step 5.—Calculate the weight change for the output layer using Eq. (5.7).
Step 6.—Calculate the weight change for the buried layer using Eq. (5.8).
Step 7.—Calculate the mean square error.
Step 8.—The preceding steps 3 through 7 are repeated if the mean square error value is greater than the threshold value.
Step 9.—If the mean square error value is less than the threshold value, declare that the network has been trained.

The suggested network design used in this research is shown in Fig. 2.

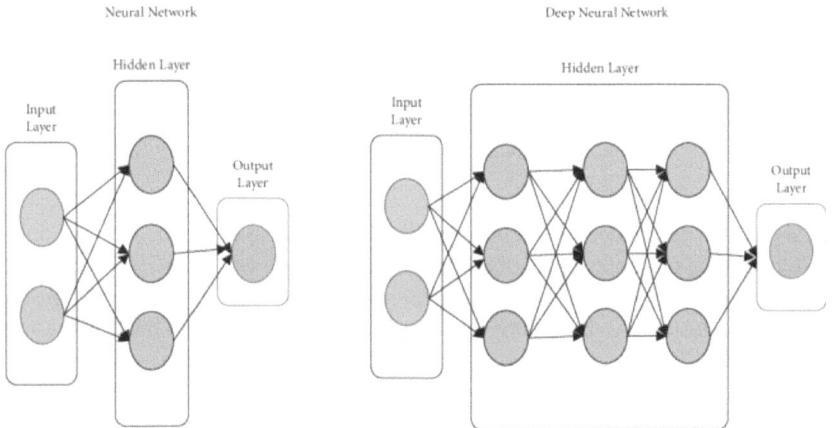

Fig. 2. Suggested network structure

5 Results and Discussion

The customized MATLAB replica was used to implement the optimal numerical model presented in this segment. The optimum supply chain information strategy in PMU was then demonstrated using this model, as seen in Fig. 3. This method screens the sign phasor across all of the PMUs that were dispersed across the matrix using the second-request Kalman channel, also referred to as the SOKF [14]. The recreation work for the suggested technique includes pushing the strategy through some substantial suffering in a range of different conditions to demonstrate the strength of the model that is being considered. This method brings the number of stage mistakes down to a manageable level. Remember that the figures below were derived from the correlation between the estimated stage in radians and the total number of tests.

The results for Supply Chain Information, as indicated by the blue line, perform better than Supply Chain Information and are close to the reference stage, demonstrating the value of Supply Chain Information for the Kalman channel. Standard stage According to

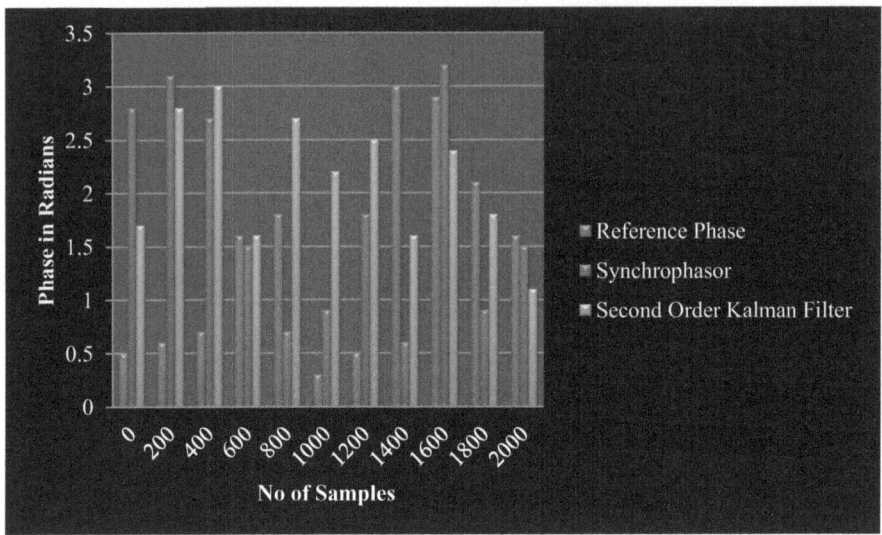

Fig. 3. Performance indicators

Fig. 2 above, the recommended method for PMU was determined to be within Gaussian noise. Comparing representations of the proposed approach with SOKF, which has been shown to have the ability to more easily screen the duration of frames in the presence of Gaussian perturbations using MSE, we find that the recommended model is shown to work very well [15]. Perturbation Evidence for this is that the provided method can easily screen the scaffold duration in the presence of Gaussian noise. Note that this increase is a staggering 21.45% increase under Gaussian noise when measured at the 5 dB rate.

However, the reproduction's outcomes were considered and are depicted in Figs. 4 and 5 when non-Gaussian commotion was present. The ability of the PMU's Supply chain information method to outsmart the Synchrophasor strategy in the face of extreme environmental conditions shows the adaptability and viability of the ideal strategy SOKF. This proves that the PMU created the optimum SOKF technique. The MSE study showed that the SOKF approach performed significantly better than previous models in the presence of non-Gaussian noise. However, this is not how the current model suggests. Therefore, we need SOKF rather than non-Gaussian perturbation to get the desired result. It is clear that adding non-Gaussian noise increases the rate centroid by 16.2 to 5 dB.

In essence, the four phase multi echelon supply chain network engineering's second arrangement of targets—TSCC and GMROI—was subjected to the multi objective assessment [7]. An exhibition inspection was carried out for each of the four help levels—half, 84%, 97%, and 99.9%—because the targets had different weights but were nevertheless equally important. The results are described in Table 1.

Table 1 shows the dynamic components associated with different assistance levels. Equal weights are assigned to each component. Thanks to the results of the multiobjective execution analysis presented earlier, we can see that his TSCC in SCN increases as the management level increases [16]. However, the GMROI decreases as the dosage increases. This is because every logistics facility requires security stock with varying

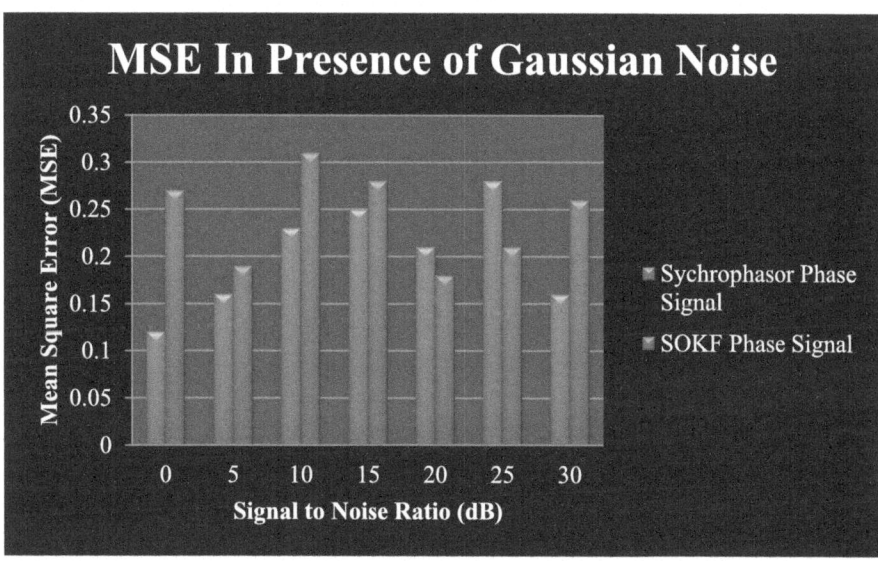

Fig. 4. MSE when a classifier is present

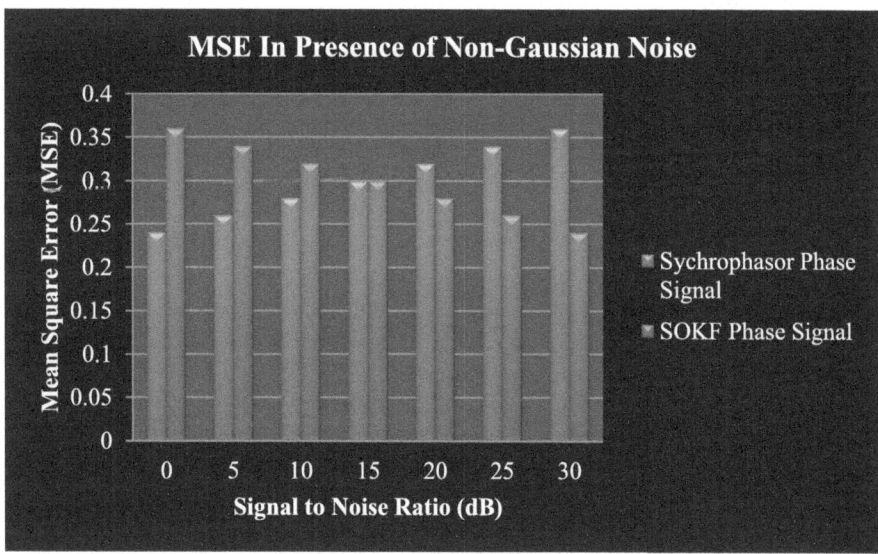

Fig. 5. MSE when non-Gaussian noise is present

amounts depending on the assistance needed. In order to meet their needs (level of management to meet customer needs) and achieve the objectives of his multi-level four-phase SCN plan, the supervisor can choose an approach that is justified in the light of his experience.

Table 1. Performance indicators for the suggested work

S. No	Weight for w1	Objectives w2	Best fitness 1	Best fitness 2	Overall objective	Objective 1 TMC/TSCC	Objective 2
1	0.1	0.9	0.985374	0.937603	0.94257	1433616	0.82232
2	0.2	0.8	0.3832778	0.942262	0.944425	1523720	0.789293
3	0.3	0.7	0.998484	0.936462	0.950268	1445483	0.822852
4	0.4	0.6	0.780337	0.953728	0.955884	1406253	0.837574
5	0.5	0.5	0.783042	0.952283	0.959832	1422026	0.834342
6	0.6	0.4	0.900076	0.832507	0.939502	1300202	0.882892
7	0.7	0.3	0.942238	0.98358	0.940289	1357783	0.84245
8	0.8	0.2	0.924707	0.808023	0.923342	1332623	0.874678
9	0.9	0.1	0.873898	0.831846	0.688693	1266372	0.874556

6 Recommendations and Conclusion

The use of computer based intelligence in various business fields has advanced thanks to technological advancements in fields like portable registration, artificial brain organizations, mechanical technology, stockpiling of enormous data on the web, cloud-based AI, and data managing calculations. Above all, improvements in artificial intelligence have assisted them in eliminating a variety of human tasks, including supply chain, combinations, and developments. Savvy Assembling is also currently a reality [17]. The pattern of computer-driven modern activity is significantly growing, which suggests that man-made intelligence is either already needed by some businesses throughout the world or is on the verge of becoming one. The violent opposition and contention among associations are growing as a result of the global interface and the changing times. Outstanding mechanical progress is being made, and businesses are racing to enter the growth and income era. We can observe that mechanical technology has been adopted in many domains to handle challenging tasks. In order to balance human and machine labor, businesses are currently embracing mechanization across all industries. However, as simulated intelligence develops, it will become more sophisticated than it is now. This new development will strengthen the collaboration between human and artificial intelligence to a significantly higher level, where it could have a historic impact on the supply chain as well as other important areas.

We recommend that the future examination plan might solve the adaptability and maintainability problems with block chains as well as the ambiguity and sporadic streamlining problems with artificial intelligence [6]. Customization as change and personalization as a result of these technological advancements can thus significantly reduce disruption in the supply chain.

References

1. Abbas, K., Afaq, M., Ahmed Khan, T. and Song, W.-C.: A block chain and machine learning-based drug supply chain management and recommendation system for smart pharmaceutical industry. Electronics **9**(5), 852 (2020). https://doi.org/10.3390/electronics9050852
2. Alomar, M.A.: Performance optimization of industrial supply chain using artificial intelligence. Comput. Intell. Neurosci. **2022**, 1 (2022). https://doi.org/10.1155/2022/9306265
3. Al-Zahrani, A., Marghalani, A.: How artificial intelligent transform business. SSRN Electron. J. (2018). https://doi.org/10.2139/ssrn.3226264
4. Tsang, Y.P., Lee, C.K.M.: Artificial intelligence in industrial design: a semi-automated literature survey. Eng. Appl. Artif. Intell. **112**, 104884 (2022). https://doi.org/10.1016/j.engappai.2022.104884
5. Ramachandran, G., Kannan, S.: Artificial intelligence and deep learning applications: a review. J. Artif. Intell. Mach. Learn. Neural Netw. (12), 10–13 (2021). https://doi.org/10.55529/jaimlnn.12.10.13
6. Ahmad, S., Kamruzzaman, M.: Optimization in apparel supply chain using artificial neural network. Eur. J. Logist. Purchasing Supply Chain Manag. **10**(1), 1–14 (2022). https://doi.org/10.37745/ejlpscm.2013/vol10no1pp.1-14
7. Toorajipour, R., Sohrabpour, V., Nazarpour, A., Oghazi, P., Fischl, M.:. Artificial intelligence in supply chain management: a systematic literature review. J. Bus. Res. **122**(1), 502–517 (2021). https://doi.org/10.1016/j.jbusres.2020.09.009
8. Wang, D., Yu, A.: Supply chain resources and economic security based on artificial intelligence and blockchain multi-channel technology. Int. J. Inf. Technol. Syst. Approach **16**(3), 1–15 (2023). https://doi.org/10.4018/ijitsa.322385
9. Mohsen, B.M.: Impact of artificial intelligence on supply chain management performance. J. Serv. Sci. Manag. **16**(01), 44–58 (2023). https://doi.org/10.4236/jssm.2023.161004
10. Mayo, R.C., Leung, J.: Artificial intelligence and deep learning – radiology's next frontier? Clin. Imaging **49**, 87–88 (2018). https://doi.org/10.1016/j.clinimag.2017.11.007
11. Kamble, S.S., Gunasekaran, A., Kumar, V., Belhadi, A., Foropon, C.: A machine learning based approach for predicting blockchain adoption in supply Chain. Technol. Forecast. Social Change, 120465 (2020). https://doi.org/10.1016/j.techfore.2020.120465
12. Hennet, J.-C., Arda, Y.: Supply chain coordination: a game-theory approach. Eng. Appl. Artif. Intell. **21**(3), 399–405 (2008). https://doi.org/10.1016/j.engappai.2007.10.003
13. Dželihodžić, A., Đonko, D.: Comparison of ensemble classification techniques and single classifiers performance for customer credit assessment. Model. Artif. Intell. (3), 140–150 (2016). https://doi.org/10.13187/mai.2016.11.140
14. Serbaya, S.H., Abualsauod, E.H., Basingab, M.S., Bukhari, H., Rizwan, A., Mehmood, M.S.: Structure and performance attributes optimization and ranking of gamma irradiated polymer hybrids for industrial application. Polymers **14**(1), 47 (2021). https://doi.org/10.3390/polym14010047
15. Taylan, O., Rizwan, A., Parsaei, H.: Optimization of engineering student learning and assessment by cognitive methods. Eng. Educ. Lett. **2017**(1), 2 (2017). https://doi.org/10.5339/eel.2017.2
16. Wang, M., Wu, Y., Chen, B., Evans, M.: Blockchain and supply chain management: a new paradigm for supply chain integration and collaboration. Oper. Supply Chain Manag. Int. J. **14**(1), 111–122 (2020). https://doi.org/10.31387/oscm0440290
17. Travé-Massuyès, L.: Bridging control and artificial intelligence theories for diagnosis: a survey. Eng. Appl. Artif. Intell. **27**, 1–16 (2014). https://doi.org/10.1016/j.engappai.2013.09.018

Prediction on Real Estate House Price Using Regression Models

S. Iniyan[1(✉)] and Siddhartha Gaba[2]

[1] Department of Computing Technologies, School of Computing, SRM Institute of Science and Technology, Kattankulathur, Chengalpattu 603203, India
iniyans@srmist.edu.in

[2] Department of Computer Science and Engineering, SRM Institute of Science and Technology, Kattankulathur, Chengalpattu 603203, India

Abstract. Machine learning has had a significant impact on image recognition, product recommendation, clinical diagnosis, and nearly every other field of technology. Every aspect of our life is surrounded by machine learning algorithms that help to improve security, public safety, medicine, transportation, and so on. Because of the growth in urbanization and the volatility of property values, there is a greater demand for a system that predicts property prices. Real estate price prediction can assist in resolving this issue and forecasting house prices so that customers can examine them. This paper reflects the effort made to solve the aforementioned challenge. This study uses machine learning algorithms as a research technique to create models for predicting house prices. We developed a combined feature selection based model for predicting housing costs, look at the precision of the models including linear regression, lasso regression, random forest and decision tree. Such models are used to create a predictive model and to select the best performing model by comparing the prediction errors derived from various models, and the analysis shows that the linear regression algorithm consistently outperforms other methods in terms of accuracy, which comes at 84.78. The authors attempt to construct a user-friendly interface design that will allow consumers to select from their options based on their needs and receive an estimated pricing for the property.

Keywords: Machine learning · real estate · regression · linear regression · lasso · decision tree regression · random forest algorithm · price prediction

1 Introduction

In any developing country or city, real estate has always been in great demand. Housing is one of the most fundamental human requirements, yet land property is more than just a need in today's society; it also signifies an individual's position and reputation. An investment in real estate seems advantageous because its value does not decrease fast. It makes sense to invest in the real estate industry. Therefore, knowing the value of land may be a big financial gain.

© The Author(s), under exclusive license to Springer Nature Switzerland AG 2025
A. Gupta et al. (Eds.): ICAIA 2023, CCIS 2308, pp. 274–285, 2025.
https://doi.org/10.1007/978-3-031-84394-5_19

In reality, purchasing a home is a stressful experience. One must pay large quantities of money and invest many hours, and there is always the question of whether it is a good bargain or not. The majority of properties are purchased through realtors. Because there exists quite a large amount of legal labor and paper-work in this process, buyers rarely buy directly from the seller. As a result, real estate agents act as a middleman, allowing them to take advantage of the consumer and raise the price of the property to whatever they want. As a result, the houses are overvalued, and it becomes vital for the consumer to have a better idea of the true value of the properties.

In this research paper, we attempted to use some of the most often used algorithms and assess their efficiency in estimating the price of properties in the city of Bengaluru, India. We attempted to establish which algorithm outperforms the others when conducting the procedure of predicting the price of a property with the parameters provided.

2 Literature Review

According to Ravikumar [2], We are in a real need for a proper prediction on properties and housing in the real-estate market that runs throughout the buying and selling of properties. That is why, [3] in recent times, predicting the value of a property has become an important area. It is a challenge for researchers to identify all of the minor features that might affect property investment and develop a prediction model that takes all of the features into account. The real estate market's coordinated expansion under the influence of digitalization was researched by [4]. In order to anticipate the real estate price, [5] looked at the relationship between supply and demand, which is the real estate price generated under the impact of social, economic, and material elements. [6] implies that every firm in today's market is working extremely hard to get a viable advantage on other businesses or their competitors. There exists a demand to make the procedure simpler for the average person while offering the significantly good outcomes. [7] proposed utilising machine learning to create an algorithm capable of predicting house values built on specific parameters The business use-case was that organisations can straightforwardly use it to estimate the costs of new properties which will be compiled by gathering several input features and determining an appropriate and fair pricing. [8] made use of the Google Colab/Jupiter IDE. Jupiter IDE is a free and open-source online application that allows us to exchange and create documents. It includes tools for data cleansing, data translation, numerical value simulation, statistical modelling, data visualisation, and machine learning. [9] built a system that will assist consumers in determining the approximate real estate cost The costs of the requested attributes will be provided to the user when they submit their specifications.

[10] uses machine learning techniques to gather information on the sales prices of various homes and plots of land. When it was built, how it was built, the materials used, and other factors all affect how much it costs. The model was created using supply-side regression approaches, Lasso regression, and SVM algorithms. According to the author, using a larger dataset will ensure that the prediction is as accurate as possible. In [11] Support Vector Machine, Least-Square SVM, and Partial-Least Squares algorithms and the related characteristics are used to examine and estimating the worth of a certain property in a particular region of Boston city. [12] estimates housing prices using the

Support vector regressor. The author finds a tradeoff between model complexity and performance. As a result, they developed a supervised machine learning model for forecasting home values. Following the removal of null samples from the initial dataset, 400 vaues are handled as training data and the rest 52 are treated as test data. The paper [13] suggested a complex house prediction system based on linear regression. They employed Linear Regression on the dataset, it produced excellent precision. Two modules—Admin and User—were used in the project to anticipate home prices. Admin has the ability to add and view locations. Admin is capable of increasing the density on the basis of per-unit-area. On the other hand, Users might examine the area location and the estimated property cost for that location. [14] creates a user-friendly interface that users can utilise while modelling in accordance with their requirements. In Bengaluru, data from the preceding few years' worth of data were used by the author. A custom-built ML model was trained and processed using the datasets, and the results are accurate.

In [15] Multiple Linear Regression is implemented, which predicts using more than one feature. This article pertains to the most updated Forecast on Research projections taking new developments into account to cope with their economics in a better way. In [16, 17] Ridge regression and Lasso Regressions are used in which Ridge standardizes the regression coefficient by establishing a size interest. It is also similar to Lasso regression with a little difference. Lasso regression is used as a model due to its practical and randomized model selection techniques [18].

3 Methodology

The Methodology is a description of the framework that is being implemented. It includes a number of milestones that must be met in order to reach the goal. The figure below, Fig. 1, depicts step-by-step tasks that must be completed.

3.1 Data Gathering and Analysis

The dataset used here in this analysis includes the prices and attributes of residential properties in Bangalore, India. This dataset contains over 13321 entries and 9 columns of house price data gathered from a variety of reliable sources. There are eight independent variables and one target variable, which is the property price. It has the following features: area_type, availability, location, size, society, total_sqft, bath, and balcony (Fig. 2).

The first stage is data collection, the dataset utilised in this example is from Kaggle. This is what the machine learning model will be trained on. The dataset gathered at this point is unstructured and raw. The costs are given in Indian Rupee, and the plot size is specified in sq. ft. The dependent variable in the dataset is the price column, while the other columns are independent variables (also called features).

3.2 Data Pre-processing

Data pre-processing is the process of cleansing our data collection. The raw dataset is organised in this stage so that a machine learning model may be trained using it. The dataset may contain missing values or the characteristic values may be skewed. These

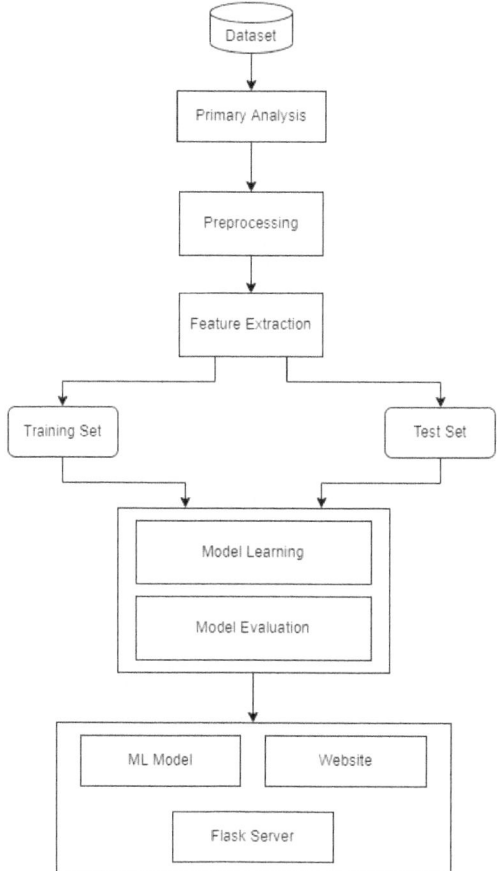

Fig. 1. Project Methodology

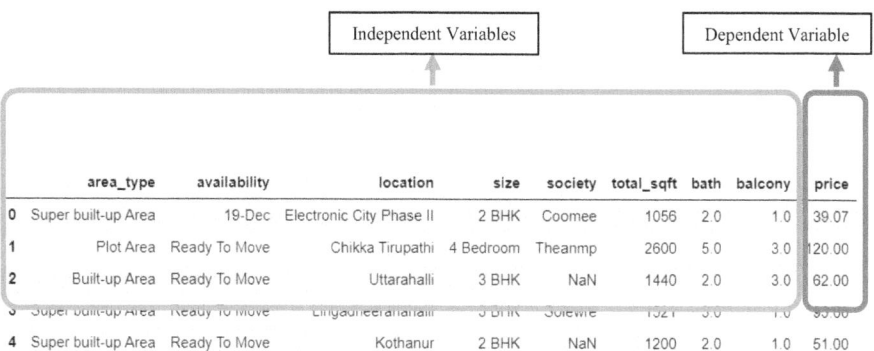

Fig. 2. Dataset

problems can be resolved via data cleaning. If a variable has null values (NaN), we will either delete them from our dataset or replace them with the average value. If there are a lot of missing values, we will also need to remove the variable altogether. Due to their high number of missing values or the fact that they weren't crucial for estimating the price of a property, certain columns were eliminated from the dataset. A new feature was created, and outliers were eliminated using several ways. For categorical data, one Hot Encoding was used.

3.3 Combine Feature Selection

The proposed work gives major attention to the ensemble approach in feature selection process, which combines the K best features of the many feature selection strategies used in the use of deep learning models to anticipate crop production. ANNOVA F-statistic Feature Selection (AFFS), Mutual Information Feature Selection (MIFS) and Recursive Feature Selection (RFS) were taking part in the Combined Feature Selection (CFS) as shown in the Fig. 3, produced better prediction results applying with machine learning models.

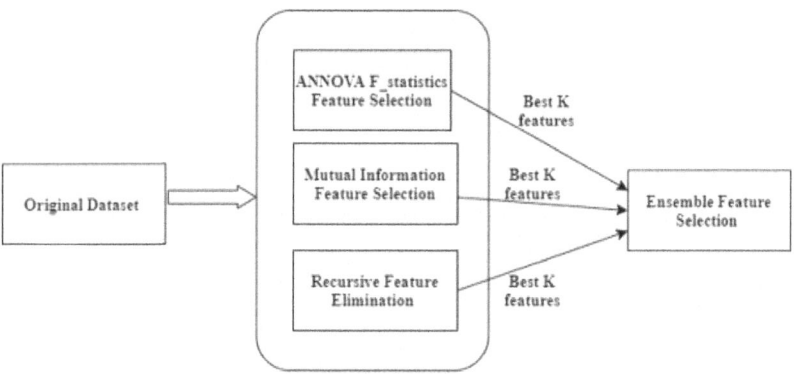

Fig. 3. Combined Feature Selection

3.4 Developing and Training the Model

The dataset we had was split into two sections: training and testing, with training using eighty percent of the data and testing using the remaining twenty percent.

For such predictions, regression is the best option. Regression is a machine learning approach that enables you to produce predictions by taking into account - from recent measurable data - the relationships between your objective parameter and numerous independent variables. To get the most accurate results possible for the suggested model, several regression techniques such as Multivariable Linear Regression, Decision Tree Regression, Lasso Regression, Randon Forest Regression were compared:

Multivariable Linear Regression Model (MLR)

An advanced form of Linear Regression, also known as Multiple Linear Regression, employs more than one independent variable to identify the dependent variable, as opposed to linear regression, which uses one independent variable to predict one dependent variable. The independent variables in MLR are not significantly connected with one another. MLR is widely utilised in econometrics and financial analysis.

Lasso Regression

This model makes use of shrinkage. The process of shrinkage involves bringing data values closer to the mean, a midpoint. Basic, sparse models are encouraged by the lasso method (i.e. models with fewer parameters). When you want to automate certain steps in the model selection process, such as variable selection and parameter removal, or when your models have a high degree of multicollinearity, this kind of regression is perfect.

Decision Tree Regression Model (DTR)

It builds regression and classification models using a tree-like structure, as the name implies. It gradually divides a dataset into smaller sections at the same time, creating a decision tree. The final result is a tree with leaf nodes and decision nodes. Each of a decision node's two or more branches corresponds to a different value for the property being thought about. A numerical target selection is reflected in a leaf node. The root node is the decision node at the top of the tree. A decision tree can handle categorical and numerical data of both types.

Random Forest Regression Model (RFR)

It is a Supervised Learning algorithm that does classification and regression. A bagging strategy, not a boosting one. There are parallel trees in random forests. There is no interaction between the trees while they are being created. It builds a lot of decision trees during training and outputs the class that represents the mean prediction (regression) of all the individual trees or the mode of the classes (classification).

4   Results and Discussion

Machine learning is used to create a model using the obtained data. To find the optimum procedure and model parameters, we will mostly employ GridSearchCV and K-fold Cross Validation.

Model training and testing are done appropriately using K-fold Cross-Validation. It uses resampling to assess ML models on a tiny sample of data.

Using K-Fold Cross Validation, the accuracy values for the various models are shown in the Figs. 4, 5, 6 and 7.

4.1   Testing and Comparison

We'll use the Root-Mean-Square Error (RMSE) approach to compare the models.

It is described as the prediction error standard deviation. The distance between the data points and the regression line is measured by prediction errors (residuals); RMSE

```
from sklearn.linear_model import LinearRegression
from sklearn.model_selection import ShuffleSplit
from sklearn.model_selection import cross_val_score

cv = ShuffleSplit(n_splits=5, test_size=0.2, random_state=0)

res = cross_val_score(LinearRegression(), X, y, cv=cv)
print("MLR result:", res)

MLR result: [0.82702546 0.86027005 0.85322178 0.8436466  0.85481502]
```

Fig. 4. K-Fold Validation for Multivariable Linear Regression

```
from sklearn.linear_model import Lasso
from sklearn.model_selection import ShuffleSplit
from sklearn.model_selection import cross_val_score

cv = ShuffleSplit(n_splits=5, test_size=0.2, random_state=0)

res = cross_val_score(Lasso(), X, y, cv=cv)
print("Lasso result:", res)

Lasso result: [0.6982002  0.73053589 0.72642204 0.7586498  0.71603286]
```

Fig. 5. K-Fold Validation for Lasso Regression

```
from sklearn.tree import DecisionTreeRegressor
from sklearn.model_selection import ShuffleSplit
from sklearn.model_selection import cross_val_score

cv = ShuffleSplit(n_splits=5, test_size=0.2, random_state=0)

res = cross_val_score(DecisionTreeRegressor(), X, y, cv=cv)
print("DTR result:", res)

DTR result: [0.82134697 0.75016687 0.5579375  0.41790934 0.77005506]
```

Fig. 6. K-Fold Validation for Decision Tree Regression

also measures how much area outside of the line these residuals words. In plainer terms, it demonstrates how tightly the data is clustered around the line of best fit. To validate experimental data, root mean square error is often employed in climatology, forecasting, and regression analysis.

$$RMSE = \sqrt{[\Sigma (P_i - O_i)^2 / n]} \tag{1}$$

where,

- P_i is the predicted value for i^{th} value
- O_i is the actual value for i^{th} value

```
from sklearn.ensemble import RandomForestRegressor
from sklearn.model_selection import ShuffleSplit
from sklearn.model_selection import cross_val_score

cv = ShuffleSplit(n_splits=5, test_size=0.2, random_state=0)

res = cross_val_score(RandomForestRegressor(), X, y, cv=cv)
print("RFR result:", res)

RFR result: [0.84020325 0.84106997 0.76792344 0.62372172 0.83971594]
```

Fig. 7. K-Fold Validation for Random Forest Regression

- N is the size of the sample

We have the following errors for different models using the Root Mean Square Error are shown in the following Figs. 8, 9, 10 and 11.

```
mlr=LinearRegression()
mlr.fit(X_train, y_train)
from sklearn.metrics import mean_squared_error as mse
err = (np.sqrt(mse(y_test, mlr.predict(X_test))))
print("RMSE:", err)

RMSE: 26.665640548312062
```

Fig. 8. RMSE for Multivariable Linear Regression

```
mlr=Lasso()
mlr.fit(X_train, y_train)
from sklearn.metrics import mean_squared_error as mse
err = (np.sqrt(mse(y_test, mlr.predict(X_test))))
print("RMSE:", err)

RMSE: 38.21281781662464
```

Fig. 9. RMSE for Lasso Regression

4.2 Evaluating the Model

To assess the model's performance, a scatter plot is generated that compares the actual prices of properties listed in the dataset to the prices predicted by the model.

```
mlr=DecisionTreeRegressor()
mlr.fit(X_train, y_train)
from sklearn.metrics import mean_squared_error as mse
err = (np.sqrt(mse(y_test, mlr.predict(X_test))))
print("RMSE:", err)
```

RMSE: 38.6517096665062

Fig. 10. RMSE for Decision Tree Regression

```
mlr=RandomForestRegressor()
mlr.fit(X_train, y_train)
from sklearn.metrics import mean_squared_error as mse
err = (np.sqrt(mse(y_test, mlr.predict(X_test))))
print("RMSE:", err)
```

RMSE: 33.84740509359653

Fig. 11. RMSE for Random Forest Regression

The following chart shows that for some datapoints, the true price is quite near to the projected price, indicating that the model is very accurate for those datapoints. However, the chart also reveals that for certain data points, the discrepancy between the real price and the predicted price is substantial, indicating that the result is less reliable. Overall, we may claim that the model is rather accurate.The Table 1 shows the accuracy and error of different models.

Table 1. Error and Accuracy of different models

Model	RMSE	Accuracy Using Cross Validation
MLR	26.66	[0.82702546, 0.86027005, 0.85322178, 0.8436466, 0.85481502]
Lasso	38.21	[0.6982002, 0.73053589, 0.72642204, 0.7586498, 0.71603286]

(*continued*)

Table 1. (*continued*)

Model	RMSE	Accuracy Using Cross Validation
DTR	38.65	[0.82134697, 0.75016687, 0.5579375, 0.41790934, 0.77005506]
RFR	33.85	[0.84020325, 0.84106997, 0.76792344, 0.62372172, 0.83971594]

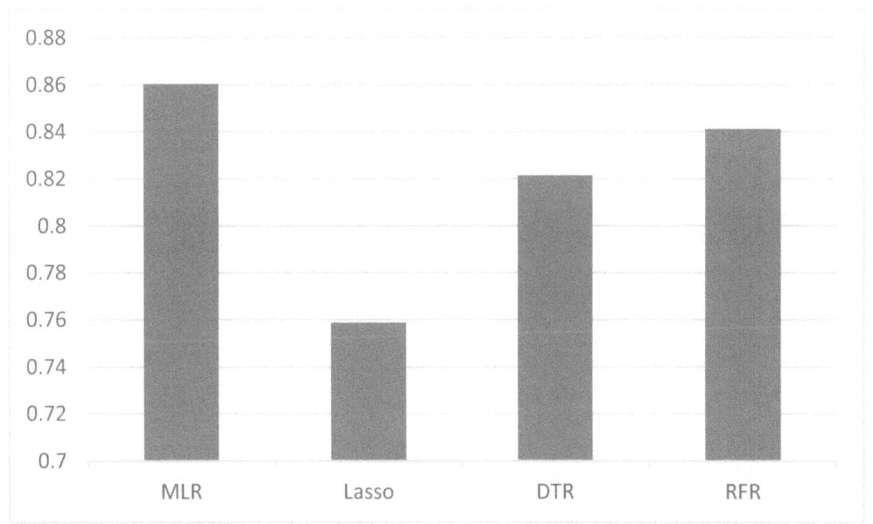

Fig. 12. Comparison of Best Accuracy Score of different models

Figures 12 and 13 shows the accuracy compariosn with other regression models and scatterplot representation between actual and predicted values. After the implementation of the model is done, and it accurately predicts the cost of a house, the model will be deployed using the Flask framework and develop a user interface in which the user shall input the intended value and the chosen model will predict what the outcome should be. This will be achieved by utilising the python library Flask to create an API.

To develop the web application and connect the Model to it, we must first extract our model into pickle and json files and generate a webpage using HTML, CSS, and JavaScript. Our model is now ready to be presented and predicted on a web application as ahown in the Fig. 14.

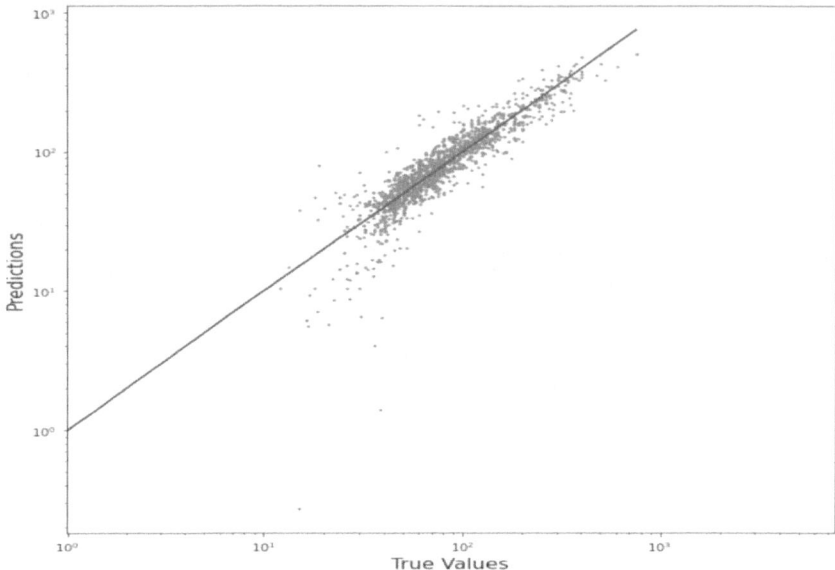

Fig. 13. Scatterplot between actual and predicted price

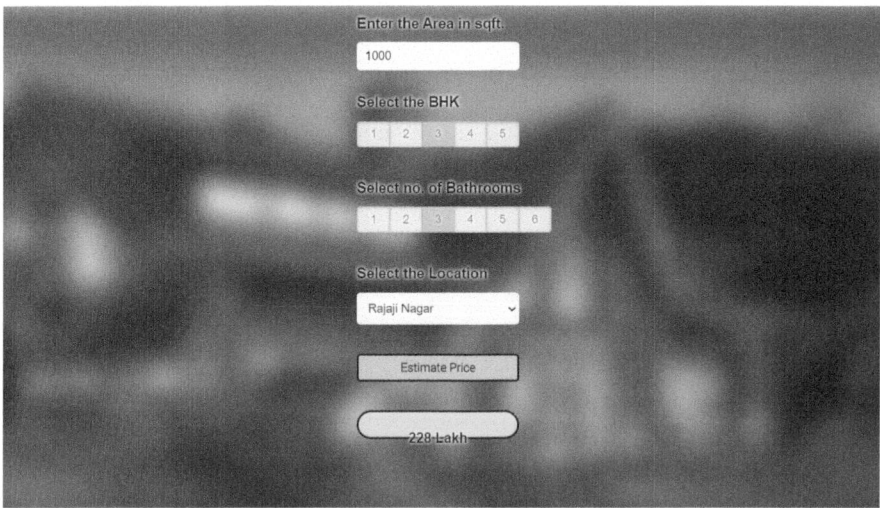

Fig. 14. Deployment of model using Flask

5 Conclusion

In this study, we employed machine learning techniques to forecast housing prices.

The data is pre-processed and input into several regression models to select the optimal model, and the accuracy of different models is determined using both K-Fold Cross-Validation and RMSE. So, based on the findings above, we can conclude that the most effective model, Multivariable Linear Regression, has the best accuracy of 84.78. Therefore, it may be a great model for estimating house prices, which is advantageous for both buyers and sellers.

A additional feature that may be added to the suggested system is to provide users with a full-fledged user interface so that users can utilise the ML model for many locations with multiple functionality. An Amazon EC2 connection will also advance the system and improve usability. Finally, the project will be completed by creating a well-integrated online application that can estimate prices anytime consumers want it to.

References

1. Weisberg, S.: Applied Linear Regression, vol. 528. John Wiley & Sons, Hoboken (2005)
2. Ravikumar, A.S.: Real Estate Price Prediction Using Machine Learning. Masters thesis, Dublin, National College of Ireland (2017)
3. Tiwari, P., Thakur, V.S., Shukla, S.S.P.: Comparison on advance regression techniques on house price prediction. IJSPR **73**(173) (2020)
4. Pirogova, O., Temnova, N.: Dynamics of coworking growth in the real estate market under digitalization. In: E3S Web of Conferences, vol. 244, no. 3, Article ID 10052 (2021)
5. Chernyshova, M., Malenkaya, A., Mezhuyeva, T.: Analysis of pricing factors in real estate market. Interexpo GEOSiberia **6**(2), 79–85 (2019)
6. Kuvalekar, A., Manchewar, S., Mahadik, S., Jawale, S.: House price forecasting using machine learning. In: Proceedings of the 3rd International Conference onAdvances in Science & Technology (ICAST) (2020)
7. Putatunda, S.: PropTech for Proactive Pricing of Houses in Classified Advertisements in the Indian Real Estate Market (2019)
8. Sivasankar, B., Ashok, A.P., Madhu, G., Fousiya, S.: House price prediction. Int. J. Comput. Sci. Eng. (IJCSE) **8**(7) (2020)
9. Naikare, R., Gahandule, G., Dumbre, A., Agrawal, K., Manka, C.: House planning and price prediction system using machine learning. Int. Eng. Res. J. **3**(3) (2019)
10. Chaturvedi, S., Ahlawat, L., Patel, T., Talha,M.: School of Computing Science and Engineering. Real Estate Price Prediction ((2021))
11. Mu, J., Wu, F., Zhang, A.: Housing value forecasting based on machine learning methods. In: Hindawi Publishing Corporation Abstract and Applied Analysis, vol. 2014 (2014)
12. Vasquez, C., Chellamuthu, V.: House price prediction with statistical analysis in support vector machine learning for regression estimation. Interdisc. J. Res. Innov. (2021)
13. Lu, S., Li, Z., Qin, Z., Yang, X., Goh, R.S.M.: A hybrid regression technique for house prices prediction. In: 2017 IEEE International Conference on Industrial Engineering and Engineering Management (IEEM), pp. 319–323 (2017). https://doi.org/10.1109/IEEM.2017.8289904
14. Abdul Razak, M.S., Anand, A., Priya, B., Neha, S., Ghanate, N.G.: House price prediction system for Bengaluru City. IJSART **7**(8) (2021)
15. Manjula, R., Jain, S., Srivastava, S., Kher, P.R.: Real estate value prediction using multivariate regression models. In: IOP Conference Series: Materials Science and Engineering (2017)
16. Hromada, E.: Mapping of real estate prices using data mining techniques. Czech Technical University, Czech Republic (2015)
17. Gupta, R.: Forecasting US real house price returns over 1831. Evidence from copula models (2013)
18. Garud, Y., Vispute, H., Bisai, N., Nashipudimath, M.: Housing price prediction using machine learning. Int. Res. J. Eng. Technol. (IRJET) **7**(5) (2020)

An Ensemble Approach for Detection and Classification of Fake News

Mohd. Maaz Khan[1], Aditya Agarwal[1], and S. Iniyan[2(✉)]

[1] Department of Computer Science and Engineering, SRM Institute of Science and Technology, Kattankulathur, Chengalpattu 603203, India
{mk8151,aa3634}@srmist.edu.in
[2] Department of Computing Technologies, School of Computing, SRM Institute of Science and Technology, Kattankulathur, Chengalpattu 603203, India
iniyans@srmist.edu.in

Abstract. Fake news is misleading information which can often damage individual's reputation and can change person's opinion. Recently, the rise of social media platforms gave a boost to circulation of fake news across the globe. Our paper presents a machine learning model using voting ensemble with equal and different weights of each base model. We used 4 different classification algorithms as base learners- 1- Logistic Regression, 2-Multinomial Naive Bayes, 3- Passive Aggressive Classification Algorithm, 4- Random forest algorithm. All algorithms chosen works best with the long textual in stream of data. We compared accuracy of each base model. After assigning different weights to each base model, voting ensemble gave an accuracy of 93.5%.

Keywords: Voting Ensemble · Base Learners · Logistic Regression · Multinomial Naive Bayes · Passive Aggresive Classifiers · Random Forest Claasifiers

1 Introduction

In today's world circulation of fake news to the readers is a frequent thing. Now a day's people are finding difficult to believe the article they are reading is real or fake. It is very easy to post fake news or some rumors related to someone on social media which gets circulated to newspapers, magazines which is read by each and every age group of the society. Due to this people start to create misconceptions on the subject and get manipulated easily and start to believe the wrong facts on the subject. Fake news makes an impact on everyone such that their opinion differs from others which sometimes lead to heated arguments or even violence. There is so much information online that it is becoming near impossible to know the true form of news. Increased used of social media sites have also been a reason for circulation of fake news, where anyone can post anything without any restrictions which gives freedom to wrong doers to spread false facts around the globe.

However, with the help of recent advances in subjects like Artificial Intelligence and Machine Learning, it is possible to detect fake articles. With the growing technology

A. Gupta et al. (Eds.): ICAIA 2023, CCIS 2308, pp. 286–297, 2025.
https://doi.org/10.1007/978-3-031-84394-5_20

we can classify news into fake or real. We took a step forward and researched on the subject. We used some of the efficient working algorithms which works best with lengthy textual paragraphs and classified the news articles as fake or real. We also tried different combination of algorithms and ensemble techniques and tried to find the most optimal result.

2 Literature Review

Detection of fake news had been an old topic for researchers to discover and find new optimal solutions. News often comes in long text paragraphs which is usually not machine understandable [1–3]. Some research used K-Nearest Neighbors (KNN) algorithm and Naive Bayes classifier for classifying the news [4]. Researchers did comparison with hybrid model and other machine learning techniques like random forest, naive bayes, KNN, decision tree, support vector machine [6]. Ensemble was also done on the data by using bagging, boosting and stacking techniques [7, 8]. Deep learning techniques and neural networks were also explored [9]. Linguistic analysis on news text was also performed in 2018 [10]. Text similarity between news articles was performed by some researchers to identify the correctness of news articles, where they used TF-idf and cosine similarity [11]. Because of the easy access social media became a place to spread Hate speeches; some papers identified articles with hate speech. They proposed 'Hate Classify' where they used CNN as a service framework [12]. A hybrid deep model was made for better classification of news [13]. A survey was performed where rumors on social media were detected, Resolution for the same was also provided [14]. Classification of news into real or fake can be classified using many machine learning models, Therefore in these papers the researchers compared those algorithms performance wised [15]. They implemented a web plugin for twitter where the posts goes credibility analysis on three levels-Social, Text, User. A front end was also included for user experience [16]. Two algorithms were used in this research Multinomial Naive Bayes and Support Vector Machine [17]. As the social media platforms popularity is increasing, There are different options available on different platforms. There are social bots which help to analyze and report different news dynamically but these bots can be exploited to spread fake propaganda. This paper focused on removing the social bots that spread false news [18]. "Fake news identification"-Being a common topic for researchers, there is a lot of research already done, this paper evaluated current approaches and measured the prediction performance [19]. A common arena for fake news is social media; this paper found some interesting facts by analyzing Facebook shares and comments on real and fake news, Also often the fake news disappears from site after few days so simply applying algorithms would be inefficient to overcome these challenges. So Different text classification methods were used to convert data into machine understandable. As we are dealing with long texted paragraphs of news articles we need to use classifiers which work best with large amount of textual data example passive aggressive classifier which many of the writers used for long in stream of textual news to classify its correctness [20]. Fake news have also its adverse effects on human psychology therefore we can state misinformation as a cyber-security threat which was analyzed in this paper.

3 Methodology

3.1 Dataset

In this paper, the dataset is collected from kaggle.com this dataset has about 20,799 rows from various articles found on the internet. After some preprocessing and data cleaning, the size of the dataset became 18285 rows. Dataset for fake news can be gathered from more than one source like different social media websites as Twitter, Facebook and others. It is not easy to distinguish the variety of news manually. Therefore, an annotator who expertizes analyzing the claims, evidence, and context from trustworthy sources is required. Another way to gather news data is from the internet itself, there are many websites which provides legit news. Journalists and workers in Press Company can also provide news data. Figure 1 shows the architectural diagram which will be followed throughout the course of our project.

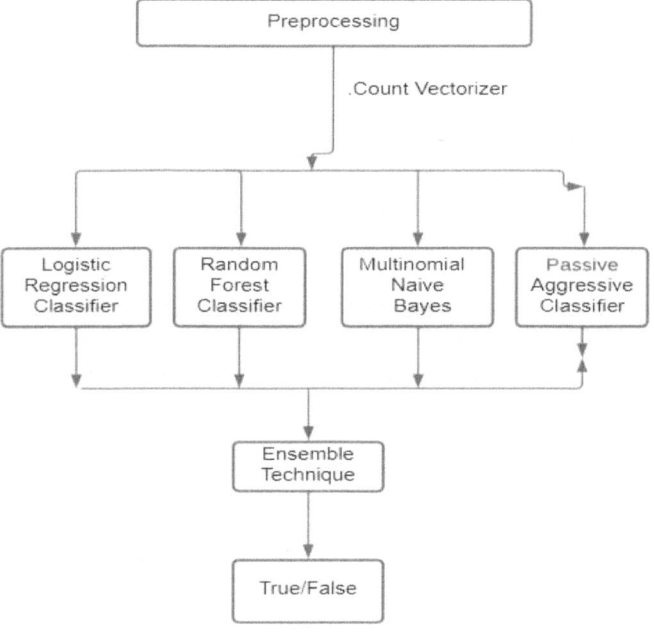

Fig. 1 .

3.2 Logistic Regression

It is a classification algorithm which predicts the probability of the final outcome which will be our target variable. It is one of the best techniques for solving classification problems. This model was treated as a base learner for the voting ensemble. It gave an accuracy of 93%. Mathematically, Logistic Regression can be represented using the following formula for P which is a function of x.

$$P = \frac{1}{(1 + e^-}(\beta 0 + \beta 1 x) \tag{1}$$

where P is the dependent variable, $\beta 0$ is the intercept, $\beta 1$ is the coefficient, x is the independent variable. The above formula can be represented in a graphical view which is shown in Fig. 2.

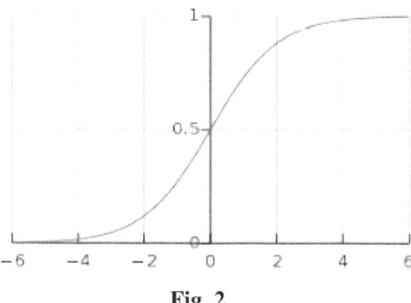

Fig. 2 .

The value of P lies between 0 and 1. If the value is greater than equal to 0.5, it will be classified into positive class i.e. True. But if the value is less than 0.5 would be negative. Figure 3 shows the confusion matrix for Logistic regression on dataset mentioned in Sect. 3.1.

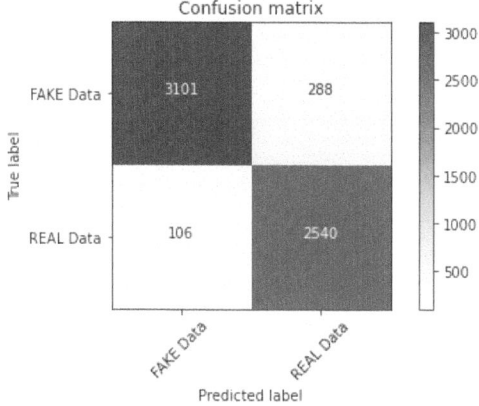

Fig. 3 .

3.3 Multinomial Naive Bayes

Multinomial Naive Bayes algorithm is a method to find probability of each feature in a given sample and produce the tag with highest probability. It is a supervised classification that is used for the analysis of the categorical text data. Before applying the algorithm we convert our textual data into features using count vectoriser which gives the frequency of each word. Naive Bayes typically use bag of words features to identify the correctness of the news, an approach commonly used in text classification.

Bayes theorem calculates probability P (F|w) where F is possible outcomes and w is the given word which we will classify.

$$P\left(\frac{F}{w}\right) = P\left(\frac{w}{F}\right) * P(F)/P(w) \tag{2}$$

Here, P denotes conditional probability. w denotes a single word in the news article.

and F denotes False/Fake news in the conditional probability.

P(F|w): It denotes probability that a news article is fake given that word w is present in the news or is true.

P(w|F): It denotes probability of word w is already in the text given that the news article is fake;

P(F) – probability that given article/news is false/fake;

P(w)- probability of occurrence of word w. It gave an accuracy of 90%.

Figure 4 shows the confusion matrix for Multinomial Naïve Bayes on dataset mentioned in Sect. 3.1.

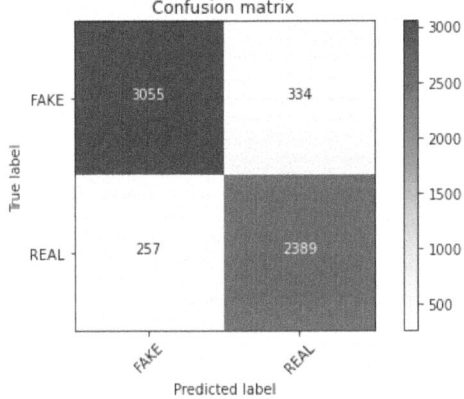

Fig. 4 .

3.4   Passive Aggressive Classification Algorithm

Passive-Aggressive algorithms belong to the category of online learning algorithms that are popularly used in big data applications used for large-scale learning. As the name suggests it responds in two ways, passive which implies true and aggressive which implies as false or any miscalculations.

The main reason behind using this classifier as a base model for ensemble is that it could detect false information on a social media websites like Facebook, Twitter where

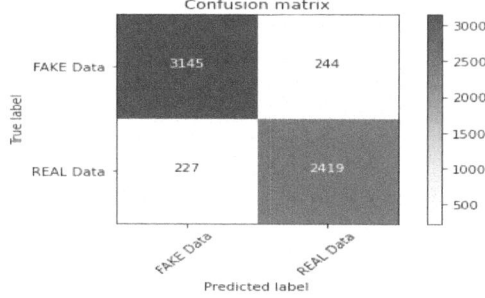

Fig. 5 .

new data is being added every second. To dynamically read the data from social media websites continuously would be a tiring job as the data would be huge and using an online-learning algorithm would be optimal choice.It gave an accuracy of 92%. Figure 5 shows the confusion matrix for Passive Aggressive Classifier on dataset mentioned in Sect. 3.1.

3.5 Random Forest Classifier

Random forest is a Supervised Machine Learning Algorithm that is used in Classification and Regression problems.As we are dealing with classification problem, where we classify the text into true or false. We will be focusing on random forest classifier which builds decision trees on different sample data sets and takes their majority vote for classification. We will get a better understanding of it using a real life example, suppose a student wants to watch an anime show, but he is not able to decide by himself so he asks his friends. He classifies the suggestions he got from his friend and the show which had maximum vote will be the one which the student chose. We will also be doing similar

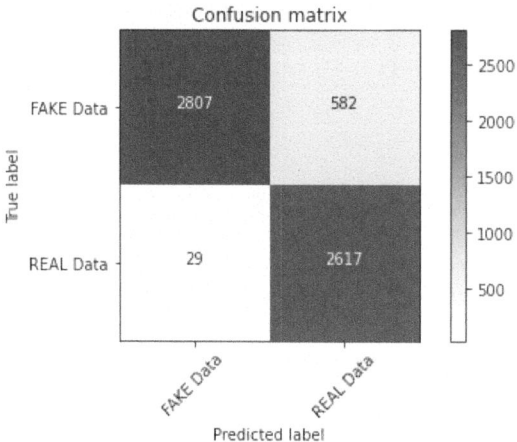

Fig. 6 .

thing, after making decision trees from our dataset. We will make variable count for fake and real news, And whichever had more votes would be the final answer.

It gave an accuracy of 90%. Figure 6 shows the confusion matrix for Random Forest Classifier on dataset mentioned in Sect. 3.1.

4   Proposed Voting Ensemble

Voting Classifier groups different classification algorithms and treat them as base learners. The model is built when we combine different models which themselves are stronger head-classifier, This way we can bring out the individuals algorithms weakness on the given data. Voting classifier can only be of two types namely Hard Voting and Soft Voting.

In this paper we worked on hard voting ensemble classifier, a hard voting ensemble involves summing the votes from other models and predicting the class with the most votes.

We can further subdivide the voting technique on the basis of weight: When equal weights have been assigned to each classifier: When vote is taken based on equal weights. Let's say there are 5 classifiers $c1, c2, c3, c4, c5$. For a particular data, the prediction for the above mentioned classifiers are $[1, 1, 0, 1, 0]$. It gave an accuracy of 93.2%. Figure 7 shows the confusion matrix for Voting based on equal weights on dataset mentioned in Sect. 3.1.

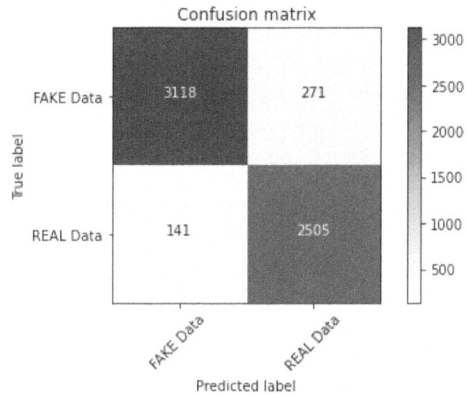

Fig. 7 .

When different weights have been assigned to each classifier: When different weights are assigned, vote is taken based on different weights; the final prediction is computed on the basis of weighted models. For example, Classifiers $c1,c2,c3$ have prediction as $[0,0,1]$ but weights as $[1, 2, 5]$. Then the final equation comes out as $[0, 0, 0, 1, 1, 1, 1, 1]$. As classifier $c3$ has weight of 5 it will overcome both $c1, c2$ thus the final result would be $[1]$. It gave an accuracy of 93.5%. Figure 8 shows the confusion matrix for Voting based on different weights on dataset mentioned in Sect. 3.1.

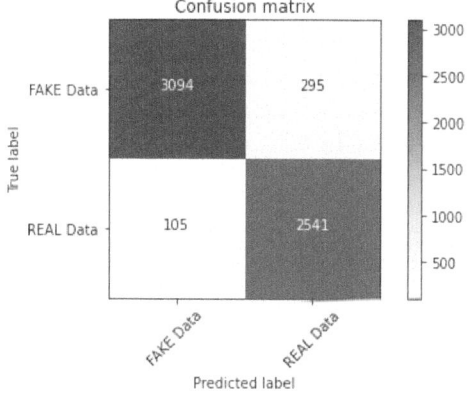

Fig. 8 .

4.1 Comparison Between Voting Based on Equal Weights and Voting Based on Different Weights

When every model was given equal weights the accuracy came to be 93.2%. But when some weight was assigned to different model the accuracy was increased to 93.5%. When Logistic Regression Model and Passive Aggressive Model was given double the weight of Multinomial Naive Bayes Model and Random Forest Model, Then the accuracy was found highest at 93.5%.

5 Results and Discussion

Using different python packages and exploring dataset from kaggle, we classified news into fake or real. We used four different classifiers and treated them as base learners for the meta-classifier which is our voting ensemble. Every model gave different accuracy among which Logistic Regression model and Passive Aggressive Model gave highest accuracy. Table 1 shows the comparison table for each model we used in our project.

Table 1. Comparison with other models

Model	Accuracy
Multinomial Naïve Bayes Algorithm	0.9
Passive Aggressive Classifier	0.92
Random Forest Algorithm	0.9
Logistic Regression	0.93
Ensemble	**0.93**

Figure 9 shows the Bar chart comparison for each model we used in our project. It is clear from the Fig. 9 that Logistic Regression and Ensemble approach gave the best results.After evaluating each model, they were treated as input for voting ensemble technique.We explored two different ways of evaluating our ensemble result. First one was when we considered the weight of every model as equal and second was when we considered weight of every model different. When giving more weight to efficient model our ensemble accuracy increased by 0.3.

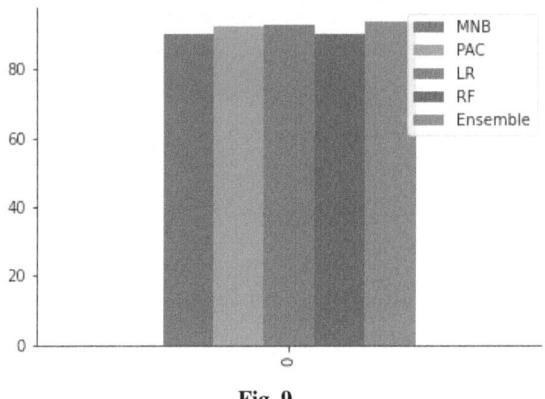

Fig. 9 .

Table 2 shows the cross validation score for each model used in our project. Figure 10 shows Box-Whisker Plot for all models used in the project. Table 3 shows Comparison of all models which is sorted in descending order of Test Accuracy. Figure 11 shows ROC Curve of all models used in the project.

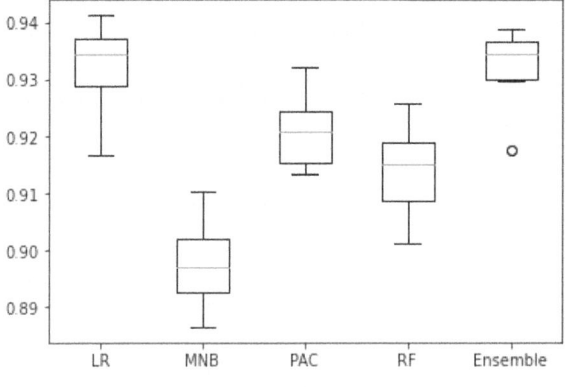

Fig. 10 .

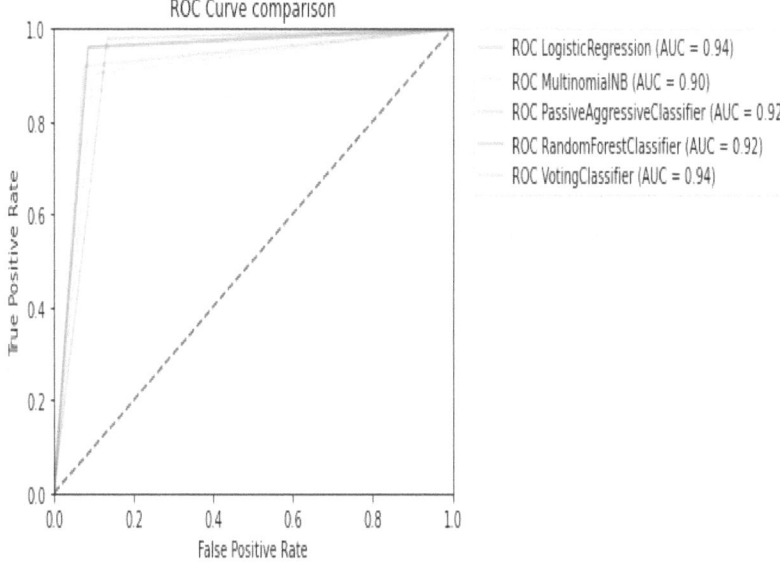

Fig. 11 .

Table 2. Cross Validation Score

```
LR| Mean=0.931755 Standard Deviation=0.007798
MNB| Mean=0.897633 Standard Deviation=0.007109
PAC| Mean=0.920816 Standard Deviation=0.005909
RF| Mean=0.913959 Standard Deviation=0.007043
Ensemble| Mean=0.931347 Standard Deviation=0.007459
```

Table 3. Comparison of Test Accuracy

Model	Train accuracy	Test accuracy	Precision	Recall
Logistic Regressioin	0.9884	0.9347	0.898161	0.959940
Voting Classifier	0.9887	0.9344	0.896965	0.960695
Passive Aggressive Classifier	1.0000	0.9246	0.905891	0.924036
Random Forest Classifier	0.9236	0.9152	0.850295	0.978836
Multinomial Naïve Bayes	0.9278	0.9021	0.877341	0.902872

6 Conclusion

The majority of tasks in the twenty-first century are practiced online. Newspapers, which were once distributed to every doorstep, are gradually being replaced by online media platforms. Because of the easy access provided by internet services and websites changing the news articles into something totally fallacious. A very important part of politics is how the news is being spread across the nation. Therefore a well written algorithm to detect the falseness of the news on online media platforms will be needed in coming future. There are no fast stream check software's available in the market for fake news identification. As more people are moving to social media platforms, the greater crowd gets exposure to misinformation. Therefore there is great need of some news filters on social media platforms which can identify and classify the news intelligently for the end-users. We used 4 different classification algorithms as base learners.All algorithms chosen works best with the long textual in stream of data. We compared accuracy of each base model. After assigning different weights to each base model, voting ensemble gave an accuracy of 93.5%.

References

1. Kesarwani, A., Chauhan, S.S., Nair, A.R.: Fake news detection on social media using k-nearest neighbor classifier. In: 2020 International Conference on Advances in Computing and Communication Engineering (ICACCE). IEEE (2020)
2. Islam, M.R., et al.: Detecting depression using k-nearest neighbors (knn) classification technique. In: 2018 International Conference on Computer, Communication, Chemical, Material and Electronic Engineering (IC4ME2). IEEE (2018)
3. Granik, M., Mesyura, V.: Fake news detection using naive Bayes classifier. In: 2017 IEEE First Ukraine Conference on Electrical and Computer Engineering (UKRCON). IEEE (2017)
4. Reddy, P.B.P., et al.: Fake data analysis and detection using ensembled hybrid algorithm. In: 2019 3rd International Conference on Computing Methodologies and Communication (ICCMC). IEEE (2019)
5. Jain, A., et al.: A smart system for fake news detection using machine learning. In: 2019 International Conference on Issues and Challenges in Intelligent Computing Techniques (ICICT), vol. 1. IEEE (2019)
6. Agarwal, A., Dixit, A.: Fake news detection: an ensemble learning approach. In: 2020 4th International Conference on Intelligent Computing and Control Systems (ICICCS). IEEE (2020)
7. Zhang, J., Dong, B., Philip, S.Y.: Fakedetector: effective fake news detection with deep diffusive neural network. In: 2020 IEEE 36th International Conference on Data Engineering (ICDE). IEEE (2020)
8. Katarya, R., Massoudi, M.: Recognizing fake news in social media with deep learning: a systematic review. In: 2020 4th International Conference on Computer, Communication and Signal Processing (ICCCSP). IEEE (2020)
9. Dey, A., et al.: Fake news pattern recognition using linguistic analysis. In: 2018 Joint 7th International Conference on Informatics, Electronics & Vision (ICIEV) and 2018 2nd International Conference on Imaging, Vision & Pattern Recognition (icIVPR). IEEE (2018)
10. Samantaray, S.D., Jodhani, G.: Fake news detection using text similarity approach. Int. J. Sci. Res. (IJSR) 8(1), 1126–1132 (2019)

11. Khan, M.U.S., Abbas, A., Rehman, A., Nawaz, R.: HateClassify: A service framework for hate speech identification on social media. IEEE Internet Comput. **25**(1), 40–49 (2021). https://doi.org/10.1109/mic.2020.3037034

12. Ruchansky, N., Seo, S., Liu, Y.: Csi: a hybrid deep model for fake news detection. In: Proceedings of the 2017 ACM on Conference on Information and Knowledge Management (2017)

13. Zubiaga, A., et al.: Detection and resolution of rumours in social media: a survey. ACM Comput. Surv. (CSUR) **51**(2), 1–36 (2018)

14. Singh, K., et al.: A comprehensive study on data-driven fake news detection methods. In: 2022 6th International Conference on Intelligent Computing and Control Systems (ICICCS). IEEE (2022)

15. Cardinale, Y., Dongo, I., Robayo, G., Cabeza, D., Aguilera, A., Medina, S.: T-CREo: a twitter credibility analysis framework. IEEE Access **9**, 32498–32516 (2021)

16. Hussain, M.G., et al.: Detection of bangla fake news using mnb and svm classifier. arXiv preprint arXiv:2005.14627 (2020)

17. Shi, P., Zhang, Z., Choo, K.K.R.: Detecting malicious social bots based on clickstream sequences. IEEE Access **7**, 28855–28862 (2019)

18. Reis, J.C.S., et al.: Supervised learning for fake news detection. IEEE Intell. Syst. **34**(2), 76–81 (2019)

19. Xu, K., et al.: Detecting fake news over online social media via domain reputations and content understanding. Tsinghua Sci. Technol. **25**(1) (2019)

20. Caramancion, K.M.: An exploration of disinformation as a cybersecurity threat. In: 2020 3rd International Conference on Information and Computer Technologies (ICICT). IEEE (2020)

Author Index

A

Agarwal, Aditya II-286
Agarwal, Nidhi II-60, II-72
Aggarwal, Jai I-166
Aggarwal, Shivani II-155
Anand, Rohit I-139
Ansal, V. II-120
Arora, Nishtha II-37
Arul, N. I-241
Askarali, K. T. I-109
Ather, Danish II-90, II-178
Awadh, Ram II-60, II-72

B

Babu, J. Chinna II-1
Babu, Yogendra II-60, II-72
Batta, Vineet I-194
Bhardwaj, Akhilesh Kumar II-245
Bharti, Vishal II-26
Biradar, Ajay II-15

C

Chaudhary, Vansh II-155
Chawla, Paras I-221
Chechar, Tejas II-15

D

Devan, K. P. K. I-182
Dhal, P. K. II-120
Divya, S. I-241

F

Fredrik, E. J. Thomson I-109

G

Gaba, Siddhartha II-274
Gaddala, Babu Rao II-136
Gayathri, M. I-194
George, Navin M. I-241
Gera, Shailja I-221
Ghosh, Nabanita I-19
Gowrishankar, R. I-66

G

Grover, Dharini I-208
Gulati, Vineeta I-150
Gupta, Anish I-122
Gupta, Ankur I-139
Gupta, Gaurav I 49
Gupta, Neeraj I-139
Gupta, Ritu II-37
Gupta, Ruchika I-122

I

Indumathy, P. I-182
Iniyan, S. I-208, II-274, II-286

J

Jadhav, Renu I-194
Jain, Kashish I-166
Jayaprakash, Lithiga II-37

K

Kait, Ramesh II-49
Kajal, Sandeep I-150
Kalidas, M. II-120
Kangan II-37
Kaur, Simarpreet I-166
Khan, Mudassir II-1
Khan, Rubina Liyakat II-90, II-178
Khan, Shakir I-49
Khanna, Shivi I-19
Kiran, Ajmeera II-1
Kulkarni, Vijay I-92
Kumar, Arhath I-139
Kumar, B. P. Santosh II-1
Kumar, Lovish I-166
Kumar, Puneet I-122
Kumar, Raj II-90, II-178
Kumar, S. Sathish II-120
Kumar, Sunita I-19
Kumar, Surender II-245
Kumar, Vikash I-80
Kushwaha, Ravindra Kumar I-80
Kutti, M. Beula I-182

The manufacturer's authorised representative in the EU is Springer
Nature Customer Service Centre GmbH, Europaplatz 3, 69115 Heidelberg,
Germany. If you have any concerns regarding our products, please
contact ProductSafety@springernature.com

Printed and bound by CPI Group (UK) Ltd, Croydon, CR0 4YY

27/04/2026

02097586-0005